Advances in Computer Vision and Pattern Recognition

For further volumes:
http://www.springer.com/series/4205

Chris Aldrich • Lidia Auret

Unsupervised Process Monitoring and Fault Diagnosis with Machine Learning Methods

Springer

Chris Aldrich
Western Australian School of Mines
Curtin University
Perth, WA, Australia

Department of Process Engineering
University of Stellenbosch
Stellenbosch, South Africa

Lidia Auret
Department of Process Engineering
University of Stellenbosch
Stellenbosch, South Africa

Series Editors
Sameer Singh
Research School of Informatics
Loughborough University
Loughborough
UK

Sing Bing Kang
Microsoft Research
Microsoft Corporation
Redmond, WA
USA

ISSN 2191-6586 ISSN 2191-6594 (electronic)
ISBN 978-1-4471-5184-5 ISBN 978-1-4471-5185-2 (eBook)
DOI 10.1007/978-1-4471-5185-2
Springer London Heidelberg New York Dordrecht

Library of Congress Control Number: 2013942259

Printed on acid-free paper

Springer is part of Springer Science+Business Media (www.springer.com)

Preface

Although this book is focused on the process industries, the methodologies discussed in the following chapters are generic and can in many instances be applied with little modification in other monitoring systems, including some of those concerned with structural health monitoring, biomedicine, environmental monitoring, the monitoring systems found in vehicles and aircraft and monitoring of computer security systems. Of course, the emphasis would differ in these other areas of interest, e.g. dynamic process monitoring and nonlinear signal processing would be more relevant to structural health analysis and brain–machine interfaces than techniques designed for steady-state systems, but the basic ideas remain intact. As a consequence, the book should also be of interest to readers outside the process engineering community, and indeed, advances in one area are often driven by application or modification of related ideas in a similar field.

In a sense, the area of process monitoring and the detection and analysis of change in technical systems are an integral part of the information revolution, as the use of data-driven methods to construct the requisite process or systems models becomes dominant over first-principle or higher knowledge approaches. This revolution has changed the world as we know it and will continue to do so in as yet unforeseen ways.

Rightly or wrongly, there is a perception that the mining engineering environment is conservative as far as research spending is concerned, reluctant to embrace future technologies that do not have an immediate proven impact on the bottom line, also as far as process automation is concerned. However, this is rapidly changing, with large mining companies investing considerable sums of money in the development of advanced process automation systems with no immediate benefit. These new automation systems will have to sense changes in their environment and be able to react to these changes, consistently, safely and economically. Apart from the development of advanced sensors, process monitoring technologies would play a central role in the success of these automated mining systems. For example, in underground mining, these systems would have to be able to differentiate between mineral and the surrounding gangue material in real time or be able to differentiate between solid rock and rock that might be on the verge of collapse in a mining tunnel. Humans have mixed success in these tasks, and current automation systems are too rudimentary to improve on this.

These new diagnostic systems would have to cope with the so-called *Big Data* phenomenon, which will inevitably also have an impact on the development and implementation of the analytical techniques underpinning them. In many ways, *Big Data* can simply be seen as more of the same, but it would be unwise to see it simply as a matter that can be resolved by using better hardware. With large complex data sets, the issues of automatically dealing with unstructured data, which may contain comparatively little useful information, become paramount. In addition, these data streams are likely to bring with them new information not presently available, in ways that are as yet unforeseen. Just like video data can simply be seen as a series of images, if taken at a sufficiently high frequency, these data can reveal information on the dynamic behaviour of the system that a discontinuous series of snapshots cannot. It is easy to see that in some cases this could make a profound difference on our understanding of the behaviour of the system.

In the same way that *Big Data* can be seen as data, just more of it, machine learning can arguably be seen as statistics, simply in a different guise, as in many ways it is without a doubt. However, looking into the future, as systems rapidly grow in complexity, the ability of machines to truly learn could also be influenced in unforeseen ways. By analogy, one could consider a novice chess player, who has learnt the rules of chess and knows how to detect direct threats to his individual pieces on the board. However, it is only by experience that he learns to recognize the unfolding of more complex patterns or emergent behaviour that would require timely action to avoid or exploit.

Perth, WA, Australia

Chris Aldrich

Acknowledgements

In many ways, this book can be regarded as a product of the Anglo American Platinum Centre for Process Monitoring and the research work of a large number of postgraduate students that have passed through the Process Systems Engineering group at Stellenbosch University over the last decade or more. The collaboration between academia and industry has been especially productive in this respect.

Our special thanks therefore to Dr. J.P. Barnard and Ms. Corné Yzelle for making available the Centre's *Process Diagnostics Toolset* software without which the methods outlined in Chap. 6 in the book could not have been implemented.

In addition, we would also like to express our sincere gratitude to Dr. Gorden Jemwa, not only for his contributions to the Process Systems Engineering group over many years but also specifically for his major contribution as main author of Chap. 8 in the book.

Finally, it may be a cliché, but it does not make it less true that a book like this does not write itself, and the authors would like to make use of this opportunity to thank their families and friends for their understanding and active support in this respect.

Chris Aldrich and Lidia Auret

Contents

Acronyms

Acronym	Description
ACF	Autocorrelation function
ADALINE	Adaptive linear element
AHPCA	Adaptive hierarchical principal component analysis
AID	Automatic interaction detection
AKM	Average kernel matrix
AMI	Average mutual information
AR	Autoregressive
ARL	Alarm run length
ARMA	Autoregressive moving average
ARMAX	Autoregressive moving average with exogenous variables
AUC	Area under curve
BDKPCA	Batch dynamic kernel principal component analysis
BDPCA	Batch dynamic principal component analysis
BZ	Belousov–Zhabotinsky
CART	Classification and regression trees
CHAID	Chi-square automatic interaction detection
COW	Correlation optimized time warping
CSTR	Continuous stirred tank reactor
CUSUM	Cumulative sum
CVA	Canonical variate analysis
DD	Detection delay
DICA	Dynamic independent component analysis
DISSIM	Dissimilarity
DKPCA	Dynamic kernel principal component analysis
DPCA	Dynamic principal component analysis
DTW	Dynamic time warping

(continued)

(continued)

Acronym	Description
EEMD	Ensemble empirical mode decomposition
ELM	Extreme learning machine
EMD	Empirical mode decomposition
EWMA	Exponentially weighted moving average
FAR	False alarm rate
FS	Feature samples
ICA	Independent component analysis
INLPCA	Inverse nonlinear principal component analysis
IOHMM	Input–output hidden Markov model
JITL	Just-in-time learning
k-DISSIM	Kernel dissimilarity
KICA	Kernel independent component analysis
KKT	Karush–Kuhn–Tucker
KPCA	Kernel principal component analysis
KPLS	Kernel partial least squares
LCL	Lower control limit
MA	Moving average
MADALINE	Multiple adaptive linear element
MAID	Multiple or modified automatic interaction detection
MAR	Missing alarm rate
MCEWMA	Moving centre exponentially weighted moving average
MEB	Minimum enclosing ball
MHMT	Multi-hidden Markov tree
MICA	Multiway independent component analysis
MKICA	Multiscale kernel independent component analysis
MPCA	Multiway principal component analysis
MPLS	Multiway partial least squares
MSDPCA	Multiscale dynamic principal component analysis
MSE	Mean square error
MSKPCA	Multiscale kernel principal component analysis
MSPC	Multivariate statistical process control
MSSA	Multichannel singular spectrum analysis
MSSPCA	Multiscale statistical process control
MSSR	Mean sum of squared residuals
MVU	Maximum variance unfolding
MVUP	Maximum variance unfolding projection
NIPS	Neural information processing systems

(continued)

(continued)

Acronym	Description
NLPCA	Nonlinear principal component analysis
NN	Neural network
NOC	Normal operating conditions
OOB	Out of bag
PAC	Probably approximately correct
PCA	Principal component analysis
PDPCA	Partial dynamic principal component analysis
PLS	Partial least squares
RBM	Restricted Boltzmann machine
RF	Random forest
ROC	Receiver operating curve
RQA	Recurrence quantification analysis
SBKM	Single batch kernel matrix
SI	Subspace identification
SOM	Self-organizing map
SPC	Statistical process control
SPE	Squared prediction error
SPM	Statistical process monitoring
SSA	Singular spectrum analysis
SSICA	State space independent component analysis
SVD	Singular value decomposition
SVDD	Support vector domain description
SVM	Support vector machine (1-SVM one class SVM)
SVR	Support vector regression
TAR	True alarm rate
THAID	Theta automatic interaction detection
TLPP	Tensor locality preserving projection
UCL	Upper control limit
VARMA	Vector autoregressive moving average
VC	Vapnik–Chervonenkis

Chapter 1
Introduction

1.1 Background

Technological advances in the process industries in recent years have resulted in increasingly complicated processes, systems and products that pose considerable challenges in their design, analysis, manufacturing and management for successful operation and use over their life cycles (Maurya et al. 2007). As a consequence, not only do the maintenance and management of complex process equipment and processes, and their integrated operation, play a crucial role in ensuring the safety of plant personnel and the environment, but they are also crucial to the timely delivery of quality products in an environmentally responsible way. Since the management of process plants remains a largely manual activity, the timely detection of abnormal events and diagnosis of its probable causes to enable appropriate supervisory control decisions and actions to bring the process back to a normal, safe operating state become all the more important. Without a doubt, there is still major scope for process improvement in all these aspects of plant operation, including safety, profitability and environmental responsibility, as discussed in more detail below.

1.1.1 Safe Process Operation

Industrial statistics show that about 70 % of industrial accidents are caused by human errors (Venkatasubramanian et al. 2003). Recent events have shown that large-scale plant accidents are not just a thing of the past. Two of the worst ever chemical plant accidents, namely, Union Carbide's, Bhopal, India, accident and Occidental Petroleum's Piper Alpha accident, happened relatively recently (1980s). Such catastrophes have a significant impact on safety, the environment and the economy. The explosion at Kuwait Petrochemical's Mina Al-Ahmedhi refinery in

C. Aldrich and L. Auret, *Unsupervised Process Monitoring and Fault Diagnosis with Machine Learning Methods*, Advances in Computer Vision and Pattern Recognition,
DOI 10.1007/978-1-4471-5185-2_1, © Springer-Verlag London 2013

June 2000 resulted in damages estimated at $400 million. Likewise, the explosion of the offshore oil platform of Petrobras, Brazil, in March 2001 resulted in losses estimated at $5 billion (Venkatasubramanian et al. 2003).

Although the occurrence of major industrial accidents such as mentioned above is not common, minor accidents are very frequent and occur almost daily, resulting in many occupational injuries and sickness and costing billions of dollars every year (Venkatasubramanian et al. 2003). This suggests that there is still a long way to go to enhance the performance of human operators to improve their diagnostic capability and good judgment.

1.1.2 Profitable Operation

Industrial processes are under increased pressure to meet the changing demands of society. For example, in the mining sector, processes have to be adapted to deal with more complex or refractory ores, as more accessible resources dwindle. The same applies in the oil industry, where the search for large repositories is increasingly focusing on deep-sea beds, as many of the world's largest fields, from Ghawar in Saudi Arabia to Prudhoe Bay in Alaska, are becoming depleted. At present, deep-sea rigs are capable of reaching down more than 12 km – twice as deep as a decade ago.

With globalization and increased competition, profit margins of companies are under pressure, and companies have to be more responsive to varying customer demands, without sacrificing product and process quality. This has led to the development of quality control management methodologies, like Six Sigma and ISO 9000, and other management programs to assist organizations in addressing some of these challenges.

In addition, modern process operations have become more complex owing to plant-wide integration and high-level automation of a large variety of process tasks (Jämsä-Jounela 2007). For example, recycling of process streams is widely established to ensure efficient material and energy usage. Process plants have become intricate virtual information networks, with significant interactions among various subsystems and components. Such interconnectivity facilitates the integration of operational tasks to achieve broader business strategic goals but invariably complicates other tasks, like planning and scheduling, supervisory control and diagnosis of process operations.

1.1.3 Environmentally Responsible Operation

More recently, regulatory frameworks have become more stringent to force better control of the environmental risks posed by industrial activities. Likewise, safety and health policies and practices are now priority issues in the modern process plants. As a result, systematic frameworks have been initiated, including process hazard analysis and abnormal event management and product life cycle management. Process hazard analysis and abnormal event management are aimed at ensuring process safety, while product life cycle management places obligatory stewardship responsibilities on an organization throughout the life cycles of its entire product range, that is, from conception to design and manufacture, service and disposal (Venkatasubramanian 2005).

1.2 Trends in Process Monitoring and Fault Diagnosis

1.2.1 Instrumentation

As a result, companies are making substantial investments in plant automation as a means to achieve their operational and business goals. This includes heavy investment in instrumentation to enable real time monitoring of process units and streams. New sensor technologies such as acoustic or vibrational signal monitoring and computer vision systems have been introduced in, among other, milling plants, multiphase processes, food processing and combustion processes (Zeng and Forssberg 1992; Das et al. 2011; Chen et al. 2012; Germain and Aguilera 2012). In large process plants, these instruments have enabled the observation of many hundreds or even thousands of process variables at high frequency (Venkatasubramanian et al. 2003). As a consequence, huge volumes of data are increasingly being generated in modern process plants. These data sets do not only contain massive numbers of samples but can also contain very large numbers of variables.

For example, in spectroscopy, data are obtained by exposing a chemical sample to an energy source and recording the resulting absorbance as a continuous trace over a range of wavelengths. Digitization of the trace at appropriate intervals (wavelengths) forms sets of variables that in pyrolytic mass spectroscopy, near-infrared spectroscopy and infrared spectroscopy yield approximately 200, 700 and 1,700 such variables for each chemical sample, respectively (Krzanowski and Marriott 1994) In these cases, the number of variables usually exceeds the number of samples by far. Similar features arise with the measurement of acoustic signals, such as may be the case in online monitoring of process equipment (Zeng and Forssberg 1992) and potentiometric measurements to monitor corrosion. Likewise, where image analysis is used to monitor particulate feeds or products in comminution systems, power plants or metallurgical furnaces, each pixel in the image could represent a variable, which could easily lead to millions of variables where high-resolution two-dimensional images are concerned.

1.2.2 Information Technology Hardware

The well-documented sustained exponential growth in computational power and communication has led to profound change in virtually all areas of technology in recent decades and will apparently continue to do so in the foreseeable future. In 1965, Gordon Moore, a co-founder of Intel, first observed that the density of components in computer chips had doubled each year since 1958, and this trend was likely to continue for at least a decade. In 1975, Dr Moore modified his prediction, observing that component density was doubling every 2 years. As a consequence, the performance of personal computers has also roughly doubled every 18 months since then, conforming to what has becoming known as Moore's law.

More recently, in what might be referred to as Koomey's law, Koomey et al. (2011) have shown that since the era of the vacuum tube, computers have also approximately doubled their electrical efficiency every 1.6 years since the mid-1940s. This trend reinforces the continued explosive growth in mobile computing, sensors and controls (Koomey et al. 2011).

The cost of computer memory is showing as pronounced a decrease as that of the other computer components, with roughly cost halving annually. For example, whereas at the beginning of the decade, 40 GB was the highest hard disk drive capacity generally available in personal computers, this has increased to 3 TB at present.

These developments have had a considerable impact on the development and maintenance of advanced process monitoring and control technologies. For example, unlike a mere decade ago, it is now possible to maintain complex instrumentation and process monitoring systems remotely via the Internet. This has led to a breakthrough in the application of instruments, such as the implementation of Blue Cube's inline diffuse reflectance spectrophotometer in remote areas, where calibration of the instrument is maintained from the company's headquarters in Stellenbosch, South Africa. The same applies to Stone Three's maintenance of their computer vision monitoring systems for particulate feeds on belts.

1.2.3 Academic Research into Fault Diagnostic Systems

Figure 1.1 shows recent trends in academic research into fault diagnosis, indicating publications associated with fault diagnosis and neural networks (ANN); expert systems (XS); kernel methods and support vector machines (SVM); multivariate methods, including principal components and latent variables (PCA); artificial immune systems and immunocomputing (AIS/IC); and others, not including the previous categories (OTHER) in the *IEEE Xplore* digital library.

Publications related to fault diagnosis and expert systems have remained more or less constant over the last two decades, since expert systems are mostly associated with qualitative fault diagnosis, while the other approaches are typically associated with data-driven fault diagnosis, which show a sharp rise, especially from 2006 to 2010. Although the publications considered here were selected to belong more or less exclusively to a particular category (e.g. SVM would indicate papers containing "support vector" *or* "kernel", but not "neural network") together with

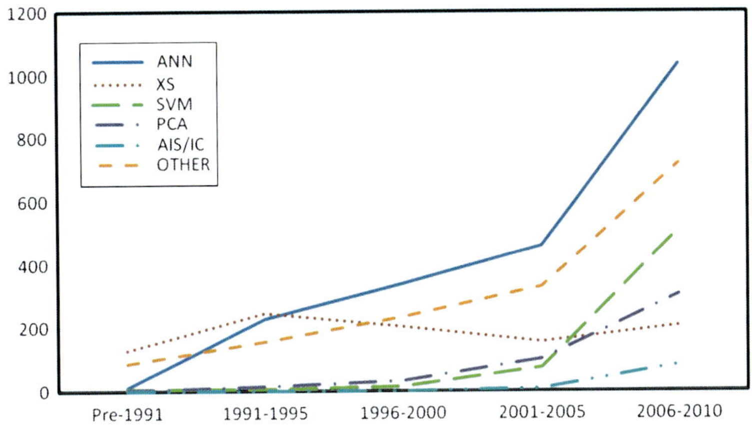

Fig. 1.1 Trends in academic research related to fault diagnosis based on number of publications in the *IEEE Xplore* digital library from 1991 to 2010

"fault diagnosis", the trends should still only be interpreted in an approximate qualitative manner, and some overlap between the categories was unavoidable. Even so, the overall trends indicate the strong growth in data-driven methods in fault diagnosis as well as the strong growth in machine learning in this area.

1.2.4 Process Analytical Technologies and Data-Driven Control Strategies

In turn, the above developments have led to further investment in advanced knowledge-based or data-driven process control strategies, collectively referred to as intelligent control systems, to enhance the information content of the data. Fortunately, advances in the information sciences have yielded data processing and analytical techniques that are very promising with respect to targeted applications in process control.

1.3 Basic Fault Detection and Diagnostic Framework

A fault can be defined as anomalous behaviour causing systems or processes to deviate unacceptably from their normal operating conditions or states. In process plants, faults can be categorized according to their sources, i.e. sensor faults affecting process measurements, actuator faults leading to errors in the operation of the plant, faults arising from erroneous operating policies or procedures as well as system component faults arising from changes in process equipment. These faults

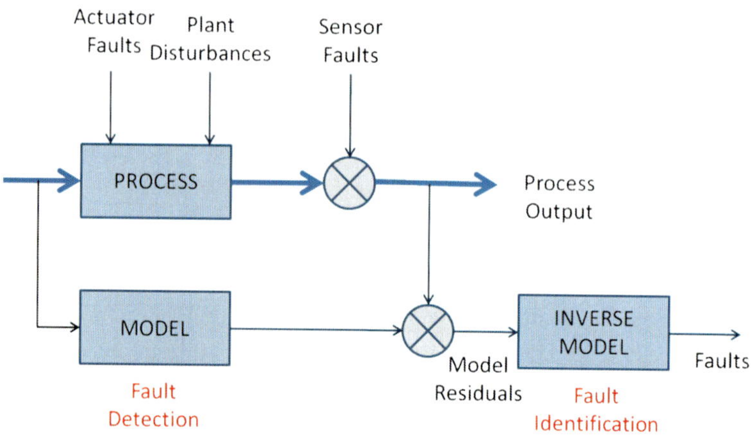

Fig. 1.2 A basic outline of the fault diagnosis problem

can arise abruptly, for example, with the sudden failure of process equipment, or faults can evolve over time, such as associated with gradual wear and tear of equipment or sensor drift.

The primary objective of fault diagnosis is the timely detection of aberrant process or system behaviour, identification of the causes of the fault and elimination of these causes with as little disruption to the process as possible. This is typically accomplished by comparing the actual behaviour of the process with a model representing normal or desirable process behaviour. The detection of process faults is based on monitoring of the deviation between the actual process behaviour and that predicted by the model, with a fault condition flagged when these deviations exceed certain predetermined limits. Once a fault is detected, identification of the root cause(s) of the problem is generally based on an inverse model. Correction of the problem depends on engineering expertise and is typically less well automated than the detection and identification problems. Figure 1.2 shows a schematic outline of the fault detection and identification problem.

1.4 Construction of Diagnostic Models

From a philosophical point of view, all fault diagnostic activities depend on models in one form or another. Models are simply compact representations of knowledge, which can either be explicit or tacit. Explicit knowledge exists in the form of documented equations, facts, rules, heuristics, etc. In contrast, tacit knowledge is more difficult to define and consists of all those things that humans know how to do, but not necessarily how to explain (Polanyi 1958). From a process perspective, it is the best practices, experience, wisdom and unrecordable intellectual property that reside within individuals and teams.

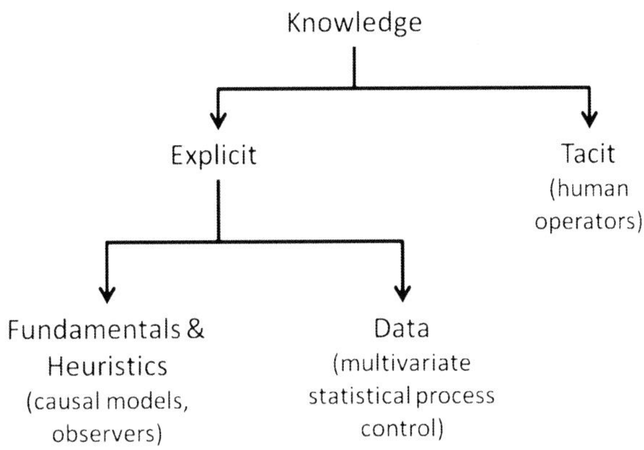

Fig. 1.3 Process fault diagnostic models as representations of process knowledge

Figure 1.3 shows a diagrammatic representation of approaches to fault diagnostic models based on different forms of knowledge. According to this diagram, process fault diagnostic methods can be categorized into models based on formal knowledge (causal models, observers), data (multivariate statistical process control) as well as manual approaches based on the tacit knowledge of human operators.

Classically, models have been derived from first principles or phenomenological models, requiring extensive knowledge of the behaviour of the process and interactions between the components of the process. These include Fickian or non-Fickian diffusion models used in the description of transport processes in leaching or adsorption process, heat conduction in warm plates, etc. Unfortunately, complete knowledge of real processes is often not available or very expensive to acquire. Under these circumstances, explicit knowledge in the form of data or process observations can be used to construct suitable models.

In some instances, tacit process knowledge or operator experience is also used to detect faults in plants. Tacit knowledge is subjective heuristic knowledge that cannot be expressed in words or numbers, often because it is context specific. For example, in froth flotation processes used in the recovery of metals, expert operators are often called upon to diagnose the condition of the process based on the appearance of the flotation froth. Similarly on food processing, the taste of the food is also sometimes used as an early indicator of the quality of the final product.

These alternative approaches to fundamental modelling based on explicit models derived from data or externalization of tacit knowledge have grown remarkably in the last half of the twentieth century based on learning from experience, such as operator knowledge or process data. Learning from data represents a paradigm shift from classical scientific inquiry in which phenomena were explained in terms of materials within a well-defined metric system. Instead, problems are cast in terms of data representation, information and knowledge. For example, a dominant theme that has emerged from the twenty-first-century computational biotechnology is the upgrade of information content in biological data, with strong parallels to the process control perspective (Aldrich 2000; Ogunnaike 1996; Venkatasubramanian 2005). Deriving knowledge from data can be achieved by statistical inferencing or planned experimental campaigns. An alternative and suitable approach that uses few or no assumptions and exploits the ever-growing volumes of process data accumulating in plant data bases is machine learning. Machine learning is concerned with developing machines and software that can discover patterns in data by learning from examples. It brings together insights and tools of mathematics, theoretical and applied computational sciences and statistics. In particular, it overlaps with many approaches that were proposed separately within the statistical community, for example, decision trees (Breiman et al. 1984; Quinlan 1986). Process fault diagnosis can also be cast as a machine learning problem, as outlined in more detail below.

1.5 Generalized Framework for Data-Driven Process Fault Diagnosis

The data-driven construction of models for process fault diagnosis can be cast in a general framework consisting of a number of elements as indicated in Fig. 1.4. These include a data matrix representing the process or system being monitored (**X**); a feature matrix extracted from the data matrix (**F**), from which diagnostic variables are derived for process monitoring and fault diagnosis; a reconstructed data matrix ($\widehat{X}$) serving as a basis for fault identification, as well as an indication of the quality of the extracted features; and, finally, a residual matrix (**E**) serving additionally as a monitoring space.

More formally, the problem can be considered given a set of sample vectors $\{\boldsymbol{x}\}_{i=1}^{N} \in \Re^{M}$, drawn from the random vector $\boldsymbol{X}$, find the mapping $\Im$: $\Re^{M} \rightarrow \Re^{q}$ and $\aleph$: $\rightarrow \Re^{M}$, such that for all $i = 1, 2, \ldots, N$, $\Im(\boldsymbol{x}_i) = \boldsymbol{f}_i$ and $\aleph(\boldsymbol{y}_i) = \widehat{\boldsymbol{x}}_i \approx \boldsymbol{x}_i$, where $\{\boldsymbol{f}\}_{i=1}^{N} \in \Re^{q}$ denote the corresponding set of reduced sample vectors or features drawn from the random vector $\boldsymbol{F}$. M and q denote the dimensionalities of the original and the feature vector or reduced latent variable space, respectively. For data visualization, $q = 2$ or 3 would be normal; otherwise, $q \ll M$.

Derivation of the mappings $\Im$ and $\aleph$ can be done by optimizing one of several possible criteria, such as the minimum mean square error or maximum likelihood criterion. For instance, with principal component analysis, the forward mapping $\Im$ is computed by eigendecomposition of the covariance matrix of the samples.

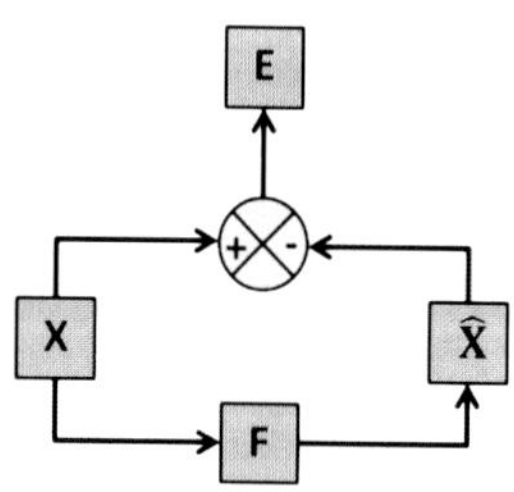

Fig. 1.4 A general framework for data-driven fault diagnosis

The reverse mapping ($\aleph$) is automatically derived from the forward mapping $\Im$. Similarly, in other linear latent variable models, such as independent component analysis, the reverse mapping is first computed from which the forward mapping can then be obtained via pseudo-inverses. Nonlinear methods can be more problematic, since mappings may not be easy to find with nonlinear transformation, and $\aleph$ is usually identified first, after which $\Im$ is defined by some projection operator.

These elements can be generated in various ways. For example, the data matrix can contain measurements of physical process variables in steady-state systems or could arise from the embedding or lagging of coordinates in dynamical systems (trajectory matrix). It could also be a component of a decomposed data matrix associated with multiscale methods.

An overview of data-driven methods to establish process fault diagnostic models is given in Chap. 2. In this book, this generalized diagnostic framework is treated from a machine learning perspective, where feature extraction is viewed as an unsupervised learning problem. Three machine learning paradigms are considered in this context, viz. neural networks, tree-based methods and kernel methods, as discussed in more detail in Chaps. 3, 4 and 5. In the remainder of the book, case studies and applications of the methodologies to different classes of fault conditions are considered.

1.6 Machine Learning

Machine learning is automatic computing based on logical and binary operations to learn tasks from examples. It can also be seen as the study of computational methods designed to improve the performance of machines by automating the acquisition of knowledge from experience or data. Different machine learning paradigms include artificial neural networks (multilayer perceptrons, self-organizing maps, radial basis function neural networks, etc.); instance-based learning (case-based reasoning, nearest neighbour methods, etc.); rule induction; genetic algorithms, where knowledge is typically represented by Boolean features, sometimes as the conditions and actions of rules; statistics; as well as analytical learning. The field of machine learning has originated from diverse technical environments and communities.

1.6.1 Supervised and Unsupervised Learning

A distinction can be made between supervised, unsupervised and reinforcement learning and combinations thereof. In supervised learning, the training data consist of a set of exemplars $\{\boldsymbol{x}, \boldsymbol{y}\}_{i=1}^{N}$, each of which is a pair comprising an input and an output vector, $\boldsymbol{x} \in \Re^{M}$, $\boldsymbol{y} \in \Re^{P}$. If the output is continuous, a regression

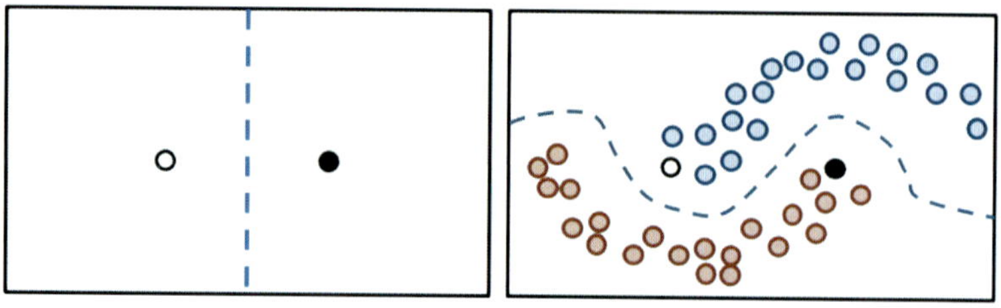

Fig. 1.5 An example of semi-supervised learning where unlabelled data (*red* and *blue markers*) are used in conjunction with labelled data (*black* and *white markers*) for learning the distribution of two classes

function is learnt; otherwise, if the output is discrete, a classification function is learnt. In unsupervised learning problems, unlabelled data are used, i.e. $\{\boldsymbol{x}\}_{i=1}^{N}$. In this case, the outputs represent the structure of the data, which is determined by a cost function to be minimized. In contrast, reinforcement learning uses a scalar reward signal to evaluate input–output pairs by trial and error to discover the optimal outputs for each input. This approach is most suited to problems where optimal input–output mappings are not available a priori, but where any given input–output pair can be evaluated. In this sense, reinforcement learning can be considered to be intermediary to supervised and unsupervised learning, since the use of a reward signal represents some form of supervision.

1.6.2 Semi-supervised Learning

Semi-supervised learning (Zhu et al. 2003; Zhao 2006) is another intermediary to supervised and unsupervised learning and comprises the solution of supervised learning tasks given labelled und unlabelled data are generated by the same distribution or underlying process.

More formally, consider a set of L independently distributed examples, $\boldsymbol{x}_1, \boldsymbol{x}_2, \ldots, \boldsymbol{x}_L \in X$, with corresponding labels $y_1, y_2, \ldots, y_L \in Y$. In addition, there are W unlabelled samples available $\boldsymbol{x}_{1+W}, \boldsymbol{x}_{2+W}, \ldots, \boldsymbol{x}_{L+W} \in X$. Semi-supervised learning attempts to make use of all the data to improve a *classification* model. This could be accomplished by clustering the unlabelled data and then labelling the clusters with the labelled data, moving the decision boundary away from high-density regions or learning an underlying manifold where the data are located, as indicated in Fig. 1.5. Semi-supervised learning is transductive, when the correct *labels* are inferred for $\boldsymbol{x}_{1+W}, \boldsymbol{x}_{2+W}, \ldots, \boldsymbol{x}_{L+W}$ only and inductive when the correct *mapping* $X \rightarrow Y$ is learnt (Vapnik 2006), as shown in Fig. 1.6.

Other approaches to machine learning are linked to the structure of the models. In deep learning, neural network models with a large multiple of layers (deep layering)

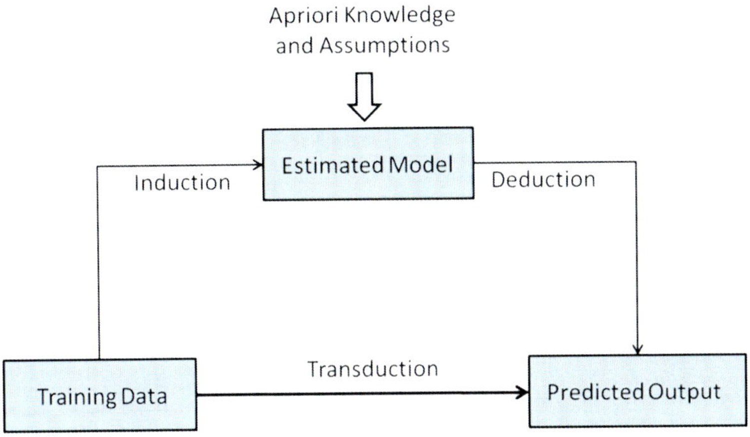

Fig. 1.6 Inductive, deductive and transductive learning

are trained in supervised or unsupervised mode, as discussed in more detail in Chap. 3. In extreme learning on the other hand, rapid learning is achieved in neural network structures with single large hidden layers by training the output layer only by means of a simple least-squares fit, as discussed in more detail in Chap. 3 as well.

1.6.3 Self-Taught or Transfer Learning

Many situations arise, where the common assumption of machine learning tasks that the training and test data are drawn from the same distribution or feature space does not hold. This could be the case with a fast changing process, where labelling of the data based on laboratory analyses becomes very expensive, or where models have to be derived for different ore systems, where characterization of the ores could again be expensive. Under these circumstances, most statistical or machine learning models need to be reconstructed using newly collected training data. In many real world situations, this may not be feasible. In these cases, the transfer of knowledge or learning between domains would be desirable.

Formally, self-taught learning or transfer learning (Pan and Yang 2010) is the solution of supervised learning tasks given labelled und unlabelled data, where the latter do not share the class labels or generative distribution of the labelled data. That is, it is not assumed that the unlabelled data can be assigned to the supervised learning task's class labels. Such data are often significantly easier to obtain than unlabelled data relevant to semi-supervised learning.

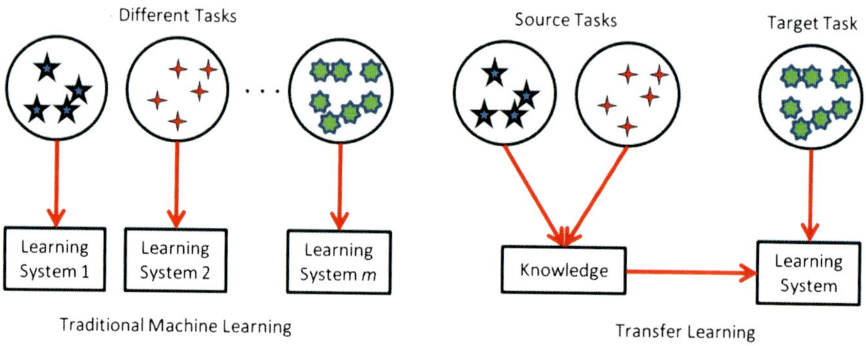

Fig. 1.7 Traditional machine learning and transfer learning (After Pan and Yang 2010)

In general, transfer learning first attempts to identify a set of features generalized across source domains, as indicated in Fig. 1.7. This is followed by adaptation, where useful features are collected to solve the problem, e.g. classification of the unlabelled or target samples (Huang et al. 2012).

1.6.4 Reinforcement Learning

Reinforcement learning considers the actions an agent should take in an environment to satisfy some notion of cumulative reward. Originally inspired by behavioural psychology, the problem is general, and it is studied in one form or another in many other disciplines, e.g. approximate dynamic programming in operations research, bounded rationality in game theory and economics, control theory and optimal control theory, statistics, information theory and evolutionary programming.

Unlike supervised learning, in reinforcement learning, input–output pairs of data are never presented to the learning machine, and suboptimal actions are not corrected explicitly. Moreover, there is a focus on finding a balance between the exploitation of current knowledge and exploration of uncharted regions in the learning space.

As indicated in Fig. 1.8, the agent or learning machine receives a stimulus or state information at time t (s_t) from its environment, which determines the action it takes (a_t). This action changes the state of the environment (s_{t+1}), the evaluation of which determines the reward (r_{t+1}) the agent receives. This reward, together with the new state information, determines the next action step of the agent. Positive rewards reinforce the behaviour of the agent, while negative rewards of penalties do the opposite.

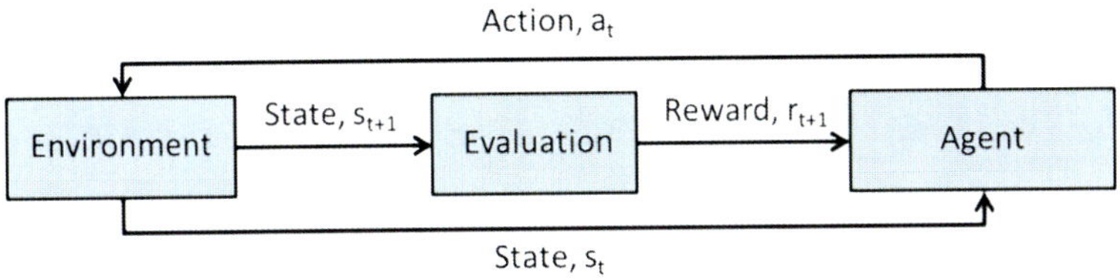

Fig. 1.8 Basic reinforcement learning scheme

1.7 Machine Learning and Process Fault Diagnosis

As indicated in Fig. 1.4, models are essential to automated process fault diagnosis. Machine learning offers significant advantages in the form of state-of-the-art models from data. This includes nonlinear modelling as well as the ability to capture data from heterogeneous sources, such as with semi-supervised learning. Traditionally, machine learning has also focused on large data sets, which is also becoming increasingly important in process monitoring.

In addition, machine learning provides flexibility in the way that learning from data can be accomplished. In many instances, target data by way of key performance measurements may not be readily available, in which case semi-supervised learning could be an option. In other instances, very few data could be available, which might make reinforcement learning an option.

As outlined above, machine learning encompasses a wide range of methods, comprehensive coverage of which is quite beyond the scope of a book such as this. As a consequence, the application of three different machine learning frameworks to process fault diagnosis will be considered, i.e. artificial neural networks, tree-based modelling approaches as well as kernel methods. As will be discussed, each of these frameworks represents a comprehensive platform for the development and implementation of all the operations required in process fault diagnosis, as outlined in Fig. 1.4, by taking full advantage of the machine learning approaches discussed above.

Therefore, the remainder of the book is organized as follows. In Chap. 2, the application of machine learning to process monitoring and fault diagnosis is reviewed. This is followed by a review of the three machine learning frameworks in Chaps. 3, 4 and 5. In Chaps. 6 and 7, the application of machine learning to steady-state and dynamic operations is considered, respectively, focusing on unsupervised learning, while the application of spectral methods to process fault diagnosis is demonstrated in Chap. 8.

References

Aldrich, C. (2000). What is AI and is it better than classical process control? *Journal of the South African Institute of Chemical Engineers, 12*(2), 27–49.

Breiman, L., Friedman, J., Stone, C. J., & Olshen, R. A. (1984). *Classification and regression trees.* Boca Raton: Chapman and Hall/CRC.

Chen, J., Chang, Y.-H., Cheng, Y.-C., & Hsu, C.-K. (2012). Design of image-based control loops for industrial combustion processes. *Applied Energy, 94*, 13–21.

Das, S. P., Das, D. P., Behera, S. K., & Mishra, B. K. (2011). Interpretation of mill vibration signal via wireless sensing. *Minerals Engineering, 24*(3–4), 245–251.

Germain, J. C., & Aguilera, J. M. (2012). Identifying industrial food foam structures by 2D surface image analysis and pattern recognition. *Journal of Food Engineering, 111*(2), 440–448.

Huang, P., Wang, G., & Qi, S. (2012). Boosting for transfer learning from multiple data sources. *Pattern Recognition Letters, 33*, 568–579.

Jämsä-Jounela, S.-L. (2007). Future trends in process automation. *Annual Reviews in Control, 31*(2), 211–220.

Koomey, J., Berard, S., Sanchez, M., & Wong, H. (2011). Implications of historical trends in the electrical efficiency of computing. *IEEE Annals of the History of Computing, 33*(3), 46–54.

Krzanowski, W. J., & Marriott, F. H. C. (1994). *Kendall's library of statistics 1. Multivariate analysis. Part I. Distributions, ordination and inference*. New York: Wiley.

Maurya, M. R., Rengaswamy, R., & Venkatasubramanian, V. (2007). Fault diagnosis using dynamic trend analysis: A review and recent developments. *Engineering Applications of Artificial Intelligence, 20*(2), 133–146.

Ogunnaike, B. A. (1996). A contemporary industrial perspective on process control theory and practice. *Annual Reviews in Control, 20*, 1–8.

Pan, S. J., & Yang, Q. (2010). A survey of transfer learning. *IEEE Transactions on Knowledge and Data Engineering, 22*(10), 1345–1359.

Polanyi, M. (1958). *Personal knowledge: Towards a post-critical philosophy*. Chicago: University of Chicago Press. ISBN 0-226-67288-3.

Quinlan, J. R. (1986). Induction of decision trees. *Machine Learning, 1*(1), 81–106.

Vapnik, V. (2006). Transductive inference and semi-supervised learning. In O. Chapelle, B. Schölkopf, & A. Zien (Eds.), *Semi-supervised learning* (pp. 453–472). Cambridge, MA: MIT Press.

Venkatasubramanian, V. (2005). Prognostic and diagnostic monitoring of complex systems for product lifecycle management: Challenges and opportunities. *Computers and Chemical Engineering, 29*(6), 1253–1263.

Venkatasubramanian, V., Rengaswamy, R., Kavuri, S. N., & Yin, K. (2003). A review of process fault detection and diagnosis Part III: Process history based methods. *Computers and Chemical Engineering, 27*(3), 327–346.

Zeng, Y., & Forssberg, K. S. E. (1992). Effects of operating parameters on vibration signal under laboratory scale ball grinding conditions. *International Journal of Mineral Processing, 35*, 273–290.

Zhao, H. (2006). Combining labeled and unlabeled data with graph embedding. *Neurocomputing, 69*, 2385–2389.

Zhu, X., Lafferty, J., & Ghabramani, Z. (2003). Semi-supervised learning using Gaussian fields and harmonic functions. In *Proceedings of the international conference on machine learning, 20* (pp. 912–919). Menlo Park: AAAI Press.

Nomenclature

Symbol	Description
$\aleph$	Reverse mapping, $\aleph : \mathbf{F} \rightarrow \widehat{\mathbf{X}}$
$\Im$	Forward mapping, $\Im : \mathbf{X} \rightarrow \mathbf{F}$
$\widehat{\mathbf{X}}$	Reconstructed data matrix, $\widehat{\mathbf{X}} \in \Re^{N \times M}$
a_t	Action taking by a reinforcement learning system at time t
$\mathbf{E}$	Residual matrix, $\mathbf{E} \in \Re^{N \times M}$
$\mathbf{F}$	Feature matrix, $\mathbf{F} \in \Re^{N \times q}$
f	Feature vector, $f \in \Re^{q_q}$
F	Feature space
L	Number of labelled samples in semi-supervised learning
M	Number of measured variables
N	Number of measured samples
p	Dimensionality of the output vector space
q	Number of features
r	Reward received by a reinforcement learning system at time t
s_t	Stimulus or state information received by a reinforcement learning system at time t
t	Time
W	Number of unlabelled samples in semi-supervised learning
$\mathbf{X}$	Data matrix, $\mathbf{X} \in \Re^{N \times M}$
$\mathbf{x}$	Measured vectors, $\mathbf{x} \in \Re^{M}$
x	Measured variable space
y	Output vector of quality or response variables, $\mathbf{y} \in \Re^{N \times p}$

Chapter 2
Overview of Process Fault Diagnosis

2.1 Background

The advent of the twenty-first century has seen the manufacturing and process industries facing stiff challenges in the form of increasing energy costs, increasingly stringent environmental regulations and global competition. As mentioned previously, although advanced control is widely recognized as essential to meeting these challenges, implementation is hindered by more complex, larger-scale circuit configurations, the tendency towards plant-wide integration and, in some cases, an increased lack of trained personnel. In these environments, where process operations are highly automated, algorithms to detect and classify abnormal trends in process measurements are critically important.

There is a plethora of literature on process fault diagnosis ranging from first-principle models on the one end of the spectrum to data-driven or statistical approaches wholly based on historical process data. It is especially the latter based on historical process data that are seen as the most cost-effective approach to dealing with the complex systems and that have seen explosive growth over the last few decades. Data-driven fault diagnosis can be traced back to control charts invented by Walter Shewhart at Bell Laboratories in the 1920s to improve the reliability of their telephony transmissions systems. In these statistical process control charts, variables of interest were plotted as time series within statistical upper and lower limits. Shewhart's methodology, described in more detail in Chap. 6, was subsequently popularized by Deming, and these statistical concepts, such as Shewhart control charts (1931), cumulative sum charts (1954) and exponentially weighted moving average charts, were well established by the 1960s (Venkatasubramanian et al. 2003).

These univariate control charts do not exploit the correlation that may exist between process variables. In the case of process data, cross-correlation is present, owing to restrictions enforced by mass and energy conservation principles, as well as the possible existence of a large number of different sensor readings on essentially the same process variable. These shortcomings have given rise to multivariate

C. Aldrich and L. Auret, *Unsupervised Process Monitoring and Fault Diagnosis with Machine Learning Methods*, Advances in Computer Vision and Pattern Recognition, DOI 10.1007/978-1-4471-5185-2_2, © Springer-Verlag London 2013

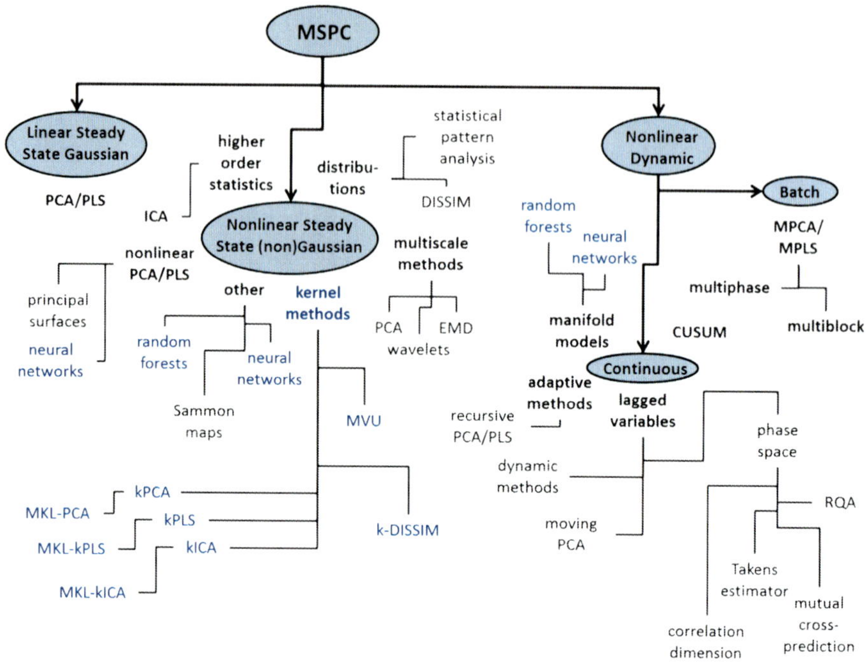

Fig. 2.1 Approaches to multivariate statistical process control (machine learning approaches are indicated in *blue*)

methods or multivariate statistical process control and related methods that have proliferated exponentially over the last number of years.

Venkatasubramanian et al. (2003) have proposed classification of these methods based on the methodology used to construct fault diagnostic models from data, drawing a distinction between statistical methods and neural networks. This seems somewhat arbitrary, as the boundaries between the different classes of methods are ill defined at best. Another way to view these approaches is on the basis of the process characteristics, as outlined in Fig. 2.1. In this diagram, a distinction is made between classical linear steady-state Gaussian processes, nonlinear steady-state stochastic processes as well as unsteady state or dynamic processes. The latter would encompass batch process monitoring, dealing with processes with a well-defined starting and end point, as well as systems dealing with continuous dynamic processes, similar to systems designed for equipment condition or structural health monitoring.

2.2 Linear Steady-State Gaussian Processes

As mentioned in the previous section, univariate control charts do not exploit the correlation that may exist between process variables, and when the assumptions of linearity, steady-state and Gaussian behaviour hold, multivariate statistical

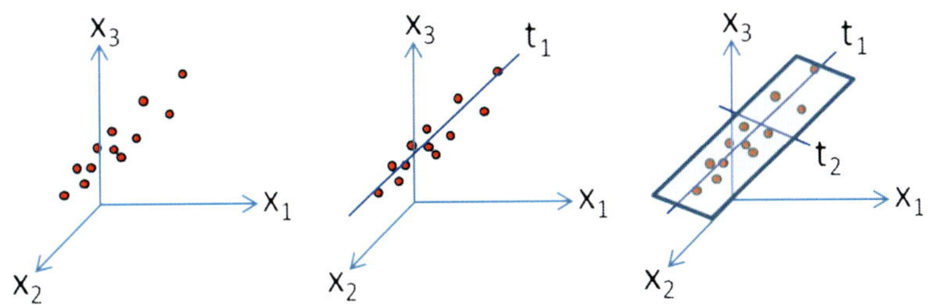

Fig. 2.2 Three-dimensional scatter plot of data (*left*), showing the first principal component (t_1) lying in the direction of maximum variation of the data (*middle*), with the second principal component (t_2) orthogonal to the first (*right*), explaining most of the remaining variance

process control based on the use of principal component analysis can be used very effectively for early detection and analysis of any abnormal plant behaviour. Since principal component analysis plays such a major role in the design of these diagnostic models, a more in-depth discussion of the methodology is in order.

2.2.1 *Principal Component Analysis*

Principal component analysis (PCA) was proposed by Karl Pearson in 1901 and developed by Hotelling in 1947 (Venkatasubramanian et al. 2003) with the aim to define a set of principal components consisting of linear combinations of the original measurement variables such that the first principal component accounts for the most variance in the data set, the second principal component for most of the remaining variance, etc. The principal components are orthogonal to each other and preserve the correlation among the process variables. As in Hotelling's T^2-statistic approach, the principal components are calculated using eigendecomposition of the covariance matrix of the process data representing normal operating conditions (NOC).

The original measurements can be located on the hyperplanes spanned by principal components. In the context of feature extraction, the score vectors obtained from projecting the process measurements onto the principal components can be seen as extracted features. The number of principal components to use in calculating the features can be determined by investigating the cumulative variance accounted for by the principal components. Figure 2.2 shows the projection of three variables (x_1, x_2 and x_3) onto a plane (t_1, t_2) oriented to show the maximum variation in the data.

In the case of two or three principal components accounting for significant total variation (e.g. more than 80 % or 90 %), the feature space can be visualized and process monitoring can consequently be based on a visual representation of the scores of the first two or three principal components. If a larger number of principal components are required, the process can still be visually summarized by diagnostic score distance and residual distance control charts.

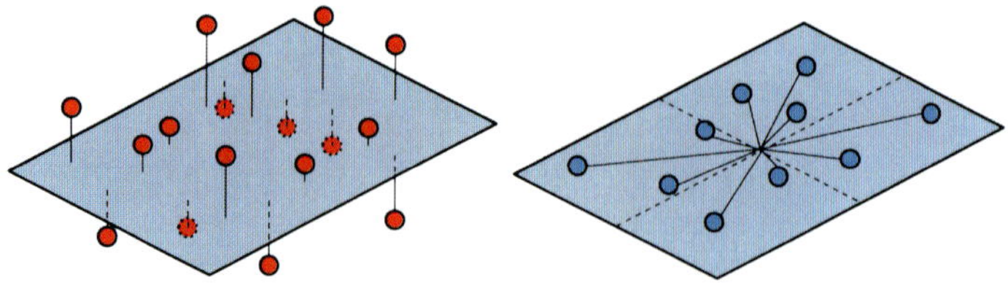

Fig. 2.3 *Q*-statistics (*left*) are based on the residuals between the observations and their projections onto the principal component plane, while T^2-statistics (*right*) are based on the deviations of the projected values from the mean on the plane

The separation of the score distance and residual distance control charts can be interpreted as the classification of abnormal variation into two parts, viz. process model variation outside its NOC control limits and variation implying a break in the expected correlation of the NOC process data (Dunia and Qin 1998).

The PCA score distance is similar to Hotelling's T^2-statistic, where only $k \ll m$ projection vectors are retained. k can be determined by an inspection of the variance decomposition of the training data by principal component number. The basis for calculation of the two statistics is illustrated in Fig. 2.3, showing the Q_i-statistics (left) as an indication of the distance of the ith sample from the principal component plane (t_1, t_2-plane in Fig. 2.2). In contrast, the T_i^2 statistic is the distance of the ith sample to the centre of the principal component plane (t_1, t_2-plane in Fig. 2.2).

$$T_i^2 = \boldsymbol{t}_i \lambda^{-1} \boldsymbol{t}_i^{\mathrm{T}} = x_i \mathbf{P} \lambda^{-1} \mathbf{P}^{\mathrm{T}} \boldsymbol{t}_i^{\mathrm{T}} \tag{2.1}$$

$$Q_i = \boldsymbol{e}_i \boldsymbol{e}_i^{\mathrm{T}} = \mathbf{x}_i \left(\mathbf{I} - \mathbf{P}_i \mathbf{P}_i^{\mathrm{T}} \right) \mathbf{x}_i^{\mathrm{T}} \tag{2.2}$$

A suggested advantage of PCA models is that the score variables obtained are linear combinations of the measurement variables and, as a consequence of the central limit theorem, should show a more normal distribution than the measurement variables themselves. From this deduction, the PCA scores should be approximately normally distributed. However, in the presence of autocorrelation, this assumption and conclusion are no longer valid, as the normal distribution assumes independently and identically distributed data (Wise and Gallagher 1996).

2.2.2 *Multivariate Statistical Process Control with PCA*

Fault identification with PCA relies on score distance and residual distance decomposition contribution plots (Russell et al. 2000a), as discussed in more detail in Chap. 6. The ease of calculation of these contributions can be attributed to the deterministic and explicit nature of PCA feature extraction. Figure 2.4 shows typical multivariate control charts based on the T^2- and Q-statistics. Samples above any one of the two confidence or control limits indicate an out-of-control process condition.

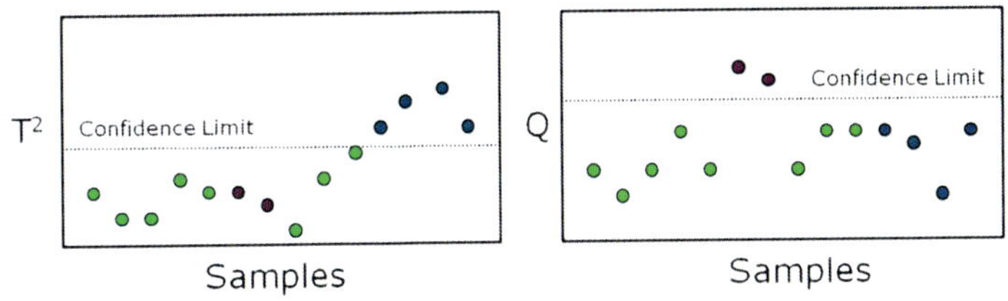

Fig. 2.4 Typical multivariate process control charts based on the T^2- and Q-statistics derived from a principal component model representing normal operating conditions

Limitations of the PCA approach include its lack of exploitation of autocorrelation (Venkatasubramanian et al. 2003) and its linear nature. The minor principal components would normally represent insignificant variance in the data for the linear case, but this cannot be said with certainty for nonlinear data. To confidently represent a nonlinear data set, more principal components have to be retained. This increases computational requirements. It is also difficult to discern which minor components capture nonlinearity and which represent insignificant variation (Dong and McAvoy 1996).

2.2.3 *Control Limits*

In classical multivariate statistical process control based on principal component analysis, the control limits required for automated process monitoring are based on the assumption that the data are normally distributed. The α upper control limit for $T^2{}_{\text{PCA}}$ is calculated from n observations based on the F-distribution, i.e.

$$\text{UCL}_{T^2{}_{\text{PCA}}} = \frac{q\,(N+1)\,(N-1)\,F_{\alpha,q,N-k}}{N\,(N-q)} \tag{2.3}$$

Then upper control limit for Q_{PCA} is calculated by means of the χ^2 distribution as

$$\text{UCL}_{Q_{\text{PCA}}} = \frac{\Lambda_1\left[1 + c_\alpha\sqrt{(2\Lambda_2\theta^2)}\,/\,\Lambda_1 + \Lambda_2\theta\,(\theta-1)\,/\,\theta\Lambda_1^2\right]}{\theta} \tag{2.4}$$

where $\Lambda_1 = \sum_{q+1}^{M} \lambda_{ji}$ (for $i = 1, 2, 3$) and $\theta = 1 - 2\Lambda_1\Lambda_3/(3\Lambda_2^2)$. c_α is the standard normal deviate corresponding to the upper $(1-\alpha)$ percentile, while m is the total number of principal components (variables). The residual Q_i is more likely to have a normal distribution than the principal component scores, since Q is a measure of the non-deterministic behaviour of the system.

The use of the deterministic variance-preserving procedure, principal component analysis (PCA), is the most widespread feature extraction fault diagnostic system applied in industrial processes (MacGregor and Kourti 1995; Kresta et al. 1991). However, the PCA assumption of stationary data with a Gaussian distribution is not always valid in chemical and other processes, partly owing to possible high sampling frequencies and time-varying behaviour (Ku et al. 1995). Apart from time-varying behaviour that will be dealt with in more detail under the rubric of dynamic process monitoring and batch processing that is inherently time variant, the following section gives a brief review of the methods dealing with non-Gaussian variables and/or nonlinear relationships between variables.

2.3 Nonlinear Steady-State (Non)Gaussian Processes

Some of the earliest approaches considered the same standard approach outlined above, with linear PCA substituted by nonlinear variants of PCA. This includes the use of autoassociative neural networks (Thissen et al. 2001), independent component analysis (Lee et al. 2004a, 2006b; Lu et al. 2008) and, more recently, kernel PCA, such as proposed by Choi et al. (2005), Jemwa and Aldrich (2006) and Khediri et al. (2011).

2.3.1 Higher-Order Statistical Methods

Several of the most prominent higher-order statistical methods proposed in recent years that are based on independent component analysis (ICA) have been proposed by Kano et al. (2003), Kano (2004), Lee et al. (2004a, b), Yoo et al. (2004), Albazzaz and Wang (2004), Wang and Shi (2010) and Zhang and Ma (2011). ICA algorithms are based on the concept of blind source separation or separation of a set of unknown mixed signals by finding unknown latent components of multivariate data (Cardoso 1998). It has seen extensively application to different signal sources, such as face recognition (Yang et al. 2004; Bartlett et al. 2002), speech signal enhancement (Ephraim and Malah 1984), image analysis (Deng and Manjunath 2001) and intrusion detection (Denning 1987).

ICA can be implemented by the use of several different algorithms, but the most popular ones are based on neural networks, higher-order statistics and minimum mutual information. The fast and robust algorithm proposed by Hyvärinen and Oja (2000) can be summarized as follows.

If M zero mean variables $\mathbf{x} = [\mathbf{x}_1, \mathbf{x}_2, \ldots, \mathbf{x}_M]$ can be expressed as a linear combination of q unknown independent components $(q < M)$, $\mathbf{s} = [\mathbf{s}_1, \mathbf{s}_2, \ldots, \mathbf{s}_q]$, then

$$\mathbf{x} = \mathbf{A}\mathbf{s}, \tag{2.5}$$

where $\mathbf{A}$ is the unknown mixing matrix, $\mathbf{A} \in \Re^{M \times q}$. The basic problem is to estimate both the mixing matrix $\mathbf{A}$ and the independent components ($\mathbf{s}$) from the measured data $\mathbf{x}$. This amounts to identification of an unmixing matrix $\mathbf{W}$ (the inverse of $\mathbf{A}$), so that the elements of the reconstructed vector $\hat{\mathbf{s}}$, given by

$$\hat{\mathbf{s}} = \mathbf{W}\mathbf{x}, \tag{2.6}$$

are as independent as possible from each other. If the independent components have a zero mean vector and are assumed to have unit variance, then the first step is to eliminate all cross-correlation between the variables. This whitening transformation can be accomplished as follows, where $\mathbf{Q}$ is the whitening matrix ($\mathbf{Q} = \mathbf{\Lambda}^{-1/2}\mathbf{U}^{\mathrm{T}}$):

$$\mathbf{z} = \mathbf{\Lambda}^{-1/2}\mathbf{U}^{\mathrm{T}}\mathbf{x} = \mathbf{Q}\mathbf{x}. \tag{2.7}$$

The orthogonal matrix of eigenvectors ($\mathbf{U}$) and the diagonal matrix of eigenvalues ($\mathbf{\Lambda}$) are obtained from the eigendecomposition of the covariance matrix $E(\mathbf{x}\mathbf{x}^{\mathrm{T}}) = \mathbf{U}\mathbf{\Lambda}\mathbf{U}^{\mathrm{T}}$. After transformation, $\mathbf{z} = \mathbf{Q}\mathbf{x} = \mathbf{Q}\mathbf{A}\mathbf{s} = \mathbf{B}\mathbf{s}$. $\mathbf{B}$ is an orthogonal matrix, since $E(\mathbf{z}\mathbf{z}^{\mathrm{T}}) = \mathbf{B}\mathrm{E}(\mathbf{s}\mathbf{s}^{\mathrm{T}})\mathbf{B}^{\mathrm{T}} = \mathbf{B}\mathbf{B}^{\mathrm{T}} = \mathbf{I}$. The independent components $\mathbf{s}$ are then estimated from $\mathbf{z}$ as follows:

$$\hat{\mathbf{s}} = \mathbf{B}^{\mathrm{T}}\mathbf{z} = \mathbf{B}^{\mathrm{T}}\mathbf{Q}\mathbf{x}, \tag{2.8}$$

and the relation between $\mathbf{W}$ and $\mathbf{B}$ is $\mathbf{W} = \mathbf{B}^{\mathrm{T}}\mathbf{Q}$.

In order to calculate $\mathbf{B}$, each column vector $\mathbf{b}$ is randomly initialized and then successively updated to ensure that the ith independent component, $\hat{\mathbf{s}}_i = (\mathbf{b}_i)^{\mathrm{T}}\mathbf{z}$, is maximally non-Gaussian. Different objective functions can be used for this purpose, including a common measure for non-Gaussianity, such as negentropy. Hyvärinen has introduced the following algorithm for fast fixed-point robust convergence approximation of negentropy:

$$\mathbf{b}_i \leftarrow E\left\{zg\left(\mathbf{b}_i^{\mathrm{T}}z\right)\right\} - E\left\{zg^*\left(\mathbf{b}_i^{\mathrm{T}}z\right)\right\}\mathbf{b}_i, \tag{2.9}$$

where g^* is the first-order derivative of g, and Hyvärinen has suggested the following functions for it:

$$g_1(u) - \tanh(a_1 u) \tag{2.10}$$

$$g_2(u) = u\exp\left(-\frac{a_{2u^2}}{2}\right) \tag{2.11}$$

$$g_3(u) = u^3. \tag{2.12}$$

ICA can decorrelate observed signals and reduce higher-order statistical dependency, and apart from the assumption of statistical independence of the components, the following assumptions and restrictions also apply (Hyvärinen 2002;

Karhunen and Joutsensalo 1994; Karhunen and Ukkonen 2007; Comon 1994; Hyvärinen and Oja 2000):

- Not more than one of the independent components can have a Gaussian distribution (the remainder should have non-Gaussian distributions).
- The source signal number should be smaller than the observed signal number.

Although ICA can be viewed as a useful extension of PCA, it has a different objective. Whereas PCA can only impose independence up to second-order statistical information (mean and variance), while constraining the direction vectors to be orthogonal, ICA has no orthogonality constraint and not only decorrelates the data (second-order statistics) but also reduces higher-order statistical dependencies (Hyvärinen and Oja 2000). Theoretically therefore, independent components reveal more useful information from observed data than principal components.

2.3.2 Nonlinear Principal Component Analysis

An obvious approach to dealing with nonlinear systems that cannot be accommodated adequately by linear multivariate methods is to use the same framework after substitution of the linear models with nonlinear models. In methods using principal component analysis, this could be accomplished by replacing linear principal component models with nonlinear models, such as principal curves and components extracted with neural networks.

Principal Curves and Surfaces

Hastie and Stuetzle (1989) have proposed principal curves as a nonlinear summary of a multivariate data set, similar in spirit to principal components. However, since this approach is non-parametric, it cannot be used directly in fault diagnostic systems, but Dong and McAvoy (1994) have shown that by making use of neural networks to represent the principal curves, the approach could be used successfully to identify fault conditions in process systems.

More formally, principal curves were defined by Hastie and Stuetzle (1989) as smooth unit speed one-dimensional manifolds in M-dimensional space $\Re^M$, satisfying the condition of self-consistency

$$f(x) = E_{\vec{Y}|g(\vec{Y})}\left\{\vec{Y}|g\left(\vec{Y}\right) = x\right\}, \forall x \in \Lambda \subseteq \Re. \tag{2.13}$$

In Eq. 2.13, E is the expectation of conditional averaging operator and $g(y)$ the projection operator given by

$$g(y) = \sup_{\lambda\in\Lambda}\left\{\lambda : \left\|\vec{Y} - f(\lambda)\right\|\right\} = \inf_{\mu\in\Lambda}\left\{\left\|\vec{Y} - f(\lambda)\right\|\right\} \tag{2.14}$$

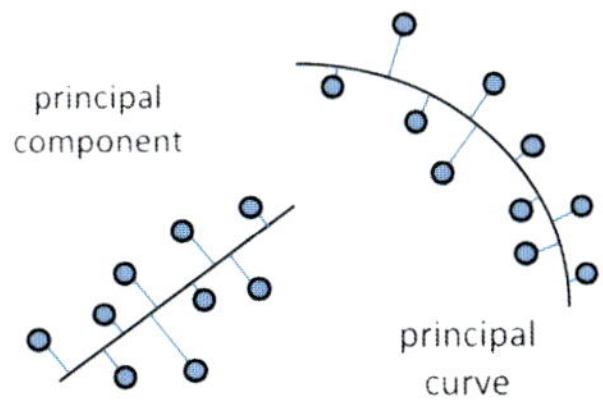

Fig. 2.5 Linear principal component (*left*) and a principal curve (*right*)

The latent variable $\boldsymbol{x}$ is usually parameterized by the arc length along $f(\boldsymbol{x})$, starting from either end, while the inf operator identifies the points on the curve f that are closest to y. The sup operator simply selects the largest coordinate among these points (Chang and Ghosh 2001). Note that theoretically, a similar approach can be followed to calculate principal (hyper)surfaces of an arbitrary dimension, but in practice this becomes intractable for large dimensional surfaces.

The principal curve approach summarizes the data with smooth curves that minimize the orthogonal deviations between the data and the curves (Fig. 2.5). If a nonlinear function can be used to represent a curve, then the function is equivalent to the principal loadings in linear PCA, while if the data are projected down onto the curve and indices are found to express the projected points, these indices can be seen as equivalent to the principal scores for linear PCA (Zhang et al. 1997).

Harkat et al. (2003) have used a similar approach with radial basis function neural networks to map the scores obtained from principal curves to the original data for use in process monitoring schemes. The NLPCA model consisted of two radial basis function neural networks. The first network, the forward model, was used to map the input data to the scores obtained with NLPCA, while the second model was used to do the reverse mapping from the scores to the input data.

Although approaches based on principal curves constitute a significant extension to linear models, principal curves themselves represent a limited class of nonlinear models in that they are based on the assumption that the nonlinear functions describing the principal curves can be approximated by linear combinations of univariate functions. This means that principal curves are restricted to the identification of nonlinear structures exhibiting additive behaviour (Jia et al. 1998).

Autoassociative Neural Networks

Autoassociative neural networks (Kramer 1991) were one of the first approaches to extend PCA-based monitoring schemes. In this sense, these neural networks are used to extract nonlinear principal components from the data based on nonlinear mappings between the original and reduced dimensional space (Antory et al. 2004, 2008), as discussed in more detail in Chap. 3. Process fault diagnostic models based on autoassociative neural networks have been described by Zhang et al. (1997) and Vedam et al. (1998). Adgar et al. (2000) have described the application of process monitoring schemes with autoassociative neural networks in surface water treatment facilities.

As discussed by Malthouse (1998), the autoassociative neural networks proposed by Kramer (1991, 1992) are unable to model curves and surfaces that intersect themselves, and cannot be used to parameterize curves with parameterizations having discontinuous jumps. Apart from this limitation, autoassociative neural networks with their multiple layers can also be difficult to train, and several authors have proposed modifications to the approach to circumvent this problem. For example, Zhang (2005) has shown that a Bayesian version of autoassociative neural networks can give more robust results than standard autoassociative neural networks.

Finally, it should be noted that autoassociative neural networks have shown promise in the compression or dimensionality reduction of hyperspectral data (Del Frate et al. 2009; Licciardi et al. 2010), as well as image data (Basso 1992), which suggest that they could also have potential use where these methods may see more extensive online application on process plants.

Input Training Neural Networks

Unlike principal curves, input training neural networks (Tan and Mavrovouniotis 1995) adjust all latent variables simultaneously and therefore represent a class of nonlinear models that can account for additive as well as multiplicative structures in data. As a consequence, input training neural networks are more efficient in capturing nonlinear behaviour than principal curves and autoassociative neural networks such as originally proposed by Kramer (1991, 1992).

Jia et al. (1998) first proposed the use of input training neural networks for multivariate statistical process control applications. Likewise, Li and Yu (2002) have used input training neural networks in conjunction with a multilayer perceptron to estimate nonlinear principal component scores, while fault diagnosis was performed by the use of statistical methods like Hotelling's T^2 and Q. The merits of the approach were demonstrated on a simulated continuous stirred tank reactor.

Shao et al. (1999) have proposed the use of input training neural networks on data denoised by wavelets, where the wavelet coefficients served as inputs to the neural networks and fault were identified by the use of control charts with non-parametric control limits. More specifically, the continuous wavelet transform can be interpreted as a correlation between the data and the time shifted and rescaled mother wavelet $\psi(x)$, as in Eq. 2.15 applied to a signal or time series data:

$$\mathrm{CWT}(a,b) = \frac{1}{\sqrt{a}} \int_{-\infty}^{\infty} f(x)\psi\left(\frac{x-b}{a}\right) \mathrm{d}x \tag{2.15}$$

In Eq. 2.15, a and b are the dilation and translation parameters, respectively, and the magnitude of the coefficient is maximized when the signal frequency matches that of the corresponding dilated wavelet. These coefficients can subsequently serve as a multiscale approximation for the data for various purposes, as, for example, discussed in a recent review by Zhu et al. (2009) in relation to tool condition monitoring. Note that since continuous wavelets are computationally inefficient, they are most frequently applied in a discrete form, typically computing the continuous wavelet at dyadic scales, $a = 2^j$ and $b = 2^j k$.

Self-Organizing and Sammon Maps

Although there is a close connection between self-organizing maps and nonlinear principal components, self-organizing maps have not been used in strictly the same way as input neural networks, principal curves or autoassociative neural networks to replace linear principal components in fault diagnostic schemes. Instead, they have been used to directly perform nonlinear mappings from high-dimensional measurement space onto a two-dimensional display which can be used to observe the system behaviour in real time (Aldrich et al. 1995a, b; Van Deventer et al. 1996; Ahola et al. 1999; Vermasvuori et al. 2002; Jämsä-Jounela et al. 2003; Frey 2008).

Aldrich (2002) has used Sammon (Sammon 1969) maps in conjunction with nonlinear models, in a similar way. The training data of the neural network were first projected onto a low-dimensional topology preserving map with the Sammon algorithm, after which the network was trained with the original data as inputs and the corresponding Sammon scores as outputs. As such, these maps were not used to diagnose faults, once detected, but this could easily be done by construction of a reverse model as well, mapping the Sammon scores back to the original data. Chemaly and Aldrich (2001) have similarly investigated the use of genetic programming models designed to construct Sammon maps of process data.

2.3.3 Monitoring Process Data Distributions

Dissimilarity Indices

An alternative approach to dealing with nonnormal process data is to monitor the distributions of the variables, instead of the variables themselves. In this spirit, statistical process monitoring based on the dissimilarity of the process data has been proposed and investigated by Kano et al. (2000, 2002) and Jian and Xie (2008). The algorithm can be summarized by considering two J-variate data sets $\mathbf{X}_1$ and $\mathbf{X}_2$ consisting of N_1 and N_2 samples, respectively. One of the data sets serves as a reference and normalized to zero mean and unit variance, while the other data set is scaled by the normalization parameters of the reference data. The covariance of the two data sets ($\mathbf{R}$) is given by Eq. 2.5:

$$\mathbf{R} = \frac{1}{N_1 + N_2}\begin{bmatrix}\mathbf{X}_1 \\ \mathbf{X}_2\end{bmatrix}^{\mathrm{T}}\begin{bmatrix}\mathbf{X}_1 \\ \mathbf{X}_2\end{bmatrix} = \frac{N_1}{N_1 + N_2}\mathbf{R}_1 + \frac{N_2}{N_1 + N_2}\mathbf{R}_2 \quad (2.16)$$

where

$$\mathbf{R}_i = \frac{1}{N_i}\mathbf{X}_i^{\mathrm{T}}\mathbf{X}_i \tag{2.17}$$

By using an eigenvalue decomposition, **R** can be diagonalized by an orthogonal matrix $\mathbf{P}_0$, where the elements of the diagonal matrix $\mathbf{\Lambda}$ are the eigenvalues of the covariance matrix **R**:

$$\mathbf{P}_0{}^{\mathrm{T}}\mathbf{R}\mathbf{P}_0 = \mathbf{\Lambda} \tag{2.18}$$

The original data matrices $\mathbf{X}_i$ can subsequently be transformed into $\mathbf{Y}_i$ as follows:

$$\mathbf{Y}_i = \sqrt{\frac{N_i}{N_1 + N_2}}\mathbf{X}_i\mathbf{P}_0\Lambda^{-1/2} = \sqrt{\frac{N_i}{N_1 + N_2}}\mathbf{X}_i\mathbf{P} \tag{2.19}$$

P is a transformation matrix defined by

$$\mathbf{P} = \mathbf{P}_0\Lambda^{-1/2}. \tag{2.20}$$

By calculating the covariance matrices $\mathbf{S}_i$ of $\mathbf{Y}_i$,

$$\mathbf{S}_i = \frac{1}{N_i}\mathbf{Y}_i{}^{\mathrm{T}}\mathbf{Y}_i, \tag{2.21}$$

the dissimilarity index D of the data sets can consequently be defined as follows:

$$D = \mathrm{dis}(\mathbf{X}_1, \mathbf{X}_2) = \frac{4}{J}\sum_{j=1}^{J}\left(\lambda_j - 0.5\right)^2. \tag{2.22}$$

In Eq. 2.22, λ_j denotes the eigenvalues of $\mathbf{S}_i$.

The index D can vary between 0 and 1, with the latter indicating very dissimilar data. For more detail on the algorithms, Kano et al. (2002) can be consulted. Although based on a different monitoring index, the dissimilarity algorithm is mathematically equivalent to PCA in that both project the data to extract and use the underlying features of the data.

Analysis of Statistical Patterns

A second approach as proposed by Wang and He (2010) is to generate diagnostic variables consisting of various statistics of the variables and monitoring these, instead of the variables themselves. These statistics calculated by means of a moving window sliding along the time axis of the variables, as indicated in Fig. 2.6, included

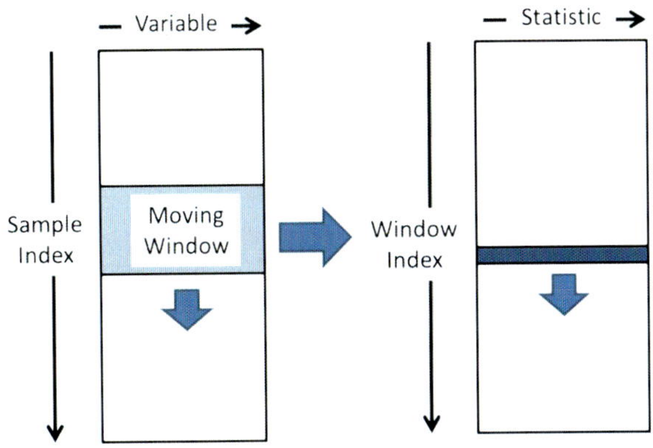

Fig. 2.6 Using a moving window (*left*) to generate statistics (*right*) from variables

the mean (first order), covariance, correlation, autocorrelation and cross-correlation (second order), as well as skewness, kurtosis and other higher-order moments.

Wang and He (2010) have scaled and normalized the new statistics pattern variables to zero mean and unit variance, prior to singular value decomposition and using the Hotelling's T^2- and Q-statistics to monitor the process. Since these variables are not generally normally distributed, control limits were determined empirically.

Data Distribution Models

A second approach as proposed by Wang and He (2010) is to generate diagnostic variables consisting of various statistics of the variables and monitoring these, instead of the variables themselves. These statistics calculated by means of a moving window sliding along the time axis of the variables, as indicated in Fig. 2.6, included the mean (first order), covariance, correlation, autocorrelation and cross-correlation (second order), as well as skewness, kurtosis and other higher-order moments.

2.3.4 *Kernel Methods*

Kernel methods have recently seen strong growth as a general framework towards nonlinear multivariate statistical analysis. They have therefore also featured prominently in more advanced process monitoring schemes designed to deal with nonlinear systems.

Kernel Principal Component Analysis (KPCA) and Kernel Partial Least Squares (KPLS)

The application of kernel versions of principal component analysis was considered by several authors (Cho et al. 2004; Stefatos and Ben Hamza 2007; Zhang 2009;

Xu et al. 2009, 2010; Alcala and Qin 2010; Xu and Hu 2010), while kernel partial least squares was investigated by Zhang et al. (2010). Although capable of capturing nonlinear relationships between variables with a high degree of accuracy, kernel principal component analysis (KPCA) suffers from a high computational cost, since it requires solution of the eigenvalue decomposition problem involving a high-dimensional kernel matrix.

Xu and Hu (2010) have considered the use of multiple kernel learning support vector machines for fault diagnosis. This was accomplished by first analysing the data with KPCA and calculating the T^2- and Q-statistics in the kernel feature space. If these statistics exceeded predefined control limits to indicate the possible presence of a fault condition, the nonlinear score vectors were further processed by a multiple kernel learning support vector machine. Simulation studies with the Tennessee Eastman problem suggested effective identification of various fault sources as well as improved speed of fault diagnosis.

In order to surmount the problem of overlearning sometimes occurring when kernel methods are used, Gan et al. (2010) have investigated the use of sparse kernel principal angles for online process monitoring. The basic idea is to first map the input space into a Hilbert feature space and then to compute the approximate basis of the feature space. Simulations on the Tennessee Eastman process indicated that this approach could be used to effectively capture nonlinear relationships in the process variables, but in a significantly simpler way than with kernel PCA.

Kernel Dissimilarity Indices (k-DISSIM)

Another variant of kernel methods, viz. kernel dissimilarity analysis, was proposed by Zhao et al. (2009) as a nonlinear extension of the dissimilarity algorithm (DISSIM). With the k-DISSIM algorithm, to facilitate analysis, the input space is first mapped into a high-dimensional kernel feature space, where the data distributions become approximately linear. The kernel-based dissimilarity analysis algorithm requires simple solution of the eigenvalue problem and avoids the demand on the specific mapping function and nonlinear optimization procedure.

Kernel Independent Component Analysis (KICA)

Diagnostic models based on kernel independent component analysis were investigated by Zhang et al. (2006), Xing et al. (2006), Zhang and Qin (2007), Zhang (2009) and Wang and Shi (2010).

Maximum Variance Unfolding

With kernel PCA, kernel functions are selected empirically, and in some cases, this may have an adverse effect on the performance of the diagnostic method, if the kernels are not suited to the structure of the data. To circumvent this problem, Shao et al. (2009) and Shao and Rong (2009) have proposed the use of a kernel function learning method to fit the kernel function to the data. Motivated by maximum variance unfolding methods (Weinberger et al. 2004), this was accomplished by optimizing the kernel function over a family of data-dependent kernels such that the data are unfolded in the kernel feature space to approximate a linear structure. MVU can be seen as a special variation of kernel PCA, whose kernel matrix is automatically learnt so that the underlying manifold structure of the training data is unfolded in the reduced space while preserving the boundary of the distribution region of the data. Simulation with the Tennessee Eastman problem suggested that this promoted more precise monitoring in the feature space.

2.3.5 *Multiscale and Multimodal Methods*

With multiscale PCA, wavelets are used to decompose the process variables under scrutiny into multiple scale representations before application of PCA to detect and identify faulty conditions in process operations. In this way autocorrelation of variables is implicitly accounted for, resulting in a more sensitive method for detecting process anomalies. Multiscale PCA constitutes a promising extension of multivariate statistical process control methods, and several authors have subsequently reported successful applications thereof, e.g. Fourie and de Vaal (2000), Rosen and Lennox (2001), Yoon and MacGregor (2004) and Lee et al. (2005).

Choi et al. (2008) have proposed nonlinear multiscale multivariate monitoring of dynamic processes based on kernel PCA. The kernel PCA model is built with reconstructed data obtained from wavelet and inverse wavelet transforms. Moreover, variable contributions to monitoring statistics were derived by calculating the derivative of the monitoring statistics with respect to the variables. Xuemin and Xiaogang (2008) have proposed an integrated multiscale approach where kernel PCA is used on measured process signals decomposed with wavelets and have also proposed a similarity factor to identify fault patterns.

Žvokelj et al. (2011) have proposed a nonlinear multivariate and multiscale statistical process monitoring and signal denoising method that combines the strengths of kernel PCA monitoring with those of ensemble empirical mode decomposition to handle multiscale system dynamics or EEMD-based multiscale KPCA (EEMD-MSKPCA). Tests on simulated and actual vibration and acoustic emission signals measured on a custom-built large-size low-speed bearing test stand indicated that the proposed EEMD-MSKPCA method provides a promising tool for dealing with nonlinear multiscale data representing a convolution of multiple events occupying different regions in the time–frequency plane. Likewise, Tianyang et al. (2011) have proposed the use of EMD to monitor flame flicker in furnaces. The flames were measured by a photo diode over the spectral band between 400 and 1,100 nm.

Yao and Gao (2007) and Ge et al. (2011a) have proposed the application of two-dimensional Bayesian methods to deal with processes in modern industries that are nonlinear and multimodal. This is based on a decomposition of the nonlinear process into different linear subspaces by means of a two-step variable selection procedure. Bayesian inferencing is subsequently used to recombine the results from the different subspaces.

Fourie and De Vaal (2000) have proposed the use of a nonlinear multiscale principal component analysis methodology for process monitoring and fault detection based on multilevel wavelet decomposition and nonlinear principal component analysis via input training neural networks. In this case, wavelets were first used to decompose the data into different scales, after which PCA was applied to the reconstituted time series data.

2.3.6 Data Density Models

A serious shortcoming of linear Gaussian models when applied to non-Gaussian data is that related to the derivation of confidence limits used to discriminate between normal and abnormal process conditions. For example, if the data distribution resembles that of three non-aligned clusters, then the assumption of a normal distribution would result in a confidence limit in the form of a large ellipse containing the data. Fault conditions associated with data in between the clusters would simply not be visible. This problem has been addressed by data density models not dependent on the assumption of Gaussian distributions, including the use of Gaussian or other mixture models (Choi et al. 2004; Yu 2012), one-class support vector machines or support vector domain descriptions (Jemwa and Aldrich 2006; Mahadevan and Shah 2009; Ge and Song 2011; Liu et al. 2011), and less frequently bag plots (Aldrich et al. 2004; Gardner et al. 2005) and convex hulls (Aldrich and Reuter 1999).

2.3.7 Other

Other methods recently considered include the use of random forests (as discussed in more detail in Chap. 6), Sammon maps (Wang 2008) and methods taking advantage of recent innovations in neural networks, such as restricted Boltzmann machines and deep learning architectures.

2.4 Continuous Dynamic Process Monitoring

Continuous dynamic systems exhibit stable dynamic behaviour or (quasi)periodic behaviour between an upper and a lower limit or more generally a bounded region in the variable space. As a result, methods designed to deal with steady-state systems

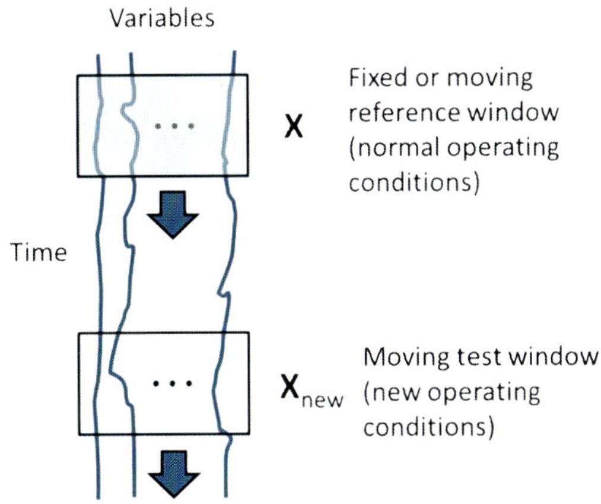

Fig. 2.7 Monitoring of continuous dynamic systems based on the generation of an augmented data matrix obtained by sliding a window along the time series variables

that do not account for this behaviour may not be able to detect changes in the process system. The problem can be dealt with in different ways, all of which comprise the construction of an augmented data matrix that contains lagged copies of the variables in the original data matrix. This can be viewed as sliding a window down the time series as indicated in Fig. 2.7. The window containing the normal operating data can be fixed or moving, while the window generating the test data captures the data to be evaluated.

More specifically, the dynamic behaviour of a process system represented by a set of time series data comprised of m variables and n observations, $\mathbf{X} = [\mathbf{x}_1, \mathbf{x}_2, \ldots, \mathbf{x}_m] \in \Re^{n \times m}$, can be accounted for by expanding the set of variables to include lagged copies of each variable, i.e. $\mathbf{X}^E = [\mathbf{x}_1(t), \mathbf{x}_1(t-k), \mathbf{x}_1(t-2k), \ldots, \mathbf{x}_1(t-(M-1)k), \mathbf{x}_2(t), \mathbf{x}_2(t-k), \mathbf{x}_2(t-2k), \ldots, \mathbf{x}_2(t-(M-1)k), \ldots, \mathbf{x}_m(t), \mathbf{x}_m(t-k), \mathbf{x}_m(t-2k), \ldots, \mathbf{x}_m(t-(M-1)k)]$. The expansion is defined by two parameters, namely, a lag k and an embedding dimension M, and optimal selection of these parameters can be done by various means (refs), depending on the purpose of the analysis to yield a sliding window of length $k \times M$ and width m (the number of variables in the original data matrix). Ideally, the embedding dimension should be sufficiently large to accommodate the largest periodicities of concern in the data. In principle, this (often substantially) expanded or aggregate lagged trajectory accounts for all the auto- and cross-correlational relationships between the variables. Moreover, in some cases denoising of the data before embedding, e.g. by use of wavelets, as proposed by Li et al. (2003), may yield more reliable estimates of embedding parameters.

2.4.1 *Determination of the Lag Parameter k*

Depending on the specific methodology, the lag parameter should be sufficiently large to ensure that the new coordinates of the system are as independent as possible, but not so large that information is lost unnecessarily. The autocorrelation function of the time series is often used as a guide in this respect, i.e. the minimum lag k at which the autocorrelation function (ACF) between the lagged variables $x(t)$ and $x(t+k)$ becomes statistically insignificant:

$$\min_k (\text{ACF}) = \sum_{k=1}^{K} \{(x(t) - \bar{x})(x(t+k) - \bar{x})\}. \tag{2.23}$$

Alternatively, mutual information is seen as a more reliable measure of independence between lag variables, and it is used in a similar way. To determine the ideal time lag, the average mutual information (AMI) between the parameter in question, $x(t)$, and a lagged version of itself, $x(t+k)$, is considered. The ideal time lag is selected to correspond to the minimum sample lag, k, at which the average mutual information reaches a minimum:

$$\min_k (\text{AMI}) = \sum_{k=1}^{K} P(x(t), x(t+k)) \log \left(\frac{P(x(t), x(t+k))}{P(x(t))\, P(x(t+k))} \right). \tag{2.24}$$

When embedding is done by the use of singular value decomposition, selection of an optimal lag value becomes less important and a trivial embedding of $k = 1$ is used, unless the sampling rate is very high, in which case higher values of can be used to reduce the number of variables in the lagged trajectory matrix.

2.4.2 Determination of the Dimension Parameter *M*

The dimensionality of the embedding should be selected such that periodic or quasiperiodic behaviour of the time series is captured in full. This can be done by the use of the false nearest neighbour algorithm, or alternatively, if singular value decomposition is used, on any of a number of methods that are normally used in the construction of reduced principal component models.

2.4.3 Multivariate Embedding

Cao et al. (1998) have proposed the embedding of all components of multidimensional observations by using an optimal Takens embedding for each component (time series). The optimal value of the embedding dimension for each component was found by minimizing the prediction error of a nearest neighbour, locally constant predictor. With this approach, each component of an observation space, $\mathbf{y} \in \Re^M$, is treated as a one-dimensional time series. Each component is embedded individually to generate a set of subspaces $[\mho_1, \mho_2, \ldots, \mho_M]$, which are then combined to form a first approximation of the attractor in $\Re^{\Lambda} = \Re^{\Pi_j^M \Lambda_j}$. Finally, the lag variables in the combined subspaces Λ are separated to reveal the structure of the attractor.

2.4.4 Recursive Methods

The use of recursive algorithms to update static models is an alternative to the dynamic modelling approaches outlined above. With these methods, the parameters of the static models are estimated incrementally, as described in more detail by Qin (1998) and others. For example, Li et al. (2000) have proposed the use of recursive principal component analysis for adaptive process monitoring, where the number of principal components is reestimated at each step, as well as the confidence limits for the Q- and T^2-statistics used with these models. Similar approaches have been considered for partial least-squares algorithms (Qin 1998; Mu et al. 2006) and kernel principal component models (Zhang et al. 2012).

2.4.5 State Space Models

If it is assumed that matrix $\mathbf{X}^{\mathrm{E}}$ contains all the dynamic information of the system, then the use of predictive models can be viewed as an attempt to remove all the dynamic information from the system to yield Gaussian residuals that can be monitored in the normal way, as indicated in Fig. 2.7.

State space models offer a principled approach to identification of the subspaces containing the data. This can be summarized as follows:

$$\mathbf{x}_{k+1} = f(\mathbf{x}_k) + \mathbf{w}_k \tag{2.25}$$

$$\mathbf{y}_k = g(\mathbf{x}_k) + \mathbf{v}_k \tag{2.26}$$

$\mathbf{x}$ and $\mathbf{y}$ are the respective state and measurement vectors of the system, and $\mathbf{w}_k$ and $\mathbf{v}_k$ are the plant disturbances and measurement errors, respectively. The plant can be approximated by a linear state space model, where $\mathbf{A}$ and $\mathbf{C}$ are unknown state and output matrices, respectively, and $\boldsymbol{\varepsilon}$ and $\boldsymbol{\eta}$ are noise vectors typically derived from independent identical Gaussian distributions:

$$\mathbf{x}_{k+1} = \mathbf{A}\mathbf{x}_k + \boldsymbol{\varepsilon}_k \tag{2.27}$$

$$\mathbf{y}_k = \mathbf{C}\mathbf{x}_k + \boldsymbol{\eta}_k \tag{2.28}$$

Without knowledge of the matrices **A** and **C**, Odiowei and Cao (2010) have used canonical variate analysis (CVA) to extract the state variables $\mathbf{x}_k$ from the process measurements, $\mathbf{y}_k$. The process measurements are stacked in the form of past ($\mathbf{y}_{p,k}$) and future spaces ($\mathbf{y}_{f,k}$), defined as follows:

$$\mathbf{y}_{p,k} = \begin{bmatrix} y_{k-1} \\ y_{k-2} \\ \vdots \\ y_{k-q} \end{bmatrix} \in \Re^{mq} \tag{2.29}$$

$$\tilde{\mathbf{y}}_{p,k} = \mathbf{y}_{p,k} - \bar{\mathbf{y}}_{p,k} \tag{2.30}$$

$$\mathbf{y}_{f,k} = \begin{bmatrix} y_{k} \\ y_{k+1} \\ \vdots \\ y_{k+q} \end{bmatrix} \in \Re^{mq} \tag{2.31}$$

$$\tilde{\mathbf{y}}_{f,k} = \mathbf{y}_{f,k} - \bar{\mathbf{y}}_{f,k} \tag{2.32}$$

$\tilde{\mathbf{y}}_{p,k}$ and $\tilde{\mathbf{y}}_{f,k}$ represent the scaled (zero mean) measurements, after subtraction of the respective sample means $\bar{\mathbf{y}}_{p,k}$ and $\bar{\mathbf{y}}_{f,k}$. With CVA, the best linear combinations of the past ($\mathbf{a}^{\mathrm{T}}\tilde{\mathbf{y}}_{p,k}$) and future values ($\mathbf{b}^{\mathrm{T}}\tilde{\mathbf{y}}_{f,k}$) are sought, so that the correlation between the combinations is maximized, i.e.

$$r(\mathbf{a}, \mathbf{b}) = \frac{\mathbf{a}^{\mathrm{T}} \boldsymbol{\Sigma}_{\mathrm{fp}} \mathbf{b}}{\sqrt{\left(\mathbf{a}^{\mathrm{T}} \boldsymbol{\Sigma}_{\mathrm{ff}} \mathbf{a}\right)\left(\mathbf{b}^{\mathrm{T}} \boldsymbol{\Sigma}_{\mathrm{pp}} \mathbf{b}\right)}} \tag{2.33}$$

where $\boldsymbol{\Sigma}_{\mathrm{ff}} = E\left(\tilde{\mathbf{y}}_{p.k}\tilde{\mathbf{y}}_{p.k}{}^{\mathrm{T}}\right)$, $\boldsymbol{\Sigma}_{\mathrm{ff}} = E\left(\tilde{\mathbf{y}}_{f.k}\tilde{\mathbf{y}}_{f.k}{}^{\mathrm{T}}\right)$ and $\boldsymbol{\Sigma}_{\mathrm{fp}} = E\left(\tilde{\mathbf{y}}_{f.k}\tilde{\mathbf{y}}_{p.k}{}^{\mathrm{T}}\right)$ are the covariance matrices Hankel matrices arising from the expanded scaled measurement vectors:

$$\boldsymbol{\Sigma}_{\mathrm{ff}} = E\left(\tilde{\mathbf{y}}_{p.k}\tilde{\mathbf{y}}_{p.k}{}^{\mathrm{T}}\right) = \frac{\mathbf{a}^{\mathrm{T}} \boldsymbol{\Sigma}_{\mathrm{fp}} \mathbf{b}}{\sqrt{\left(\mathbf{a}^{\mathrm{T}} \boldsymbol{\Sigma}_{\mathrm{ff}} \mathbf{a}\right)\left(\mathbf{b}^{\mathrm{T}} \boldsymbol{\Sigma}_{\mathrm{pp}} \mathbf{b}\right)}} \tag{2.34}$$

The solution can be found by singular value decomposition of the Hankel matrices, i.e.

$$\mathbf{z}_k = \begin{bmatrix} \mathbf{b}_1^{\mathrm{T}} \\ \mathbf{b}_2^{\mathrm{T}} \\ \vdots \\ \mathbf{b}_{mq}^{\mathrm{T}} \end{bmatrix} \tilde{\mathbf{y}}_{p,k} = \mathbf{J}\tilde{\mathbf{y}}_{p,k}, \tag{2.35}$$

where **J** is a transformation matrix transforming mq dimensional past measurements into an mq dimensional canonical variate space. This space can be separated into an nth order state and residual spaces, $\mathbf{z}_k = [\mathbf{x}_k \mathbf{e}_k]$, i.e. the canonical variate space is spanned by the state variable and residual variable space. Odiowei and Cao (2009b) reported that these state variables obtained by means of CVA provide a better basis for process monitoring than that based on direct monitoring of the observations of the dynamic system.

State space modelling with CVA has been considered by several authors, e.g. Simoglou et al. (2001, 2002, 2005), Lee et al. (2006a) and Russel et al. (2000b). Other state space approaches have extended these linear models in various ways. Odiowei and Cao (2010) have constructed a linearized state space model of dynamic processes from normal operating condition data and used independent component analysis to detect faulty behaviour in the plant. Negiz and Cinar (1997) used a vector autoregressive moving average model (VARMA) with parameters estimated by canonical variate analysis on a milk pasteurization process. The residuals of this model were monitored based on T^2- and Q-statistics derived with PCA.

Kruger et al. (2004) claimed that the integration of ARMA filters into multivariate statistical process control framework improved the monitoring of large-scale industrial processes by removing autocorrelation in the variables. Dong et al. (2010) have proposed a method based on empirical mode decomposition and a vector autoregressive moving average (VARMA) model for the detection of structural damage. A damage index was defined based on the model coefficients, which could be used to detect changes in signals.

Likewise, neural networks have been used extensively in novelty detection (Markou and Singh 2003). Of these, multilayer perceptrons are the most widely used to generate residuals that can be monitored by the use of simple thresholding and using the training data of the neural network as reference (Ryan et al. 1998; Augusteijn and Folkert 2002).

Stubbs et al. (2009) have proposed a simplified state space model for process monitoring based on a modified definition of the past vector of inputs and outputs to estimate a reduced set of state space matrices. Odiowei and Cao (2009a) have investigated canonical variate analysis (CVA) with control limits derived through kernel density estimation. To improve the performance of nonlinear dynamic process monitoring, Odiowei and Cao (2010) have proposed a state space approach based on independent component analysis (SSICA). Unlike conventional ICA, the algorithm uses canonical variate analysis to construct a state space, from which statistically independent components can be extracted for process monitoring. SSICA could detect faults earlier than other methods, such as the DICA and the CVA, in the simulated Tennessee Eastman process plant case study. Karoui et al. (2010) have developed the use of rectangular hybrid automata for dynamic process monitoring that takes the behaviour of the system and the evolution of its parameters into account. Khediri et al. (2010) have proposed the use of support vector regression to construct several control charts to monitor multivariate nonlinear autocorrelated processes.

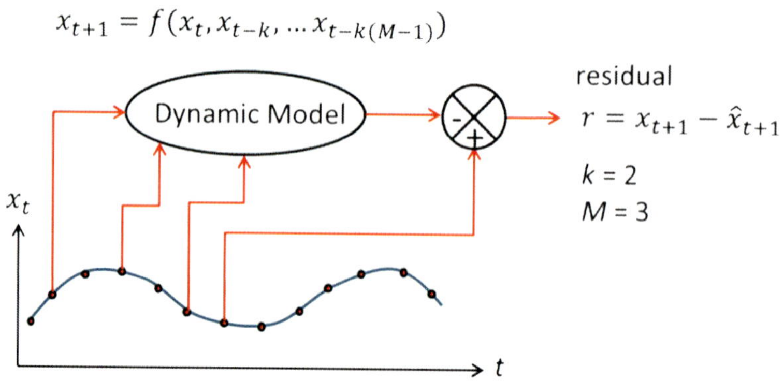

Fig. 2.8 The use of dynamic models to generate residuals for process monitoring

2.4.6 *Subspace Modelling*

The lagged or embedded variables in the windows can subsequently be used as an expanded system of variables amenable to all the analytical methods available for steady-state systems. As indicated in Fig. 2.7, with dynamic process monitoring, it is possible to use a fixed or a relative reference. In fixed reference systems, $\mathbf{X}^{\mathrm{E}}$ represents a fixed window on the behaviour of the system, while in relative reference systems, $\mathbf{X}^{\mathrm{E}}$ represents a moving window on the behaviour of the system. With the moving window approach, the embedding parameters k and M usually need to be reestimated, making it a computationally expensive process. Apart from these parameters, the size of the test window and the distance between the reference window and a test window may also need to be defined, as discussed in more detail in Chap. 7.

Predictive Model Residuals

With predictive models, the evolving behaviour of the system is predicted by some model based on the present and past, i.e. $\mathbf{x}(t+k) = f_M[\mathbf{x}(t), \mathbf{x}(t-k), \dots, \mathbf{x}(t-k(M-1))]$. If the model is fitted to data representative of normal operating conditions, then the model residuals are expected to change when the process starts to develop abnormal behaviour. The basic idea behind these methods is to capture and remove the process dynamics reflected by the data in order to enable the application of steady-state approaches to process monitoring and fault diagnosis, as indicated in Fig. 2.8.

Hill and Minsker (2010) have proposed an anomaly detection scheme based on the use of a univariate autoregressive model to predict environmental sensor data one time step ahead. In this approach, a moving window is used as input to the model. Four models were investigated, namely, a naïve predictor, a nearest cluster model, a single-layer linear network as well as a multilayer perceptron. Of these, the neural network yielded the best results.

Chen and Liao (2002) have integrated neural network (NN) and principal component analysis (PCA) for process monitoring. NN was used to represent operating process information with a nonlinear dynamic model, while PCA was used to generate simple monitoring charts for the residuals derived from the difference between process measurements and neural network predictions.

Subspace Models

Projection of the data to a lower-dimensional subspace has been investigated by various authors. Ku et al. (1995) have used PCA to accomplish this, generating T^2- and Q-statistics to monitor the data. Owing to their ability to extract features from data via unsupervised learning, autoassociative neural networks have been used as a nonlinear equivalent of principal component analysis in process monitoring schemes by a number of authors, including Marseguerra and Zoia (2005), Lopes and Menezes (2004) and Shimizu et al. (1997). Kano et al. (2001) have made use of a method they have referred to as moving principal component analysis, in which changes in the directions of the principal components are monitored in response to new data.

Observing that industrial process variables may consist of a mixture of approximately Gaussian and non-Gaussian variables, Zhang (2009) has considered a nonlinear dynamic approach combining the use of KPCA and KICA. In contrast, Zhang and Qin (2007) have considered the same idea but evaluated an improved version of KPCA and KICA, which compensated for some of the drawbacks of these methods, such as the redundancy of data mapped into the feature space and increased computational cost.

Choi and Lee (2004) have proposed nonlinear dynamic process monitoring by the use of dynamic kernel principal component analysis (DKPCA). DKPCA enabled the authors to monitor an arbitrary process with severe nonlinearity and (or) dynamics. In comparison with PCA, dynamic PCA and KPCA in terms of type I error rate, type II error rate and detection delay, the proposed methodology yielded the best performance with regard to low missing alarms and small detection delay.

Cheng and Chiu (2005) have proposed monitoring of nonlinear static or dynamic systems by the use of just-in-time learning and principal component analysis. In the JITL–PCA framework, JITL is used as the process model to remove nonlinear or dynamic information from the raw process data, while the resulting residuals between the actual process outputs and JITL's predicted outputs are used in the PCA analysis to draw the monitoring charts. Simulation results indicated that the JITL–PCA approach could outperform both PCA and dynamic PCA in the monitoring of nonlinear static or dynamic systems.

Xie et al. (2006) have investigated the influence of auto- and cross-correlation on statistical process control with Monte Carlo experiments. They have concluded that dynamic PCA and ARMA–PCA are inefficient in removing the influence of auto- and cross-correlation, unlike subspace identification-based PCA (SI-PCA). Li and Rong (2006) have addressed the weakness of principal component analysis in fault isolation by the use of partial dynamic principal component analysis (PDPCA) to obtain structured residuals and enhance the ability of dynamic process monitoring to isolate faults. Simulations on continuous stirred tank reactor have demonstrated the effectiveness of the proposed method.

Stefatos and Ben Hamza (2010) have introduced a diagnostic method using a dynamic independent component analysis approach capable of accurately detecting and isolating the root causes of individual faults. Hsu et al. (2010) have proposed a process monitoring scheme to compensate for the shortcomings of conventional ICA-based monitoring methods. This was done by first augmenting the observed data matrix to take the process dynamics into consideration.

Maximum Variance Unfolding

Lieftucht et al. (2009) have analysed subspace monitoring to identify and isolate fault conditions, and based on the assumption of a deterministic fault signature, while the monitored variables are stochastic, they have introduced a regression-based reconstruction technique. The method was demonstrated on a simulation example and the analysis of experimental data from an industrial reactive distillation unit. Since kernel PCA incurs a high computational cost owing to its dense expansion in terms of kernel functions in the online monitoring stage, Shao and Rong (2009) have proposed process monitoring based on maximum variance unfolding projections (MVUP). MVUP applies maximum variance unfolding (MVU) on training samples and can be seen as a special variation of kernel PCA, where the kernel matrix is automatically learnt such that the underlying manifold structure of training samples is unfolded in the reduced space, leading to preservation of the boundary of the distribution region of the training samples. This is followed by MVUP using linear regression to find the projection that best approximates the implicit mapping from training samples to their lower-dimensional MVU embedding. Simulation results on the benchmark Tennessee Eastman process indicated that MVUP-based process monitoring method is a good alternative to kernel PCA monitoring.

2.4.7 *Data Density Models*

Process monitoring can also be accomplished by identifying the boundaries of the data associated with normal operating conditions and monitoring the location of new data with respect to these boundaries, as discussed in Sect. 2.3.3 for steady-state systems. Support vector domain description (SVDD), as applied by Bovolo et al. (2010), has been used to detect changes in image data on this basis. With SVDD, the change detection problem was formulated as a minimum enclosing ball problem (MEB) with changed pixels (images) as target objects. The MEB problem was solved in a high-dimensional Hilbert space, and once the minimum volume hypersphere was computed, it was mapped back into the original feature space to give a nonlinear flexible boundary around the target objects (Tax and Duin 1999, 2004).

Schölkopf et al. (2001) have proposed an alternative to the approach used by the previous authors. Instead of trying to identify a hypersphere with minimal radius to fit the data, they have tried to separate the surface region containing data from empty regions. This is done by constructing a hyperplane maximally distant from the origin of the data and all the data located on the opposite side of the origin, so that the margin is positive. The disadvantage of this approach is that the origin plays a crucial role in that effectively acts as a prior for where the class of anomalies is assumed to be located.

Using similar ideas, Cui et al. (2008) have improved kernel principal component analysis for fault detection by using a feature vector selection scheme based on geometrical considerations to reduce the computational complexity of KPCA when the number of samples becomes large. Secondly, KPCA was used in conjunction with Fisher discriminant analysis to improve the performance of KPCA.

2.4.8 Chaos-Theoretical Approaches

When the columns of $\mathbf{X}^{\mathrm{E}} \in \Re^{n-k(M-1)\times mk(M-1)}$ are orthogonal, they represent the coordinates of a phase space. If not, $\mathbf{X}^{\mathrm{E}}$ can usually be orthogonalized by eigenvector decomposition. The scores of the phase space variables represent an orbit or attractor with some geometrical structure, depending on the frequencies with which different regions of the phase space are visited.

The topology of this attractor is a direct result of the underlying dynamics of the system being observed, and changes in the topology are usually an indication of a change in the parameters or structure of the system dynamics. Therefore, descriptors of the attractor geometry can serve as sensitive diagnostic variables to monitor abnormal system behaviour. The use of chaos-theoretical approaches in the analysis of corrosion based on electrochemical noise measurements has included descriptors such as the correlation dimension (Legat and Dolecek 1995; Aldrich et al. 2006; Xia et al. 2012), recurrence plots (Cazares-Ibáñez et al. 2005; Acuña-González et al. 2008), Lyapunov exponents and information entropy (Legat and Dolecek 1995). Applications of chaos-theoretical approaches in other areas include the analysis of biosignals (Übeyli and Güler 2004), arc welding (He et al. 2013) as well as structural health monitoring (Casciati and Casciati 2006).

2.4.9 Other

Lopez and Sarigul-Klijn (2009) have proposed a structural health monitoring scheme based on distance similarity matrices of dimensionally reduced data. They have considered an ensemble method of dimension reduction, which performed better than single-feature extraction methods.

Guh and Shiue (2008) have used decision trees to detect the mean shifts in multivariate control charts. Experimental results using simulation showed that the proposed model could both detect mean shifts and identify the variables that have deviated from their original means. The feature of fast learning makes the proposed DT learning-based model more adaptable to a dynamic process monitoring scenario, in which constant model redesigning and relearning is required.

Alabi et al. (2005) have described an extension to previous research into dynamic process performance monitoring based on the use of a generic dissimilarity measure and wavelets that are used to decorrelate autocorrelated process data. A comparative study with three other methods, standard PCA, dynamic PCA and multiscale PCA, has indicated that the proposed approach could yield better results in terms of false alarm rate and the time to fault detection.

More recently, Auret and Aldrich (2010) have considered the application of random forests to change point detection problems. This approach was based on the embedding of multivariate time series data associated with normal process conditions, followed by the extraction of features from the resulting lagged trajectory matrix by recasting the data into a binary classification problem, as discussed in more detail in Chap. 5. The classification problem can be solved with a random forest model, from which a proximity matrix can be calculated. Features extracted from this matrix representing the trajectory of the system in phase space are then used as a basis for monitoring the system.

2.5 Batch Process Monitoring

Construction of batch process models from first principles is difficult, owing to the high dimensionality and complexity of the processes involved, as well as the short product-to-market times often required in contemporary industrial environments. As a consequence, data-driven approaches, such as multiway principal component analysis (MPCA) and multiway partial least squares (MPLS), which extend the application of PCA and PLS from continuous processes to batch processes, have been investigated intensely over the last few decades (Nomikos and MacGregor 1995a, b). With multiway methods, process trajectories can be projected to lower-dimensional subspaces to greatly expedite the analysis and prediction of the performance of the process.

Batch data are typically represented by three-way matrices (I batches $\times$ J variables $\times$ K records) which are subsequently unfolded and rearranged into two-dimensional arrays (Nomikos and MacGregor 1994, 1995a, b; Van Sprang et al. 2002). Mean centring and unit variance scaling are performed to remove most of the possible nonlinearities of the batch trajectories.

3D arrays can be unfolded in three different ways, but only two of these are sensible. The first direction is in the batch dimension, i.e. construction of a matrix with dimensions $I \times JK$, as indicated in the top left in Fig. 2.9. The second direction is in the variable dimension, yielding a matrix with dimensions $IK \times J$, as indicated in the top right in Fig. 2.9. Other unfolding configurations are also possible if lagged measurement vectors are incorporated, as indicated in the bottom left and bottom right of Fig. 2.9.

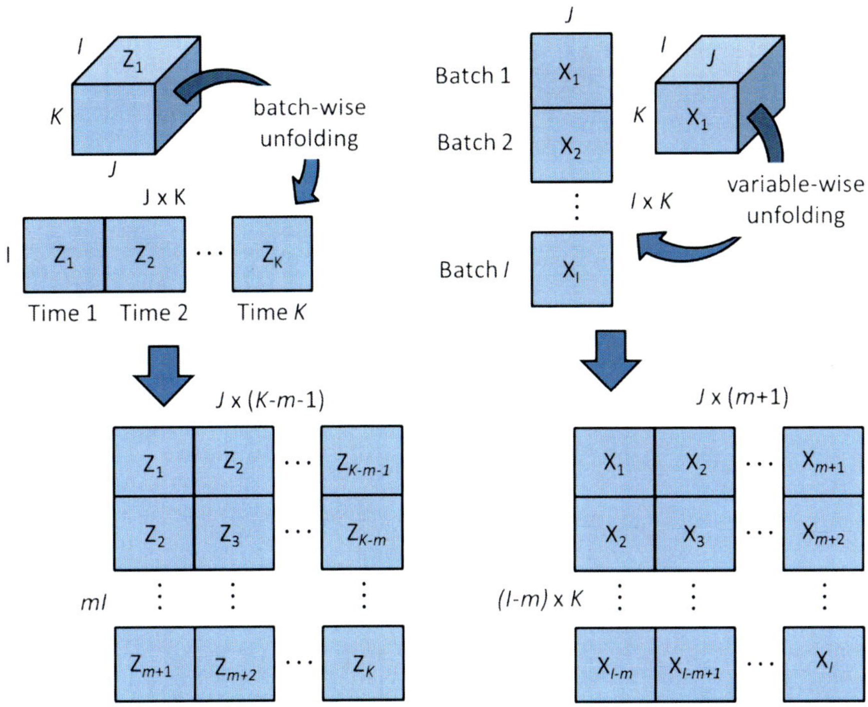

Fig. 2.9 Unfolding of three-way batch data arrays

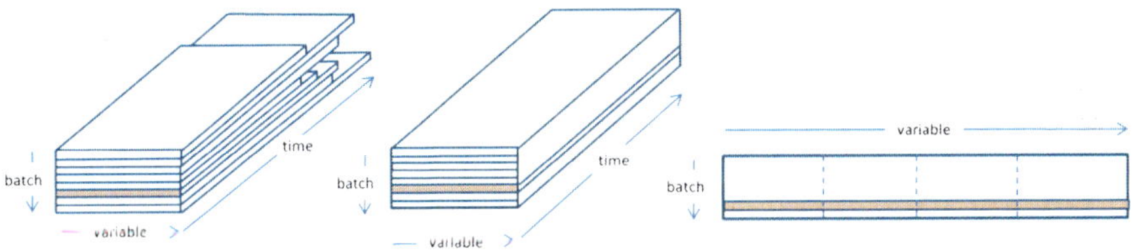

Fig. 2.10 Data preprocessing associated with batch monitoring, showing original batch records of unequal length (*left*), 3D array after trajectory synchronization (*middle*) and 2D matrix after trajectory unfolding (*right*)

Unfolding of the batch data requires all the batches to be of equal duration, and if this is not so, as is usually the case, then PCA cannot be applied directly to the raw data. As indicated in Fig. 2.10, preprocessing of the data is then required to synchronize the batches, i.e. ensuring that all batches are observed from the same evolutionary point.

This can be accomplished by different means, of which dynamic time warping (DTW) is the most popular. The classic dynamic time warping algorithm aligns two signals or time series optimally searching a grid of distances between the samples in the two time series (Kassidas et al. 1998).

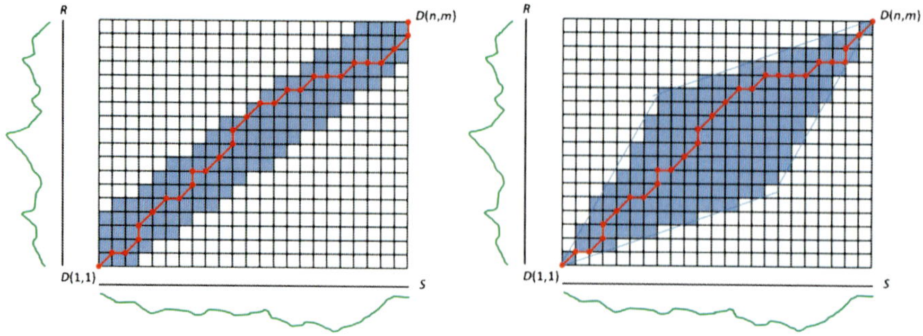

Fig. 2.11 Dynamic time warping of two signals R and S in a reduced search area with Sakoe–Chiba band constraints (*left*) and Itakura constraints (*right*)

2.5.1 Dynamic Time Warping (DTW)

More formally, if two time series are denoted by $R = \{r_1, r_2, \ldots, r_n\} \in \Re$ and $S = \{s_1, s_2, \ldots, s_n\} \in \Re$, then they can be arranged to form an $n \times m$ grid, where each grid point corresponds to an alignment between the elements of $r_i \in \Re$ and $s_j \in \Re$. A warp path can subsequently be defined as a sequence of grid points $W = \{w_1, w_2, \ldots, w_n\}$, where each w_k corresponds to a point $(i, j)_k$. W maps elements of the sequences R and S and is typically constrained as follows:

(a) *Boundary conditions*: $w_1 = (1,1)$ and $w_K = (n, m)$, i.e. the path starts at the first point of both sequences and ends at the last point of both sequences.
(b) *Continuity*: Let $w_k = (r, s)$, then $w_k - 1 = (r', s')$, where $r - r' \leq 1$ and $s - s' \leq 1$, that is, allowable steps in the warp path are restricted to adjacent cells in the grid.
(c) *Monotonicity*: Let $w_k = (r, s)$, then $w_k - 1 = (r', s')$, where $r - r' \geq 0$ and $s - s' \geq 0$, forcing points to be monotonically spaced in time.
(d) The optimal warp path can be found by

$$\mathrm{DTW}(R, S) = \min_W \left[\sum_{k=1}^{k} d\,(w_k) \right] / K. \tag{2.36}$$

In Eq. 2.22, $d(w_k)$ is the (usually Euclidean) distance between elements of the time series, and K is a denominator compensating for different warp paths having different lengths. Dynamic programming can be used to find the optimal warp path by using the following recursive equations: $D(0,0) = 0$ and $D(i,j) = \min\{D(i-1, j), D(i, j-1), D(i-1, j-1)\} + d(r_i, s_j)$ (Fig. 2.11).

The $O(N^2)$ time complexity of the DTW algorithm makes it efficient for relatively short time series only, and many different approaches to alleviate the problem have been proposed, either by the imposition of constraints or data abstraction.

Constraints limit the search for a warp path by reducing the search space or allowed warp along the time coordinate. The most commonly used constraints are Sakoe–Chiba bands (Sakoe and Chiba 1978) and Itakura parallelograms (Itakura 1975), as indicated in Fig. 2.11.

After construction of the optimal path, *synchronization* can be done, symmetrically or asymmetrically. In the symmetric version, the indices of all K points in the optimal path are used to generate the warped signal. During asymmetric synchronization, the vertical transitions in the optimal path are treated differently. The signal R on the horizontal axis serves as reference, so that when more than one point of the warped signal is aligned with the same point on the reference signal (vertical transition), then the average of the points is calculated and aligned with the corresponding point in the reference signal.

2.5.2 Correlation Optimized Warping (COW)

With correlation optimized warping, two signals are aligned by piecewise compression and stretching so that the correlation between fragments of the signals is maximized. If two signals are to be aligned, one is designated as the target (R) and the other as signal to be aligned with the target is designated as signal S. The unaligned signal with length L_S is divided into n' sections, each of length m'. The target signals with length L_R are likewise divided into m' segments. Each segment in signal S is subsequently stretched or compressed via linear interpolation, so that the warped signal (now designated by S*) has the same length as the target signal R. To accomplish this, the difference in the lengths of segments of the two signals is considered, i.e.

$$\Delta = \left(\frac{L_R}{n'}\right) - m'. \tag{2.37}$$

For each section, a finite number of possible warpings is investigated, as determined by an (integer) value or slack variable s set by the user. For example, if $s = 5$, then the end point can assume five different values (i.e. -2, -1, 0, 1, 2). Correlation optimized warping is therefore a piecewise or segmented data preprocessing method that operates on one sample record at a time to align a sample data vector with a reference vector by allowing limited changes to the sample segment lengths. The different segment lengths on the sample vector are selected to optimize the overall correlation between the sample and the reference, and in so doing, the problem is cast as a segment-wise correlation optimization problem that can be solved by the use of dynamic programming (Nielsen et al. 1998).

For practical purposes, any reasonable choice for segment length and slack will give an indication of the anticipated synchronization performance. Furthermore, due to the relatively small search space, a trial-and-error approach for finding the best settings is feasible even on a modest computer system. In this respect, COW may be more suitable than DTW, as the number of possible paths is much smaller and, consequently, the memory requirements are significantly reduced (Tomasi et al. 2004).

Applications of the COW algorithm include those of Fransson and Folestad (2006) who have used the approach to align batch process data as well as Skov et al. (2006) who have used it to align chromatographic data.

2.5.3 PCA/PLS Models

Before fitting PCA or related models, the average trajectory of each variable is subtracted from the batches so that the model can focus on the variability around the batch data, while scaling of the variables to unit variance may also be necessary (Camacho et al. 2009). The synchronized and preprocessed data in the unfolded matrix **X** is modelled with PCA as follows:

$$\mathbf{X} = \mathbf{T}\mathbf{P}^{\mathrm{T}} + \mathbf{E} \tag{2.38}$$

with **P** the loading matrix, **T** the score matrix and **E** the residual matrix in the PCA model, as before. Once the model has been fitted to the normal operating condition (NOC) data, the monitoring system can be constructed based on the T^2- and Q-statistics computed from the scores, and control limits can be established at certain confidence levels, while the contributions of variables when abnormalities are detected can be assessed with contribution plots.

Nomikos and MacGregor (1995a) have used multiway partial least squares (MPLS) to extract information from measured process variable trajectories relevant to the final product quality. They have proposed simple statistical process control charts to monitor new batches. Apart from monitoring batch operations, online prediction of the final product qualities was also possible. The approach was illustrated by the use of a simulation study of a styrene–butadiene batch reactor.

Kourti et al. (1995) have extended multiway PCA and PLS procedures to allow the use of not only the measured trajectory data on all the process variables and information on key performance indicators but also information on initial conditions for batches, including raw material properties, initial ingredient charges and discrete operating conditions. The approach was illustrated with data from two industrial batch polymerization processes. Likewise, Ramaker et al. (2002) have considered improvements to the conventional approach by incorporating external information into the model, viz. batch run-specific and process-specific information.

Russell et al. (2000b) have reduced batch quality monitoring to a problem of state estimation for batch and semi-batch processes. They did this with online smoothing of the initial conditions to reduce the effects of the initial uncertainty resulting from feed disturbances.

Lane et al. (2001) have presented an extension to principal component analysis which enables the simultaneous monitoring of a number of product grades or recipes. The method is based on the existence of a common eigenvector subspace for the sample covariance matrices of the individual products and estimation of the principal component loadings of the multigroup model from the pooled sample covariance matrix of the individual products. Industrial application suggested that the detection and diagnostic capabilities of the multigroup model are comparable to those of separate statistical representations of the individual products.

The online batch process monitoring scheme developed by Chen and Liu (2001) to reduce the variations of product quality was based on integration of the time-lagged windows of process dynamic behaviour with principal component analysis and partial least square for online batch monitoring. Like traditional MPCA and MPLS approaches, the only information required to set up control charts is historical data collected from past successful batches, and no expensive computations are done to anticipate future measurements. Gurden et al. (2002) have discussed the role of spectroscopy in batch process monitoring with an emphasis on dealing with the measured data as complementary to process variable measurements.

An MPLS modelling technique suitable for online real-time process monitoring was proposed by Ündey et al. (2004), and they have extended the approach to include predictions of end-of-batch quality measurements during the progress of a batch run. The process monitoring scheme was embedded into a real-time knowledge-based system, and multivariate charts were automatically interpreted through a generic rule base for efficient alarm handling.

Lee et al. (2005) have considered multiscale PCA for fault detection and diagnosis of batch processes. Adaptive multiway PCA (MPCA) models were developed to update the covariance structure at each scale to better deal with changing process conditions. The proposed method was successfully applied to a pilot-scale sequencing batch reactor for biological wastewater treatment. Hybrid or grey models combining the strengths of first-principle and data-driven models were investigated by Van Sprang et al. (2005).

Kulkarni et al. (2004) have used a formalism integrating PCA and generalized regression neural networks for modelling and monitoring of batch processes. With this PCA–GRNN hybrid methodology, process outputs could be predicted accurately, even when nonlinearly related to the input variables.

Yao and Gao (2008b) have applied two-dimensional dynamic principal component analysis to model intra- and inter-batch dynamics, in which subspace identification was combined with 2D-DPCA. This approach recognizes the fact that batch-wise dynamics may be caused by slow response variables, slowly varying feed stocks, drift of process characteristics or run-to-run adjustment in process controllers. The state space model of a 2D batch process could be identified with canonical variate analysis. They have found that the use of state variables, instead of lagged process variables, could reduce the number of variables required in the analysis and provided clearer contribution plots for fault diagnosis.

2.5.4 ICA Models

Yoo et al. (2004) have considered in-control data of nonstationary processes that contain inherent non-Gaussian data due owing to ramp changes, step changes and autocorrelated variables. They have shown that improved process monitoring of batch processes with non-Gaussian data is possible with methods based on multiway independent component analysis (MICA). The proposed method was applied to the online monitoring of a fed-batch penicillin production.

Albazzaz and Wang (2006) have extended the use of independent component analysis to develop statistical monitoring charts for batch processes by introducing time lag shifts to include process dynamics in the ICA model. This has yielded better results than what could be obtained with static ICA, static principal component analysis and dynamic PCA.

Monitoring of batch processes based on multiway PCA may perform poorly, owing to the need for estimates of future values when used in online monitoring and assumptions of Gaussian data distribution. To counter this, Ge and Song (2007, 2008) have proposed an approach based on multi-model independent component analysis (ICA) and PCA. The scheme uses ICA to monitor non-Gaussian information of the process, after which PCA is applied to the rest. The proposed method does not require prediction of the future values, since sub-models are constructed for every sample time of the batch, which means it can also be used for batch processes in which the batch lengths vary.

Tian et al. (2009) proposed a nonlinear monitoring technique using multiway kernel independent component analysis based on feature samples (FS-MKICA). With this approach, the three-way data set of a batch process was first unfolded into a two-way data set from which representative feature samples were selected. This nonlinear feature space abstracted from the unfolded two-way data space was subsequently transformed into a high-dimensional linear space via kernel independent component analysis. The small subset of samples could significantly reduce the computational cost of the kernel ICA model. Matero et al. (2009) have studied the granulation end point in a pharmaceutical mixture with multiway methods.

2.5.5 Fisher Discriminant Analysis

A number of authors have considered monitoring schemes based on feature extraction with Fisher discriminant analysis (Zhao and Shao 2006; Zhao et al. 2006). The similarity of features of the current and the reference batch was calculated online, and a contribution plot of weights in the feature direction is calculated for fault diagnosis. The approach surmounts the need for estimating or filling in the unknown portion of the process variable trajectories from the current time to the end of the batch. Marjanovic et al. (2006) described the development of a real-time monitoring system for a batch process operated by Aroma and Fine Chemicals Limited. The work was aimed at batch end-point identification. Their approach based upon multivariate statistical techniques could provide a soft sensor that could estimate the product quality throughout the batch and provide a long-term estimate of the likely cycle time.

Zhang et al. (2007) have made use of kernel Fisher discriminant analysis to model blocked reference and new data, which proved effective in the monitoring of fed-batch penicillin fermentation. Zhao et al. (2007) have successfully extended the approach based on a dissimilarity measure (DISSIM) for application to batch processes (EDISSIM). This included the use of contribution plots derived from the dissimilarity index to identify variables contributing significantly to out-of-control process states.

2.5.6 *Other Modelling Approaches*

Lee et al. (2004b) have proposed a new statistical batch monitoring approach based on variable-wise unfolding and time-varying score covariance structures to overcome the drawbacks of conventional MPCA. The proposed method does not require prediction of the future values, while the dynamic relations of data are preserved by using time-varying score covariance structures and can be used to monitor batch processes in which the batch length varies.

Chen and Chen (2006) have considered online batch process monitoring with a wavelet-based multi-hidden Markov model tree (MHMT) model. This allowed them not only to analyse the measurements at multiple scales in time and frequency but also to better capture the structure of the data than could be done with multiway PCA.

Choi et al. (2008) have described an integrated framework consisting of a multivariate autoregressive (AR) model and multiway principal component analysis (MPCA) to monitoring the performance of batch processes. After preprocessing the data, the data are filtered using an AR model to remove the auto- and cross-correlation inherent within the preprocessed batch data. MPCA was consequently applied to the residuals from the AR model. The main advantage of the proposed approach is that it can monitor batch dynamics along the mean trajectory without the requirement to estimate future observed values.

Hu and Yuan (2008) have proposed a multivariate statistical process control method using dynamic multiway neighbourhood preserving embedding (DMNPE) for fed-batch process monitoring. They have found that the neighbourhood preserving property facilitated the capture of more information than was possible by multiway PCA. An industrial cephalosporin fed-batch fermentation process was used to demonstrate the performance of the approach.

Hu and Yuan (2008) used tensor factorization tensor locality preserving projections (TLPP), instead of principal component analysis on an unfolded matrix. TLPP preserves local neighbourhood information better than PCA, which preserves the global Euclidean structure. This may have certain advantages in batch monitoring. Gunther et al. (2009) have proposed evolving online partial least-squares approach and compared it to a global PLS method. A variety of faults could be recognized during online monitoring. Faggian et al. (2009) have presented an industrial case study where the challenges related to the real-time estimation of the required time to manufacture a resin and to the instantaneous product quality estimation were addressed using multivariate statistical techniques. They have shown that stage and batch lengths can be estimated in real time with an average error not larger than 20 % of the inherent batch-to-batch variability. In contrast, quality estimations could be provided within the accuracy of the hardware instrumentation, but many times faster.

Yao et al. (2010) have integrated Gaussian mixture models with dynamic principal component analysis to construct better online control limits in batch process monitoring schemes. These Gaussian mixture models were integrated with 2D-DPCA, and control limits were estimated from joint probability density functions that were estimated to address the non-Gaussian nature of 2D dynamic batch process monitoring. A two-phase fed-batch fermentation process for penicillin production was used to verify the effectiveness of the proposed method. Chen et al. (2010) have proposed the use of an integrated framework called IOHMM–MPLS to monitor the performance of batch processes. The methodology is based on the use of an input–output hidden Markov model (IOHMM) and a multiway partial least-squares (MPLS) method. The sequence of the process variables and the product quality variables are decomposed into linear outer relations, which is handled by MPLS, as well as simple inner dynamic sequence relations, which can be accommodated by single-input – single-output IOHMM models. MPLS is used to solve the problem of high dimensionality and collinearity, while IOHMM is used to capture the transition probability of the dynamic information.

Jia et al. (2010) have proposed a nonlinear batch process monitoring method integrating kernel PCA and ARMAX time series models through estimating the average kernel matrix (AKM) of all batch runs. AKM is an average of I, the batch number, single batch kernel matrices (SBKM). The AKM contains the information of the stochastic variations and deviations among batches. This information will be very useful for the BDKPCA model to characterize the batch process in detail. Chen and Wang (2010) have used a linear time-variant system for batch processes. They have also modified their approach to enable it to differentiate between optimal and suboptimal conditions and not just normal and abnormal operating conditions. This was done by generating performance bounds based on data from batch runs operated at optimum conditions. Alvarez et al. (2010) have proposed an MSPC strategy for batch process monitoring operating in the space of the original variables. Moreover, the technique uses only the T^2-statistic for detection and identification.

More recently, Zhang and Hu (2011) have proposed the use of hierarchical kernel partial least squares (HKPLS) to monitor batch processes. Apart from capturing more nonlinear information compared to hierarchical partial least squares (HPLS) and multiway PLS (MPLS), monitoring of new batch processes using HKPLS does not need to estimate unknown parts of the process variable trajectories. Monroy et al. (2011) conducted a comparative study between multiway principal component analysis (MPCA) and batch-dynamic principal component analysis (BDPCA) of a penicillin production process. The study indicated that BDPCA is better than MPCA in terms of diagnosis. Zhao et al. (2011) have presented a method for handling uneven-length multiphase batch processes. This was based on classifying irregular

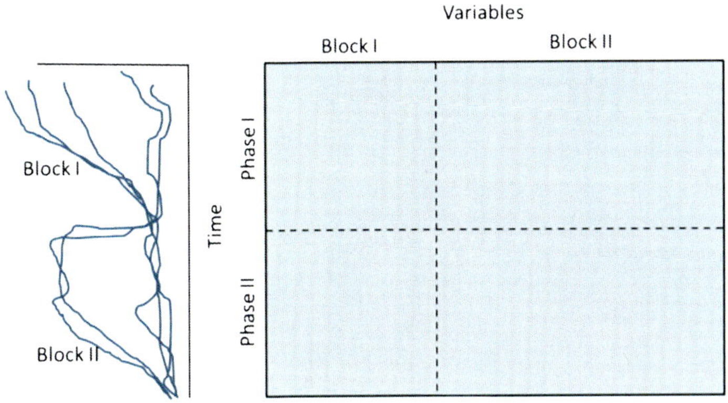

Fig. 2.12 A multiblock (variables) and multiphase (time) approach to an unfolded batch data matrix

batches into different uneven-length groups based on changes in their underlying characteristics. This was followed by creating two different subspaces to model the group common and group-specific information and constructing corresponding confidence regions by searching similar patterns. He and Wang (2011) have proposed a statistical framework for monitoring batch processes in the semiconductor industry. This approach was based on analysis of batch statistics, including higher-order statistics, such as the skewness and kurtosis of batch variables.

2.5.7 *Multiblock, Multiphase and Multistage Batch Processes*

Many batch processes have important multiphase[1] and multistage[2] characteristics that are not readily accommodated by multiway methods, as outlined above, since process dynamics may differ considerably between different phases and stages (Fig. 2.12). The relationships between these differing dynamics are not considered with MPCA/MPLS, where all the data are treated as a single object. This can severely compromise understanding of the process as well as monitoring thereof.

In recent years, a range of approaches have been proposed to deal with multistage and multiphase processes, and a good overview of these methods is given by Yao and Gao (2009a). Flores-Cerrillo and MacGregor (2004) have considered extension of the multiblock MPCA/MPLS approach to explicitly incorporate batch-to-batch trajectory information summarized by the scores of previous batches while retaining the advantages and monitoring statistics of traditional MPCA/MPLS methods. The main advantage of this approach was that it could be useful for detecting problems when monitoring new batches in the early stages of operation.

---

[1] Multiphase refers to a batch process with a single processing unit, but multiple operating regimes.

[2] Multistage refers to a batch process with multiple processing units.

2.5.8 Phase Segmentation

Three major approaches to identify different phases in batch processes have emerged recently. These include the use of expert knowledge, process analysis and automated data-based approaches (Yao and Gao 2009b). Camacho and Picó (2006a) proposed a new strategy, designed for online monitoring based on four steps, which included subtraction of the mean batch trajectory and autoscaling, variable-wise unfolding, addition of lagged variables to account for process dynamics and multiphase modelling.

Simoglou et al. (2005) have described statistical process control tools that have been applied for the online monitoring of an industrial fed-batch sugar crystallization process, characterized by distinct operating phases during operation and the presence of strong nonlinear, dynamic relationships between the variables (Table 2.1).

2.5.9 Multiblock Methods

Since conventional MPCA/MPLS models do not consider the unique characteristics of each phase in multistage or multiphase batch processes, more reasonable models can be obtained by using two different groups of techniques. The first group includes various multiblock PCA/PLS modelling methods (Smilde et al. 2003; Westerhuis et al. 1998) that separate process variables in different operating stages and/or phases into different blocks. The variable correlations within each block are then modelled together with the correlations among blocks. With phase-based methods, each phase is considered separately. These modelling approaches have been applied to process analysis and online monitoring, online quality prediction and online quality control. They are portrayed diagrammatically in Fig. 2.13.

PCA/PLS

With multiblock methods, the process variables are grouped into meaningful blocks, after which both the relationships within and between blocks are considered (MacGregor et al. 1994; Kourti et al. 1995; Westerhuis and Coenegracht 1997; Lee and Vanrolleghem 2003). Process monitoring and fault diagnosis can be based on the overall T^2- and Q-statistics as well as the same statistics and corresponding contribution plots for each block (Qin et al. 2001; Choi and Lee 2005).

Although multiblock models are easier to interpret than global models, they do not provide methods for phase division. In addition, the estimation of future unavailable measurements is still required in online monitoring.

Table 2.1 Comparison of phase division methods (After Yao and Gao 2009b)

Approach	Methodology	Prior knowledge required	References	Additional comments
Expert knowledge	Locations of phase division points[a]	Process knowledge	Dong and McAvoy (1996), Kosanovich et al. (1994), Rainikainen and Höskuldsson (2007), Liu and Wong (2008), Smilde et al. (2003), Westerhuis et al. (1998)	The phase division corresponding to operating stages/phases may not reflect variable correlation changes
Process analysis	DTW[b]	Prototype cycle with known phase division	Gollmer and Posten (1996)	
	Multivariate rules[b]	Process features associated with phases	Muthuswamy and Srinivasan (2003)	
	Landmarks in indicator variables[b]	Indicator variables with significant landmarks	Ündey and Cinar (2002), Facco et al. (2007), Doan and Srinivasan (2008)	
	Variances explained by principal components[b]	Nil	Kosanovich et al. (1996)	Only gives some rough indications
Automated database methods	Local correlation analysis[c]	Nil	Lu et al. (2004)	Can be applied to different processes
	PCA predictions[c]	Nil	Camacho and Picó (2006a, b), Camacho et al. (2008), Zhao et al. (2008)	
	ISODATA dynamic clustering	Nil	Chen et al. (2010)	

[a]Phases and operating phases usually correspond
[b]Phases and operating phases may noy correspond
[c]J process variables measured over K sampling points yielding a data matrix of dimensions $J \times K$ from each batch run. Therefore, a set of I normal batch runs result in a three-way process data matrix, $X(I \times J \times K)$, which is the most popular data form for batch processes. The horizontal slice $X(J \times K)$ is the data matrix from each batch run, while the vertical slice $\tilde{X}\ (I \times J)$ is a time-sliced matrix that is used to obtain the process correlation at sampling time k

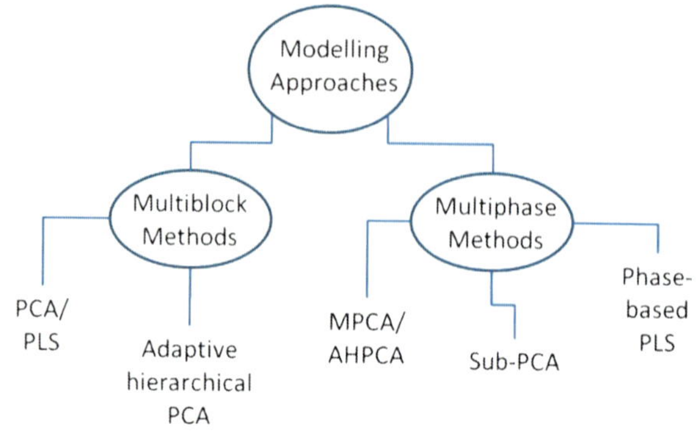

Fig. 2.13 Modelling approaches to block- and phase-based batch processes

Adaptive Hierarchical PCA

Ranner et al. (1998) have proposed the use of adaptive hierarchical principal component analysis (AHPCA), where the blocks are time-sliced data matrices. Models are then built sequentially for each time interval, and weighting factors are used to make the models adaptive. AHPCA has a high computational and storage burden, and in some situations, it may be difficult to find a proper weight for the entire batch run. A similar concept was considered by Ramaker et al. (2005), who have used local models for batch process monitoring.

2.5.10 Multiphase Methods

Camacho et al. (2010) compared batch-wise, variable-wise, batch-dynamic, local and multiphase approaches using simulated data. They came to the conclusion that the best monitoring approach depends more on the type of fault than on the process dynamics, that parsimonious models and the model structure are important and that it may be advisable to combine several unfolding methods to improve the detection ability of predefined types of faults.

MPCA/AHPCA

Phase-based modelling can be accomplished by direct representation of the phases with separate MPCA or AHPCA models. As with conventional MPCA and AHPCA, phase-based MPCA and AHPCA suffer from shortcomings associated with high computational cost and future data estimation.

Yao and Gao (2009b) have modelled each phase separately with PCA and have proposed an index based on the angles between different principal components to quantify the similarities between PCA models. Yao and Gao (2009b) considered multiple operational phases within a batch and proposed an approach to deal with these phases simultaneously.

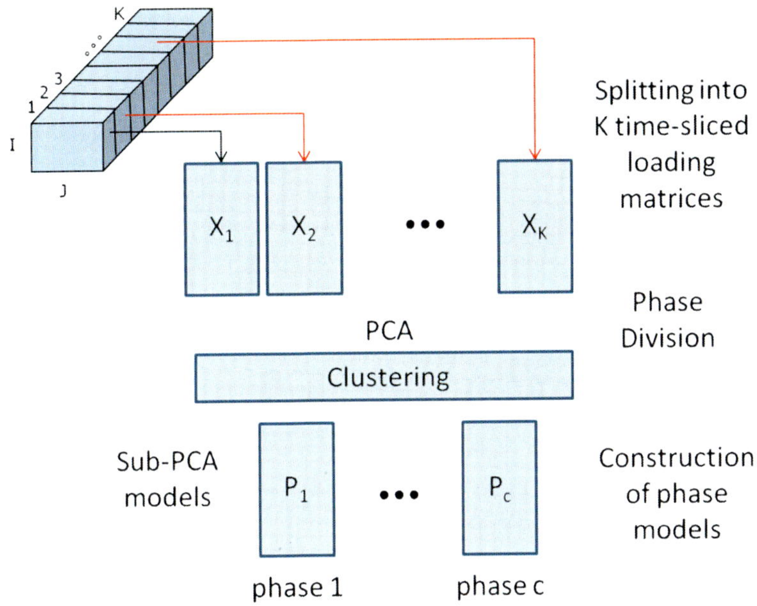

Fig. 2.14 Batch process phase division and phase-based sub-PCA modelling

More recently, Ge et al. (2011b) have considered batch process monitoring as a one-class classification problem. They have used support vector data description, which are not dependent on assumptions regarding the distribution of the data and have extended this to multiphase, multimodal systems.

Sub-PCA Modelling

With phase-based sub-PCA modelling (Lu et al. 2004), the phase changes in a batch process are typically indicated by changes in the correlational structure of the variables. By modelling each phase separately, the unique characteristics of each phase can be captured more accurately, and fault detection can consequently be done more reliably. As indicated in Fig. 2.14, for a normalized $I \times J \times K$ batch process,[3] each vertical slice ($I \times J$) is a time-sliced matrix on which PCA is performed.

The K loading matrices are subsequently transformed into a composite loading matrix by weighting the loadings of each matrix as follows:

---

[3] J process variables measured over K sampling points yielding a data matrix of dimensions $J \times K$ from each batch run. Therefore, a set of I normal batch runs result in a three-way process data matrix, X ($I \times J \times K$), which is the most popular data form for batch processes. The horizontal slice $X(J \times K)$ is the data matrix from each batch run, while the vertical slice $\tilde{X}$ ($I \times J$) is a time-sliced matrix that is used to obtain the process correlation at sampling time k.

$$\mathbf{P}^k = \left[w_1^k p_1^k; w_2^k p_2^k \ldots w_j^k p_j^k\right] \tag{2.39}$$

where λ_j^k is the jth eigenvalue of the covariance matrix $(\mathbf{X}_k)^{\mathrm{T}}(\mathbf{X}_k)$, and the weights are defined as follows:

$$w_j^k = \frac{\lambda_j^k}{\sum_{j=1}^{j} \lambda_j^k} \tag{2.40}$$

Phase division is accomplished by the use of a clustering method, such as k-means clustering of the time-sliced matrices, based on the composite loading matrix.

Phase-Based PLS

Phase division can be accomplished in a way similar to that for sub-PCA models. The time-sliced data matrices are regressed to the quality data matrix with PLS. k-Means clustering is subsequently used for phase division based on the regression coefficients of the time-sliced PLS models.

Lu and Gao (2005) have developed a phase-based approach to compensate for the loss in prediction accuracy that may result from the inclusion of irrelevant information in phases that are not critical to product quality. More recently, Qi et al. (2011) have investigated batch process monitoring and quality prediction with multiphase dynamic partial least squares. The batch process data from a fed-batch penicillin fermentation production system were divided into several phases by means of a Gaussian mixture model and preprocesses by dynamic time warping. This allowed the construction of a dynamic PLS model for each phase, which yielded better results than a single model.

2.6 Conclusions

As Qin (2012) has pointed out, the combination of basic models or building blocks, such as multiscale methods, multiway approaches, kernel methods or independent component analysis, can give rise to a large variety of new methods. This could include recursive multimodal multiway kernel principal component analysis, but these models need to be justified by clearly demonstrating their relative merits over other approaches, since they tend to be more complex, which carries a penalty in itself.

With machine learning methods, the scope for advances in fault diagnostic systems is even larger. Not only will progress be driven by advances in machine learning methods themselves, e.g. more advanced deep learning methods for neural networks, or further advances in tree-based modelling, but also by the fact that machine learning can be extended by a variety of other algorithms. For example, algorithms associated with manifold learning, such as Sammon maps (Sammon

1969), Isomaps (Tenenbaum et al. 2000), locally linear embeddings (Roweis and Saul 2000) and Laplacian eigenmaps (Belkin and Niyogi 2003), could readily be incorporated in forward and reverse models of the fault diagnostic framework discussed in this book. This may have certain advantages when complex data have to be interpreted, as could arise from spectral- or image-based monitoring of systems that may exhibit complex dynamics.

References

Acuña-González, N., Garcia-Ochoa, E., & González-Sanchez, J. (2008). Assessment of the dynamics of corrosion fatigue crack initiation applying recurrence plots to the analysis of electrochemical noise data. *International Journal of Fatigue, 30*, 1211–1219.

Adgar, A., Cox, C. S., & Bohme, T. J. (2000). Performance improvements at surface water treatment works using ANN-based automation schemes. *Transactions of the Institute for Chemical Engineers Part A, 78*, 1026–1039.

Ahola, J., Alhoniemi, E., & Simula, O. (1999). Monitoring industrial processes using the self-organizing map. In *IEEE midnight-sun workshop on soft computing methods in industrial applications* (pp. 22–27). Piscataway: IEEE. Available at: http://ieeexplore.ieee.org/lpdocs/epic03/wrapper.htm?arnumber=782702. Accessed 25 Dec 2011.

Alabi, S., Morris, A., & Martin, E. (2005). On-line dynamic process monitoring using wavelet-based generic dissimilarity measure. *Chemical Engineering Research and Design, 83*, 698–705.

Albazzaz, H., & Wang, X. Z. (2004). Statistical process control charts for batch operations based on independent component analysis. *Industrial and Engineering Chemistry Research, 43*, 6731–6741.

Albazzaz, H., & Wang, X. Z. (2006). Multivariate statistical batch process monitoring using dynamic independent component analysis. In *Computer aided chemical engineering* (pp. 1341–1346). Amsterdam: Elsevier. Available at: http://linkinghub.elsevier.com/retrieve/pii/S1570794606802336. Accessed 27 Nov 2011.

Alcala, C. F., & Qin, S. J. (2010). Reconstruction-based contribution for process monitoring with kernel principal component analysis. *Industrial and Engineering Chemistry Research, 49*(17), 7849–7857.

Aldrich, C. (2002). *Exploratory analysis of metallurgical process data with neural networks and related methods*. Amsterdam: Elsevier.

Aldrich, C., & Reuter, M. A. (1999). Monitoring of metallurgical reactors by the use of topographic mapping of process data. *Minerals Engineering, 12*(11), 1301–1312.

Aldrich, C., Moolman, D. W., & Van Deventer, J. S. J. (1995a). Monitoring and control of hydrometallurgical processes with self-organizing and adaptive neural net systems. *Computers and Chemical Engineering, 19*(S1), 803–808.

Aldrich, C., Moolman, D. W., Eksteen, J. J., & Van Deventer, J. S. J. (1995b). Characterization of flotation processes with self-organizing neural nets. *Chemical Engineering Communications, 139*, 25–39.

Aldrich, C., Gardner, S., & Le Roux, N. J. (2004). Monitoring of metallurgical process plants by use of biplots. *AICHE Journal, 50*(9), 2167–2186.

Aldrich, C., Qi, B. C., & Botha, P. J. (2006). Analysis of electrochemical noise with phase space methods. *Minerals Engineering, 19*(14), 1402–1409.

Alvarez, C. R., Brandolin, A., & Sánchez, M. C. (2010). Batch process monitoring in the original measurement's space. *Journal of Process Control, 20*(6), 716–725.

Antory, D., Kruger, U., Irwin, G., & McCullough, G. (2004). Industrial process monitoring using nonlinear principal component models. In *2nd international conference on intelligent systems* (pp. 293–298). Piscataway: IEEE. Available at: http://ieeexplore.ieee.org/lpdocs/epic03/wrapper.htm?arnumber=1344685. Accessed 22 Dec 2011.

Antory, D., Irwin, G., Kruger, U., & McCullough, G. (2008). Improved process monitoring using nonlinear principal component models. *International Journal of Intelligent Systems, 23*(5), 520–544.

Augusteijn, M. F., & Folkert, B. A. (2002). Neural network classification and novelty detection. *International Journal of Remote Sensing, 23*(14), 2891–2902.

Auret, L., & Aldrich, C. (2010). Change point detection in time series data with random forests. *Control Engineering Practice, 18*(8), 990–1002.

Bartlett, M. S., Movellan, J. R., & Sejnowski, T. J. (2002). Face recognition by independent component analysis. *IEEE Transactions on Neural Networks, 13*(6), 1450–1464.

Basso, A. (1992). Autoassociative neural networks for image compression: A massively parallel implementation. In *Proceedings of the IEEE-SP Workshop* (pp. 373–381). Piscataway: IEEE. Available at: http://ieeexplore.ieee.org/lpdocs/epic03/wrapper.htm?arnumber=253675. Accessed 24 Dec 2011.

Belkin, M., & Niyogi, P. (2003). Laplacian eigenmaps for dimensionality reduction and data representation. *Neural Computation, 15*(6), 1373–1396.

Bovolo, F., Camps-Valls, G., & Bruzzone, L. (2010). A support vector domain method for change detection in multitemporal images. *Pattern Recognition Letters, 31*(10), 1148–1154.

Camacho, J., & Picó, J. (2006a). Online monitoring of batch processes using multi-phase principal component analysis. *Journal of Process Control, 16*(10), 1021–1035.

Camacho, J., & Picó, J. (2006b). Multi-phase principal component analysis for batch processes modelling. *Chemometrics and Intelligent Laboratory Systems, 81*(2), 127–136.

Camacho, J., Picó, J., & Ferrer, A. (2008). Multi-phase analysis framework for handling batch process data. *Journal of Chemometrics, 22*(11–12), 632–643.

Camacho, J., Picó, J., & Ferrer, A. (2009). The best approaches in the on-line monitoring of batch processes based on PCA: Does the modelling structure matter? *Analytica Chimica Acta, 642*, 59–68.

Camacho, J., Picó, J., & Ferrer, A. (2010). Data understanding with PCA: Structural and variance information plots. *Chemometrics and Intelligent Laboratory Systems, 100*(1), 48–56.

Cao, L., Mees, A., & Judd, K. (1998). Dynamics from multivariate time series. *Physica D, 121*, 75–88.

Cardoso, J.-F. (1998). Blind signal separation: Statistical principles. *Proceedings of the IEEE, 86*(10), 2009–2025.

Casciati, F., & Casciati, S. (2006). Structural health monitoring by Lyapunov exponents of nonlinear time series. *Structural Control and Health Monitoring, 13*(1), 132–146.

Cazares-Ibáñez, E., Vázquez-Coutiño, A. G., & García-Ochoa, E. (2005). Application of recurrence plots as a new tool in the analysis of electrochemical oscillations of copper. *Journal of Electroanalytical Chemistry, 583*(1), 17–33.

Chang, K.-Y., & Ghosh, J. (2001). A unified model for probabilistic principal surfaces. *IEEE Transactions on Pattern Analysis and Machine Intelligence, 23*(1), 22–41.

Chemaly, T. P., & Aldrich, C. (2001). Visualization of process data by use of evolutionary computation. *Computers and Chemical Engineering, 25*, 1341–1349.

Chen, X., Gao, X. Zhang, Y, & Qi, Y. (2010a). Enhanced batch process monitoring and quality prediction based on multi-phase multi-way partial least squares. In *International conference on Intelligent Computing and Intelligent Systems (ICIS)* (pp. 32–36). Piscataway: IEEE. Available at: http://ieeexplore.ieee.org/lpdocs/epic03/wrapper.htm?arnumber=5658834. Accessed 28 Dec 2011.

Chen, J., & Chen, H.-H. (2006). On-line batch process monitoring using MHMT-based MPCA. *Chemical Engineering Science, 61*(10), 3223–3239.

Chen, J., & Liao, C.-M. (2002). Dynamic process fault monitoring based on neural network and PCA. *Journal of Process Control, 12*(2), 277–289.

Chen, J., & Liu, K. C. (2001). Derivation of function space analysis based PCA control charts for batch process monitoring. *Chemical Engineering Science, 56*(10), 3289–3304.

Chen, J., & Wang, W.-Y. (2010). Performance monitoring of MPCA-based control for multivariable batch control processes. *Journal of the Taiwan Institute of Chemical Engineers, 41*(4), 465–474.

Chen, J., Song, C.-M., & Hsu, T.-Y. (2010). Online monitoring of batch processes using IOHMM based MPLS. *Industrial and Engineering Chemistry Research, 49*(6), 2800–2811.

Cheng, C., & Chiu, M. (2005). Nonlinear process monitoring using JITL-PCA. *Chemometrics and Intelligent Laboratory Systems, 76*, 1–13.

Cho, J.-H., Lee, J.-M., Choi, S. B., Lee, D., & Lee, I.-B. (2004). Sensor fault identification based on kernel principal component analysis. In *International conference on Control Applications* (pp. 1223–1228). Taipei: IEEE. Available at: http://ieeexplore.ieee.org/lpdocs/epic03/wrapper.htm?arnumber=1387540. Accessed 25 Dec 2011.

Choi, S. W., & Lee, I.-B. (2004). Nonlinear dynamic process monitoring based on dynamic kernel PCA. *Chemical Engineering Science, 59*(24), 5897–5908.

Choi, S. W., & Lee, I.-B. (2005). Multiblock PLS-based localized process diagnosis. *Journal of Process Control, 15*(3), 295–306.

Choi, S. W., Park, J. H., & Lee, I.-B. (2004). Process monitoring using a Gaussian mixture model via principal component analysis and discriminant analysis. *Computers and Chemical Engineering, 28*(8), 1377–1387.

Choi, S. W., Lee, C., Lee, J.-M., Park, J. H., & Lee, I.-B. (2005). Fault detection and identification of nonlinear processes based on kernel PCA. *Chemometrics and Intelligent Laboratory Systems, 75*(1), 55–67.

Choi, S., Morris, J., & Lee, I. (2008). Dynamic model-based batch process monitoring. *Chemical Engineering Science, 63*, 622–636.

Comon, P. (1994). Independent component analysis, a new concept? *Signal Processing, 36*, 287–314.

Cui, P., Li, J., & Wang, G. (2008). Improved kernel principal component analysis for fault detection. *Expert Systems with Applications, 34*, 1210–1219.

Del Frate, F., Licciardi, G., & Duca, R. (2009). Autoassociative neural networks for features reduction of hyperspectral data. In *First Workshop on Image and Signal Processing: Evolution in Remote Sensing, WHISPERS '09* (pp. 1–4). Piscataway: IEEE. Available at: http://ieeexplore.ieee.org/lpdocs/epic03/wrapper.htm?arnumber=5288997. Accessed 24 Dec 2011.

Deng, Y., & Manjunath, B. S. (2001). Unsupervised segmentation of color-texture regions in images and video. *IEEE Transactions on Pattern Analysis and Machine Intelligence, 23*(8), 800–810.

Denning, D. E. (1987). An intrusion-detection model. *IEEE Transactions on Software Engineering, SE-13*(2), 222–232.

Doan, X., & Srinivasan, R. (2008). Online monitoring of multi-phase batch processes using phase-based multivariate statistical process control. *Computers and Chemical Engineering, 32*(1–2), 230–243.

Dong, D., & McAvoy, T. J. (1994). Nonlinear principal component analysis – Based on nonlinear principal curves and neural networks. In *Proceedings of the American Control Conference* (pp. 1284–1288). American Control Conference, Baltimore, MD, USA.

Dong, D., & McAvoy, T. J. (1996). Batch tracking via nonlinear principal component analysis. *AICHE Journal, 42*(8), 2199–2208.

Dong, Y., Li, Y., & Lai, M. (2010). Structural damage detection using empirical-mode decomposition and vector autoregressive moving average model. *Soil Dynamics and Earthquake Engineering, 30*(3), 133–145.

Dunia, R., & Qin, S. J. (1998). Joint diagnosis of process and sensor faults using principal component analysis. *Control Engineering Practice, 6*(4), 457–469.

Ephraim, Y., & Malah, D. (1984). Speech enhancement using a minimum-mean square error short-time spectral amplitude estimator. *IEEE Transactions on Acoustics, Speech, and Signal Processing, 32*(6), 1109–1121.

Facco, P., Olivi, M., Rebuscini, C., Bezzo, F., & Barolo, M. (2007). Multivariate statistical estimation of product quality in the industrial batch production of resin. In *8th International Symposium on Dynamics and Control of Process Systems (Dycops)* (pp. 93–98).

Faggian, A., Facco, P., Doplicher, F., Bezzo, F., & Barolo, M. (2009). Multivariate statistical real-time monitoring of an industrial fed-batch process for the production of specialty chemicals. *Chemical Engineering Research and Design, 87*(3), 325–334.

Flores-Cerrillo, J., & MacGregor, J. F. (2004). Multivariate monitoring of batch processes using batch-to-batch information. *AICHE Journal, 50*(6), 1219–1228.

Fourie, S. H., & De Vaal, P. L. (2000). Advanced process monitoring using an on-line non-linear multiscale principal component analysis methodology. *Computers and Chemical Engineering, 24*(2–7), 755–760.

Fransson, M., & Folestad, S. (2006). Real-time alignment of batch process data using COW for on-line process monitoring. *Chemometrics and Intelligent Laboratory Systems, 84*(1–2), 56–61.

Frey, C. W. (2008, July 14–16). Diagnosis and monitoring of complex industrial processes based on self-organizing maps and watershed transformations. In *Proceedings of the IEEE International Conference on Computational Intelligence for Measurement Systems and Applications (CIMSA)* (pp. 87–92). Available at: http://ieeexplore.ieee.org/lpdocs/epic03/wrapper.htm?arnumber=4595839. Accessed 25 Dec 2011.

Gan, L., Liu, H., & Shen, X. (2010). Sparse kernel principal angles for online process monitoring. *Journal of Computational Information Systems, 6*(5), 1601–1608.

Gardner, S., Le Roux, N. J., & Aldrich, C. (2005). Process data visualization with biplots. *Minerals Engineering, 18*(9), 955–968.

Ge, Z., & Song, Z. (2007). Process monitoring based on Independent Component Analysis – Principal Component Analysis (ICA – PCA) and similarity factors. *Industrial and Engineering Chemistry Research, 46*(7), 2054–2063.

Ge, Z., & Song, Z. (2008). Online batch process monitoring based on multi-model ICA-PCA method. In *7th World Congress on Intelligent Control and Automation, WCICA 2008* (pp. 260–264). Piscataway: IEEE. Available at: http://ieeexplore.ieee.org/lpdocs/epic03/wrapper.htm?arnumber=4594430. Accessed 26 Dec 2011.

Ge, Z., & Song, Z. (2011). A distribution free method for process monitoring. *Expert Systems with Applications, 38*(8), 9821–9829.

Ge, Z., Gao, F., & Song, Z. (2011a). Two-dimensional Bayesian monitoring method for nonlinear multimode processes. *Chemical Engineering Science, 66*(21), 5173–5183.

Ge, Z., Gao, F., & Song, Z. (2011b). Batch process monitoring based on support vector data description method. *Journal of Process Control, 21*, 949–959.

Gollmer, K., & Posten, C. (1996). Supervision of bioprocesses using a dynamic time warping algorithm. *Control Engineering Practice, 4*, 1287–1295.

Guh, R., & Shiue, Y. (2008). An effective application of decision tree learning for on-line detection of mean shifts in multivariate control charts. *Computers and Industrial Engineering, 55*(2), 475–493.

Gunther, J. C., Conner, J. S., & Seborg, D. E. (2009). Process monitoring and quality variable prediction utilizing PLS in industrial fed-batch cell culture. *Journal of Process Control, 19*, 914–921.

Gurden, S. P., Westerhuis, J. A., & Smilde, A. K. (2002). Monitoring of batch processes using spectroscopy. *AICHE Journal, 48*(10), 2283–2297.

Harkat, M.F., Mourot, G., Ragot, J. (2003). Nonlinear PCA combining principal curves and RBF-networks for process monitoring. In *Proceedings of the 42nd IEEE conference on Decision and Control* (pp. 1956–1961), Maui, Hawaii, USA.

Hastie, T., & Stuetzle, W. (1989). Principal curves. *Journal of the American Statistical Association, 84*(406), 502–516.

He, Q. P., & Wang, J. (2011). Statistics pattern analysis: A new process monitoring framework and its application to semiconductor batch processes. *AICHE Journal, 57*(1), 107–121.

He, K., Li, Q., & Chen, J. (2013). An arc stability evaluation approach for SW AC SAW based on Lyapunov exponent of welding current. *Measurement, 46*(1), 272–278.

Hill, D. J., & Minsker, B. S. (2010). Anomaly detection in streaming environmental sensor data: A data-driven modeling approach. *Environmental Modelling and Software, 25*(9), 1014–1022.

Hsu, C.-C., Chen, M.-C., & Chen, L.-S. (2010). A novel process monitoring approach with dynamic independent component analysis. *Control Engineering Practice, 18*, 242–253.

Hu, K., & Yuan, J. (2008). Multivariate statistical process control based on multiway locality preserving projections. *Journal of Process Control, 18*(7–8), 797–807.

Hyvärinen, A. (2002). An alternative approach to infomax and independent component analysis. *Neurocomputing, 44–46*, 1089–1097.

Hyvärinen, A., & Oja, E. (2000). Independent component analysis: Algorithms and applications. *Neural Networks, 13*(4–5), 411–430.

Itakura, F. (1975). Minimum prediction residual principle applied to speech recognition. *IEEE Transactions on Acoustics, Speech, and Signal Processing, 23*(1), 67–72.

Jämsä-Jounela, S.-L., Vermasvuair, M., Endén, P., & Haavisto, S. (2003). A process monitoring system based on the Kohonen self-organizing maps. *Control Engineering Practice, 11*, 83–92.

Jemwa, G. T., & Aldrich, C. (2006). Kernel-based fault diagnosis on mineral processing plants. *Minerals Engineering, 19*(11), 1149–1162.

Jia, F., Martin, E. B., & Morris, A. J. (1998). Non-linear principal components analysis for process fault detection. *Computers and Chemical Engineering, 22*, S851–S854.

Jia, M., Chu, F., Wang, F., & Wang, W. (2010). On-line batch process monitoring using batch dynamic kernel principal component analysis. *Chemometrics and Intelligent Laboratory Systems, 101*(2), 110–122.

Jiang, L., & Xie, L. (2008). Fault detection for batch process based on dissimilarity index. In *International conference on Systems, Man and Cybernetics* (pp. 3415–3419). Piscataway: IEEE. Available at: http://ieeexplore.ieee.org/lpdocs/epic03/wrapper.htm?arnumber=1571675. Accessed 23 Dec 2011.

Kano, M. (2004). Evolution of multivariate statistical process control: application of independent component analysis and external analysis. *Computers and Chemical Engineering, 28*, 1157–1166.

Kano, M., Nagao, K., Hasebe, S., Hashimoto, I., Ohno, H., Strauss, R., & Bahshi, B. (2000). Comparison of statistical process monitoring methods: Application to the Eastman challenge problem. *Computers and Chemical Engineering, 24*, 175–181.

Kano, M., Hasebe, S., Hashimoto, I., & Ohno, H. (2001). A new multivariate statistical process monitoring method using principal component analysis. *Computers and Chemical Engineering, 25*(7–8), 1103–1113.

Kano, M., Nagao, K., Hasebe, S., Hashimoto, I., Ohno, H., Strauss, R., & Bahshi, B. (2002). Comparison of multivariate statistical process monitoring methods with applications to the Eastman challenge problem. *Computers and Chemical Engineering, 26*(2), 161–174.

Kano, M., Tanaka, S., Hasebe, S., & Hashimoto, I. (2003). Monitoring independent components for fault detection. *AICHE Journal, 49*(4), 969–976.

Karhunen, J., & Joutsensalo, J. (1994). Representation and separation of signals using nonlinear PCA type learning. *Neural Networks, 7*(1), 113–127.

Karhunen, J., & Ukkonen, T. (2007). Extending ICA for finding jointly dependent components from two related data sets. *Neurocomputing, 70*(16–18), 2969–2979.

Karoui, M. F., Alla, H., & Chatti, A. (2010). Monitoring of dynamic processes by rectangular hybrid automata. *Nonlinear Analysis: Hybrid Systems, 4*(4), 766–774.

Kassidas, A., MacGregor, J. F., & Taylor, P. (1998). Synchronization of batch trajectories using dynamic time warping. *AICHE Journal, 44*, 864–875.

Khediri, I. B., Weihs, C., & Limam, M. (2010). Support vector regression control charts for multivariate nonlinear autocorrelated processes. *Chemometrics and Intelligent Laboratory Systems, 103*, 76–81.

Khediri, I. B., Limam, M., & Weihs, C. (2011). Variable window adaptive Kernel Principal Component Analysis for nonlinear nonstationary process monitoring. *Computers and Industrial Engineering, 61*(3), 437–446.

Kosanovich, K. A., Piovoso, M. J., & Dahl, K. S. (1994). Multi-way PCA applied to an industrial batch process. In *Proceedings of the American Control Conference* (pp. 1294–1298). American Control Conference. The Stouffer Harborplace Hotel, Baltimore, MD, USA.

Kosanovich, K. A., Dahl, K. S., & Piovoso, M. J. (1996). Improved process understanding using multiway principal component analysis. *Industrial and Engineering Chemistry Research, 35*, 138–146.

Kourti, T., Nomikos, P., & MacGregor, J. F. (1995). Analysis, monitoring and fault diagnosis of batch processes using multiblock and multiway PLS. *Journal of Process Control, 5*, 277–284.

Kramer, M. A. (1991). Nonlinear principal component analysis using autoassociative neural networks. *AICHE Journal, 37*(2), 233–243.

Kramer, M. A. (1992). Autoassociative neural networks. *Computers and Chemical Engineering, 16*(4), 313–328.

Kresta, J. V., MacGregor, J. F., & Marlin, T. E. (1991). Multivariate statistical monitoring of process operating performance. *Canadian Journal of Chemical Engineering, 69*(1), 35–47.

Kruger, U., Zhou, Y., & Irwin, G. W. (2004). Improved principal component monitoring of large-scale processes. *Journal of Process Control, 14*(8), 879–888.

Ku, W., Storer, R. H., & Georgakis, C. (1995). Disturbance detection and isolation by dynamic principal component analysis. *Chemometrics and Intelligent Laboratory Systems, 30*(1), 179–196.

Kulkarni, S. G., et al. (2004). Modeling and monitoring of batch processes using principal component analysis (PCA) assisted generalized regression neural networks (GRNN). *Biochemical Engineering Journal, 18*, 193–210.

Lane, S., Martin, E. B., Kooijmans, R., & Morris, A. J. (2001). Performance monitoring of a multi-product semi-batch process. *Journal of Process Control, 11*, 1–11.

Lee, D. S., & Vanrolleghem, P. A. (2003). Monitoring of a sequencing batch reactor using adaptive multiblock principal component analysis. *Biotechnology and Bioengineering, 82*, 489–497.

Lee, J.-M., Yoo, C., & Lee, I.-B. (2004a). Statistical process monitoring with independent component analysis. *Journal of Process Control, 14*(5), 467–485.

Lee, J.-M., Yoo, C. K., & Lee, I.-B. (2004b). Enhanced process monitoring of fed-batch penicillin cultivation using time-varying and multivariate statistical analysis. *Journal of Biotechnology, 110*(2), 119–136.

Lee, D. S., Park, J. M., & Vanrolleghem, P. A. (2005). Adaptive multiscale principal component analysis for on-line monitoring of a sequencing batch reactor. *Journal of Biotechnology, 116*(2), 195–210.

Lee, C., Choi, S. W., & Li, I.-B. (2006a). Adaptive monitoring statistics based on state space updating using canonical variate analysis. *Computer Aided Chemical Engineering, 21*, 1545–1550.

Lee, J.-M., Qin, S. J., & Lee, I.-B. (2006b). Fault detection and diagnosis based on modified independent component analysis. *AICHE Journal, 52*(10), 3501–3514.

Legat, A., & Dolecek, V. (1995). Chaotic analysis of electrochemical noise measured on stainless steel. *Journal of the Electrochemical Society, 142*(6), 1851–1858.

Li, R., & Rong, G. (2006). Fault isolation by partial dynamic principal component analysis in dynamic process. *Chinese Journal of Chemical Engineering, 14*(4), 486–493.

Li, E., & Yu J. (2002). An input-training neural network based nonlinear principal component analysis approach for fault diagnosis. In *Proceedings of the 4th World Congress on Intelligent Control and Automation* (pp. 2755–2759). Piscataway: IEEE. Available at: http://ieeexplore.ieee.org/lpdocs/epic03/wrapper.htm?arnumber=1020023. Accessed 22 Dec 2011.

Li, W., Yu, H. H., Valle-Cervantes, S., & Qin, S. J. (2000). Recursive PCA for adaptive process monitoring. *Journal of Process Control, 10*(5), 471–486.

Li, X., Yu, Q., & Wang, J. (2003). Process monitoring based on wavelet packet principal component analysis. *Computer Aided Chemical Engineering, 14*, 455–460.

Licciardi, G., Del Frate, F., Schiavon, G., & Solimini, D. (2010). Dimensionality reduction of hyperspectral data: Assessing the performance of autoassociative neural networks. In *International Geoscience and Remote Sensing Symposium (IGARSS)* (pp. 4377–4380). IEEE. Available at: http://ieeexplore.ieee.org/lpdocs/epic03/wrapper.htm?arnumber=5652586. Accessed 24 Dec 2011.

Lieftucht, D., Völker, M., Sonntag, C., Kruger, U., Irwin, G. W., & Engell, S. (2009). Improved fault diagnosis in multivariate systems using regression-based reconstruction. *Control Engineering Practice, 17*, 478–493.

Liu, J., & Wong, D. S. H. (2008). Fault detection and classification for a two-stage batch process. *Journal of Chemometrics, 22*(6), 385–398.

Liu, X., Li, K., McAfee, M., & Irwin, G. W. (2011). Improved nonlinear PCA for process monitoring using support vector data description. *Journal of Process Control, 21*, 1306–1317.

Lopes, J., & Menezes, J. (2004). Multivariate monitoring of fermentation processes with non-linear modelling methods. *Analytica Chimica Acta, 515*(1), 101–108.

Lopez, I., & Sarigul-Klijn, N. (2009). Distance similarity matrix using ensemble of dimensional data reduction techniques: Vibration and aerocoustic case studies. *Mechanical Systems and Signal Processing, 23*(7), 2287–2300.

Lu, N., & Gao, F. (2005). Stage-based process analysis and quality prediction for batch processes. *Industrial and Engineering Chemistry Research, 44*(10), 3547–3555.

Lu, N., Gao, F., Yang, Y., & Wang, F. (2004). PCA based modeling and on-line monitoring strategy for uneven length batch processes. *Industrial and Engineering Chemistry Research, 43*, 3343–3352.

Lu, C.-T., Lee, T.-S., & Chin, C.-C. (2008). *Statistical process monitoring using independent component analysis based disturbance separation scheme* (pp. 232–237). IEEE. Available at: http://ieeexplore.ieee.org/lpdocs/epic03/wrapper.htm?arnumber=4633795. Accessed 22 Dec 2011.

MacGregor, J. F., & Kourti, T. (1995). Statistical process control of multivariate processes. *Control Engineering Practice, 3*(3), 403–414.

MacGregor, J. F., Jaeckle, C., Kiparessides, C., & Koutoudi, M. (1994). Processing monitoring and diagnosis by multiblock PLS methods. *AICHE Journal, 40*, 826–838.

Mahadevan, S., & Shah, S. L. (2009). Fault detection and diagnosis in process data using one-class support vector machines. *Journal of Process Control, 19*(10), 1627–1639.

Malthouse, E. C. (1998). Limitations of nonlinear PCA as performed with generic neural networks. *Neural Networks, IEEE Transactions on, 9*(1), 165–173.

Marjanovic, O., Lennox, B., Sandoz, D., Smith, K., & Crofts, M. (2006). Real-time monitoring of an industrial batch process. *Computers and Chemical Engineering, 30*(10–12), 1476–1481.

Markou, M., & Singh, S. (2003). Novelty detection: A review—Part 2: Neural network based approaches. *Signal Processing, 83*(12), 2499–2521.

Marseguerra, M., & Zoia, A. (2005). The autoassociative neural network in signal analysis: II. Application to on-line monitoring of a simulated BWR component. *Annals of Nuclear Energy, 32*(11), 1207–1223.

Matero, S., Poutiainen, S., Leskinen, J., Reinikainen, S.-P., Ketolainen, J., Järvinen, K., & Poso, A. (2009). Monitoring the wetting phase of fluidized bed granulation process using multi-way methods: The separation of successful from unsuccessful batches. *Chemometrics and Intelligent Laboratory Systems, 96*(1), 88–93.

Monroy, I., Villez, K., Graells, M., & Venkatasubramanian, V. (2011). Dynamic process monitoring and fault detection in a batch fermentation process. In *Computer aided chemical engineering* (pp. 1371–1375). Amsterdam: Elsevier. Available at: http://linkinghub.elsevier.com/retrieve/pii/B9780444542984500532. Accessed 26 Dec 2011.

Mu, S., Zeng, Y., Liu, R., Wu, P., Su, H., & Chu, J. (2006). Online dual updating with recursive PLS model and its application in predicting crystal size of purified terephthalic acid (PTA) process. *Journal of Process Control, 16*(6), 557–566.

Muthuswamy, K., & Srinivasan, R. (2003). Phase-based supervisory control for fermentation process development. *Journal of Process Control, 13*, 367–382.

Negiz, A., & Cinar, A. (1997). PLS, balanced, and canonical variate realization techniques for identifying VARMA models in state space. *Chemometrics and Intelligent Laboratory Systems, 38*(2), 209–221.

Nielsen, N. P. V., Carstensen, J. M., & Smedsgaard, J. (1998). Aligning of single and multiple wavelength chromatographic profiles for chemometric data analysis using correlation optimised warping. *Journal of Chromatography. A, 805*, 17–35.

Nomikos, P., & MacGregor, J. F. (1994). Monitoring batch processes using multiway principal component analysis. *AICHE Journal, 40*, 1361–1375.

Nomikos, P., & MacGregor, J. F. (1995a). Multiway partial least squares in monitoring batch processes. *Chemometrics and Intelligent Laboratory Systems, 30*, 97–108.

Nomikos, P., & MacGregor, J. F. (1995b). Multivariate SPC charts for monitoring batch processes. *Technometrics, 37*(1), 41–59.

Odiowei, P. P., & Cao, Y. (2009a). Nonlinear dynamic process monitoring using canonical variate analysis and kernel density estimations. *Computer Aided Chemical Engineering, 27*, 1557–1562.

Odiowei, P. P., & Cao, Y. (2009b). Nonlinear dynamic process monitoring using canonical variate analysis and kernel density estimations. *IEEE Transactions on Industrial Informatics, 6*(1), 36–45.

Odiowei, P. P., & Cao, Y. (2010). State-space independent component analysis for nonlinear dynamic process monitoring. *Chemometrics and Intelligent Laboratory Systems, 103*, 59–65.

Qi, Y., Wang, P., & Gao, X. (2011). Enhanced batch process monitoring and quality prediction using multi-phase dynamic PLS. In *Proceedings of the 30th Chinese Control Conference, CCC 2011* (pp. 5258–5263). Piscataway: IEEE.

Qin, S. J. (1998). Recursive PLS algorithms for data adaptive modelling. *Computers and Chemical Engineering, 22*(4), 503–514.

Qin, S. J. (2012). Survey on data-driven industrial process monitoring and diagnosis. *Annual Reviews in Control, 36*, 220–234.

Qin, S. J., Valle, S., & Piovoso, M. J. (2001). On unifying multiblock analysis with application to decentralized process monitoring. *Journal of Chemometrics, 15*(9), 715–742.

Rainikainen, S. P., & Höskuldsson, A. (2007). Multivariate statistical analysis of a multistep industrial process. *Analytica Chimica Acta, 595*, 248–256.

Ramaker, H.-J., Van Sprang, E. N. M., Gurden, S. P., Westerhuis, J. A., & Smilde, A. K. (2002). Improved monitoring of batch processes by incorporating external information. *Journal of Process Control, 12*, 569–576.

Ramaker, H.-J., Van Sprang, E. N. M., Westerhuis, J. A., & Smilde, A. K. (2005). Fault detection properties of global, local and time evolving models for batch process monitoring. *Journal of Process Control, 15*(7), 799–805.

Ranner, S., MacGregor, J. F., & Wold, S. (1998). Adaptive batch monitoring using hierarchical PCA. *Chemometrics and Intelligent Laboratory Systems, 73–81*.

Rosen, C., & Lennox, J. A. (2001). Multivariate and multiscale monitoring of wastewater treatment operation. *Water Research, 35*(14), 3402–3410.

Roweis, S. T., & Saul, L. K. (2000). Nonlinear dimensionality reduction by locally linear embedding. *Science, 290*, 2323–2326.

Russell, E. L., Chiang, L. H., & Braatz, R. D. (2000a). *Data-driven techniques for fault detection and diagnosis in chemical processes*. London: Springer.

Russell, E. L., Chiang, L. H., & Braatz, R. D. (2000b). Faut detection in industrial processes using canonical variate analysis and dynamic principal component analysis. *Chemometrics and Intelligent Laboratory Systems, 51*, 81–93.

Ryan, J., Lin, M., & Mikkulainen, R. (1998). Intrusion detection with neural networks. In *Advances in neural information processing systems* (Vol. 10). Cambridge, MA: MIT Press.

Sakoe, H., & Chiba, S. (1978). Dynamic programming algorithm optimization for spoken word recognition. *IEEE Transactions on Acoustics, Speech, and Signal Processing, 26*(1), 43–49.

Sammon, J. W. (1969). A nonlinear mapping for data structure analysis. *IEEE Transactions on Computers, C-18*(5), 401–409.

Schölkopf, B., Platt, J. C., Shawe-Taylor, J., Smola, A. J., & Williamson, R. C. (2001). Estimating the support of a high-dimensional distribution. *Neural Computation, 13*(7), 1443–1471.

Shao, J.-D., & Rong, G. (2009). Nonlinear process monitoring based on maximum variance unfolding projections. *Expert Systems with Applications, 36*(8), 11332–11340.

Shao, R., Jia, F., Martin, E. B., & Morris, A. J. (1999). Wavelets and non-linear principal components analysis for process monitoring. *Control Engineering Practice, 7*, 865–879.

Shao, J.-D., Rong, G., & Lee, J. M. (2009). Generalized orthogonal locality preserving projections for nonlinear fault detection and diagnosis. *Chemometrics and Intelligent Laboratory Systems, 96*(1), 75–83.

Shimizu, H., Yasuoka, K., Uchiyama, K., & Shioya, S. (1997). On-line fault diagnosis for optimal rice a-amylase production process of a temperature-sensitive mutant of Saccharomyces cerevisiae by an autoassociative neural network. *Journal of Fermentation and Bioengineering, 83*(5), 435–442.

Simoglou, A., Argyropoulos, P., Martin, E. B., Scott, K., Morris, A. J., & Taam, W. M. (2001). Dynamic modelling of the voltage response of direct methanol fuel cells and stacks Part I: Model development and validation. *Chemical Engineering Science, 56*, 6761–6772.

Simoglou, A., Martin, E. B., & Morris, A. J. (2002). Statistical performance monitoring of dynamic multivariate processes using state space modelling. *Computers and Chemical Engineering, 26*, 909–920.

Simoglou, A., Georgieva, P., Martin, E. B., Morris, A. J., & Feyo de Azevedo, S. (2005). On-line monitoring of a sugar crystallization process. *Computers and Chemical Engineering, 29*, 1411–1422.

Skov, T., van den Berg, F., Tomasi, G., & Bro, R. (2006). Automatic alignment of chromatographic data. *Journal of Chemometrics, 20*(11–12), 484–497.

Smilde, A. K., Westerhuis, J. A., & de Jong, S. (2003). A framework for sequential multiblock methods. *Journal of Chemometrics, 17*, 323–337.

Stefatos, G., & Ben Hamza, A. (2007). Statistical process control using kernel PCA. In *Mediterranean conference on Control and Automation* (pp. 1–6). Piscataway: IEEE. Available at: http://ieeexplore.ieee.org/lpdocs/epic03/wrapper.htm?arnumber=4433899. Accessed 25 Dec 2011.

Stefatos, G., & Ben Hamza, A. (2010). Dynamic independent component analysis approach for fault detection and diagnosis. *Expert Systems with Applications, 37*, 8606–8617.

Stubbs, S., Zhang, J., & Morris, A. J. (2009). Fault detection of dynamic processes using a simplified monitoring-specific CVA state space approach. *Computer Aided Chemical Engineering, 26*, 339–344.

Tan, S., & Mavrovouniotis, M. L. (1995). Reducing data dimensionality through optimising neural network inputs. *AICHE Journal, 41*(6), 1471–1480.

Tax, D. M. J., & Duin, R. P. W. (1999). Support vector domain description. *Pattern Recognition Letters, 20*(11–13), 1191–1199.

Tax, D. M. J., & Duin, R. P. W. (2004). Support vector data description. *Machine Learning, 54*(1), 45–66.

Tenenbaum, J., Silva, V., & Langford, J. (2000). A global geometric framework for nonlinear dimensionality reduction. *Science, 290*, 2319.

Thissen, U., Melssen, W. J., & Buydens, L. M. C. (2001). Nonlinear process monitoring using bottle-neck neural networks. *Analytica Chimica Acta, 446*, 371–383.

Tian, X., Zhang, X., Deng, X., & Chen, S. (2009). Multiway kernel independent component analysis based on feature samples for batch process monitoring. *Neurocomputing, 72*(7–9), 1584–1596.

Tianyang, C., Huaibo, Z., & Qingfeng, Y. (2011). A method for flame flicker frequency calculation with the empirical mode decomposition. In *3rd International Conference on Measuring Technology and Mechatronics Automation (ICMTMA)* (Vol. 1, pp. 104–106). Piscataway: IEEE. Available at: http://ieeexplore.ieee.org/lpdocs/epic03/wrapper.htm?arnumber=5720732. Accessed 22 Dec 2011.

Tomasi, G., van den Berg, F., & Andersson, C. (2004). Correlation optimized warping and dynamic time warping as preprocessing methods for chromatographic data. *Journal of Chemometrics, 18*, 231–241. doi:10.1002/cem.859.

Übeyli, E. D., & Güler, U. (2004). Detection of electrocardiographic changes in partial epileptic patients using Lyapunov exponents with multilayer perceptron neural networks. *Engineering Applications of Artificial Intelligence, 17*(6), 567–576.

Ündey, C., & Cinar, A. (2002). Statistical monitoring of multistage, multiphase batch processes. *IEEE Control Systems Magazine, 22*(5), 40–52.

Ündey, C., Ertunc, S., Tatara, E., Teymour, F., & Cinar, A. (2004). Batch process monitoring and its application to polymerization systems. *Macromolecular Symposia, 206*(1), 121–134.

Van Deventer, J. S. J., Aldrich, C., & Moolman, D. W. (1996). Visualisation of plant disturbances using self-organising maps. *Computers and Chemical Engineering, 20*, S1095–S1100.

Van Sprang, E. N. M., Ramaker, H.-J., Westerhuis, J. A., Gurden, S. P., & Smilde, A. K. (2002). Critical evaluation of approaches for on-line batch process monitoring. *Chemical Engineering Science, 57*(18), 3979–3991.

Van Sprang, E. N. M., Ramaker, H.-J., Westerhuis, J. A., Smilde, A. K., & Wienke, D. (2005). Statistical batch process monitoring using gray models. *AICHE Journal, 51*, 931–945.

Vedam, H., Venkatasubramanian, V., & Bhalodia, M. (1998). A B-spline based method for data compression, process monitoring and diagnosis. *Computers and Chemical Engineering, 22*((Supplement 1)), S827–S830.

Venkatasubramanian, V., Rengaswamy, R., Kavuri, S. N., & Yin, K. (2003). A review of process fault detection and diagnosis Part III: Process history based methods. *Computers and Chemical Engineering, 27*(3), 327–346.

Vermasvuori, M., Enden, P., Haavisto, S., & Jamsa-Jounela, S.-L. (2002). The use of Kohonen self-organizing maps in process monitoring. In *First international IEEE symposium on Intelligent Systems* (pp. 2–7). Piscataway: IEEE. Available at: http://ieeexplore.ieee.org/lpdocs/epic03/wrapper.htm?arnumber=1042576. Accessed 22 Dec 2011.

Wang, Q. (2008). *Use of topographic methods to monitor process systems*. M.Sc. thesis, University of Stellenbosch, Stellenbosch, South Africa.

Wang, J., & He, Q. P. (2010). Multivariate statistical process monitoring based on statistics pattern analysis. *Industrial and Engineering Chemistry Research, 49*(17), 7858–7869.

Wang, L., & Shi, H. (2010). Multivariate statistical process monitoring using an improved independent component analysis. *Chemical Engineering Research and Design, 88*(4), 403–414.

Weinberger, K. Q., Sha, F., & Saul, L. K. (2004). Learning a kernel matrix for nonlinear dimensionality reduction. In *Proceedings of the 21st International Conference on Machine Learning (ICML-04)* (pp. 839–846). Banff: ACM Press.

Westerhuis, J. A., & Coenegracht, P. M. J. (1997). Multivariate modelling of the pharmaceutical two-step process of wet granulation and tableting with multiblock partial least squares. *Journal of Chemometrics, 11*, 379–392.

Westerhuis, J. A., Kourti, T., & MacGregor, J. F. (1998). Analysis of multiblock and hierarchical PCA and PLS models. *Journal of Chemometrics, 12*, 301–321.

Wise, B. M., & Gallagher, N. B. (1996). The process chemometrics approach to process monitoring and fault detection. *Journal of Process Control, 6*(6), 329–348.

Xia, D., Song, S., Wang, J., Shi, J., Bi, H., & Gao, Z. (2012). Determination of corrosion types from electrochemical noise by phase space reconstruction theory. *Electrochemistry Communications, 15*(1), 88–92.

Xie, L., Zhang, J., & Wang, S. (2006). Investigation of dynamic multivariate chemical process monitoring. *Chinese Journal of Chemical Engineering, 14*(5), 559–568.

Xing, R., Zhang, S., & Xie, L. (2006). Nonlinear process monitoring based on improved kernel ICA. In *International conference on Computational Intelligence and Security* (pp. 1742–1746). Piscataway: IEEE. Availaxiaorg/lpdocs/epic03/wrapper.htm?arnumber=4076265. Accessed 25 Dec 2011.

Xu, J., & Hu, S. (2010). Nonlinear process monitoring and fault diagnosis based on KPCA and MKL-SVM. In *International conference on Artificial Intelligence and Computational Intelligence* (pp. 233–237). Piscataway: IEEE. Available at: http://ieeexplore.ieee.org/lpdocs/epic03/wrapper.htm?arnumber=5656754. Accessed 25 Dec 2011.

Xu, J., Hu, S., & Shen, Z. (2009). Fault detection for process monitoring using improved kernel principal component analysis. In *International conference on Artificial Intelligence and Computational Intelligence, component analysis (AICI '09)* (pp. 334–338). Piscataway: IEEE. Available at: http://ieeexplore.ieee.org/lpdocs/epic03/wrapper.htm?arnumber=5375775. Accessed 25 Dec 2011.

Xu, J., Hu, S., & Shen, Z. (2010). Combining KPCA with Sparse SVM for nonlinear process monitoring. In *Asia Pacific Power and Engineering Conference (APPEEC)* (pp. 1–4). Piscataway: IEEE. Available at: http://ieeexplore.ieee.org/lpdocs/epic03/wrapper.htm?arnumber=5448914. Accessed 25 Dec 2011.

Xuemin, T., & Xiaogang, D. (2008). A fault detection method using multi-scale kernel principal component analysis. In *Proceedings of the 27th Chinese Control Conference*, Kunming, Yunnan, China.

Yang, J., Zhang, D., Frangi, A. F., & Yang, J.-Y. (2004). Two-dimensional PCA: A new approach to appearance-based face representation and recognition. *IEEE Transactions on Pattern Analysis and Machine Intelligence, 26*(1), 131–137.

Yao, Y., & Gao, F. (2007). Batch process monitoring in score space of two-dimensional dynamic Principal Component Analysis (PCA). *Industrial and Engineering Chemistry Research, 46*(24), 8033–8043.

Yao, Y., & Gao, F. (2008a). Stage-oriented statistical batch process monitoring, quality prediction and improvement. In M. J. Chung & P. Misra (Eds.), *Proceedings of the IFAC World Congress, 17*(1), 4499–4510.

Yao, Y., & Gao, F. (2008b). Subspace identification for two-dimensional dynamic batch process statistical monitoring. *Chemical Engineering Science, 63*(13), 3411–3418.

Yao, Y., & Gao, F. (2009a). A survey on multistage/multiphase statistical modeling methods for batch processes. *Annual Reviews in Control, 33*(2), 172–183.

Yao, Y., & Gao, F. (2009b). Multivariate statistical monitoring of multiphase two-dimensional dynamic batch processes. *Journal of Process Control, 19*, 1716–1724.

Yao, Y., Chen, T., & Gao, F. (2010). Multivariate statistical monitoring of two-dimensional dynamic batch processes utilizing non-Gaussian information. *Journal of Process Control, 20*(10), 1188–1197.

Yoo, C. K., Lee, J.-M., Vanrolleghem, P. A., & Lee, I.-B. (2004). On-line monitoring of batch processes using multiway independent component analysis. *Chemometrics and Intelligent Laboratory Systems, 71*(2), 151–163.

Yoon, S., & MacGregor, J. F. (2004). Principal-component analysis of multiscale data for process monitoring and fault diagnosis. *AICHE Journal, 50*(11), 2891–2903.

Yu, J. (2012). A nonlinear kernel Gaussian mixture model based inferential monitoring approach for fault detection and diagnosis of chemical processes. *Chemical Engineering Science, 68*(10), 506–519.

Zhang, F. (2005). Bayesian neural networks for nonlinear multivariate manufacturing process monitoring. In *Proceedings of the International Joint Conference on Neural Networks, IJCNN '05* (pp. 2308–2312). IEEE . Available at: http://ieeexplore.ieee.org/lpdocs/epic03/wrapper.htm?arnumber=1556261. Accessed 24 Dec 2011.

Zhang, Y. (2009). Enhanced statistical analysis of nonlinear processes using KPCA, KICA and SVM. *Chemical Engineering Science, 64*(5), 801–811.

Zhang, Y., & Hu, Z. (2011). Multivariate process monitoring and analysis based on multi-scale KPLS. *Chemical Engineering Research and Design*. Available at: http://linkinghub.elsevier.com/retrieve/pii/S0263876211001857

Zhang, Y., & Ma, C. (2011). Decentralized fault diagnosis using multiblock kernel independent component analysis. *Chemical Engineering Research and Design*. Available at: http://linkinghub.elsevier.com/retrieve/pii/S0263876211003479. Accessed 21 Dec 2011.

Zhang, Y., & Qin, S. J. (2007). Fault detection of nonlinear processes using multiway kernel independent component analysis. *Industrial and Engineering Chemistry Research, 46*(23), 7780–7787.

Zhang, J., Martin, E. B., & Morris, A. J. (1997). Process monitoring using non-linear statistical techniques. *Chemical Engineering Journal, 67*(3), 181–189.

Zhang, X., Yan, W., Zhao, X., & Shao, H. (2006). Nonlinear on-line process monitoring and fault detection based on kernel ICA. In *IEEE international conference on Information and Automation* (pp. 222–227). Piscataway: IEEE. Available at: http://ieeexplore.ieee.org/lpdocs/epic03/wrapper.htm?arnumber=4250206. Accessed 25 Dec 2011.

Zhang, X., Yan, W., Zhao, X., & Shao, H. (2007). Nonlinear biological batch process monitoring and fault identification based on kernel Fisher discriminant analysis. *Process Biochemistry, 42*, 1200–1210.

Zhang, Y., Zhou, H., Qin, S. J., & Chai, T. (2010). Decentralized fault diagnosis of large-scale processes using multiblock kernel partial least squares. *IEEE Transactions on Industrial Informatics, 6*(1), 3–10.

Zhang, Y., Li, S., & Hu, Z. (2012). Improved multi-scale kernel principal component analysis and its application for fault detection. *Chemical Engineering Research and Design, 90*(9), 1271–1280.

Zhao, X., & Shao, H.-H. (2006). On-line batch process monitoring and diagnosing based on Fisher discriminant analysis. *Journal of Shanghai Jiaotong University, 11E*(3), 307–312.

Zhao, X., Yan, W., & Shao, H. (2006). Monitoring and fault diagnosis for batch process based on feature extract in Fisher subspace. *Chinese Journal of Chemical Engineering, 14*(6), 759–764.

Zhao, C., Wang, F., & Jia, M. (2007). Dissimilarity analysis based batch process monitoring using moving windows. *AICHE Journal, 53*, 1267–1277.

Zhao, C., Wang, F., Mao, Z., Lu, N., & Jia, M. (2008). Adaptive monitoring based on independent component analysis for multiphase batch processes with limited modeling data. *Industrial and Engineering Chemistry Research, 47*(9), 3104–3113.

Zhao, C., Wang, F., & Zhang, Y. (2009). Nonlinear process monitoring based on kernel dissimilarity analysis. *Control Engineering Practice, 17*(1), 221–230.

Zhao, C., Mo, S., Gao, F., Lu, N., & Yao, Y. (2011). Statistical analysis and online monitoring for handling multiphase batch processes with varying durations. *Journal of Process Control, 21*(6), 817–829.

Zhu, K., Wong, Y. S., & Hong, G. S. (2009). Wavelet analysis of sensor signals for tool condition monitoring: A review and some new results. *International Journal of Machine Tools and Manufacture, 49*(7–8), 537–553.

Žvokelj, M., Zupan, S., & Prebil, I. (2011). Non-linear multivariate and multiscale monitoring and signal denoising strategy using kernel principal component analysis combined with ensemble empirical mode decomposition method. *Mechanical Systems and Signal Processing, 25*(7), 2631–2653.

Nomenclature

Symbol	Description
λ	Eigenvalue
Λ	Scalar composite
θ	Composite variable, function of Λ_1, Λ_1 and Λ_3
$\mathbf{\Lambda}$	Diagonal matrix of eigenvalues
Λ	Aggregated subspace
Δ	Difference in length between two signal segments
χ^2	Chi-square statistic
λ_j^k	jth eigenvalue of kth data matrix
$\boldsymbol{\eta}_k$	kth state noise variable
$\sum_{\mathrm{ff}}$	Covariance matrix of zero mean scaled future process measurements, $\tilde{\mathbf{y}}_{\mathrm{f},k}$
$\sum_{\mathrm{fp}}$	Covariance matrix of zero mean called past and future process measurements, $\tilde{\mathbf{y}}_{\mathrm{p},k}$ and $\tilde{\mathbf{y}}_{\mathrm{f},k}$
Λ_j	jth subspace

(continued)

(continued)

Symbol	Description
$\boldsymbol{\varepsilon}_k$	kth measurement noise variable
$\sum_{pp}$	Covariance matrix of zero mean scaled past process measurements, $\tilde{\mathbf{y}}_{p,k}$
$\tilde{\mathbf{y}}_{f,k}$	kth zero mean scaled future process measurement
$\bar{\mathbf{y}}_{f,k}$	Mean future process measurement
$\tilde{\mathbf{y}}_{p,k}$	kth zero mean scaled past process measurement$\tilde{\mathbf{y}}_{f,k}$
$\bar{\mathbf{y}}_{p,k}$	kth mean past process measurement
$F_{\alpha,k,n-k}$	F statistic at α confidence level and degrees of freedom q and $N-q$; feature space
L_R	Length of target signal R
L_S	Length of unaligned signal S
P_j^k	jth loading of kth data matrix
P^k	Composite loading matrix
S_i	Covariance matrix of $\mathbf{Y}_i$
w_j^k	jth weight associated with kth loading
$\bar{x}$	Mean of variable x
$\vec{Y}$	Vector resulting from mapping of variables onto principal curve
$\hat{\mathbf{s}}$	Reconstructed vector of independent components, $\hat{\mathbf{s}} \in \Re^q$
$\mathbf{A}$	State matrix in a state space model; mixing matrix, $\mathbf{A} \in \Re^{M \times q}$
$\mathbf{a}$	CVA model coefficients associated with past measurements
a_1, a_2, a_3	Parameters of $g(\cdot)$
$\mathbf{B}$	Orthogonal matrix
$\mathbf{b}$	CVA model coefficients associated with future measurements; column vector of matrix $\mathbf{B}$
b_i	ith element of column vector $\mathbf{b}$
$\mathbf{C}$	Output matrix in state space model
c_α	Standard normal deviate corresponding to upper $(1-\alpha)$ percentile
D	Dissimilarity index between two data matrices; warp path
$d(\cdot)$	(Euclidean) distance measure
$\mathbf{e}$	Residual vector, $\mathbf{e} \in \Re^N$
$E(\cdot)$	Expectation operator
$\mathbf{e}_k$	Residual vector of canonical variate model
$f(\cdot)$	One-dimensional function defining principal curve; function in general
$g(\cdot)$	Objective functions representing non-Gaussianity; projection operator; function in general
g^*	First-order derivative of g
$\mathbf{I}$	Unity matrix
I, J, K	Dimensions of a batch array $I \times J \times K =$ (batches × variables × records)
J	Number of variables
$\mathbf{J}$	Transformation matrix
k	Lag parameter
M	Embedding dimension; number of variables
m'	Number of segments in target signal
N	Number of samples or observations
n'	Number of segments in unaligned signal
N_1, N_2	Number of samples in each of two different data sets
$\mathbf{P}$	Transformation matrix; loading matrix of principal component model, $\mathbf{P} \in \Re^{N \times q}$

(continued)

(continued)

Symbol	Description
$P(\cdot)$	Probability
$\mathbf{P}_0$	Orthogonal matrix
Q	Statistic of principal component model
$\mathbf{Q}$	Whitening matrix
$\mathbf{R}$	Covariance matrix of two data sets
r	Correlation between two variables
R,S	Two time series
$\mathbf{R}_i$	Covariance matrix of ith data set
$\mathbf{s}$	Vector of independent components, $\mathbf{s} \in \Re^q$
$\boldsymbol{t}$	Principal component score vector
t	Time
T^2	Hotelling's statistic of principal component model
$\mathbf{U}$	Matrix of eigenvectors
u	Arbitrary variable
$\mathbf{v}_k$	Measurement errors in state space model
$\mathbf{W}$	Unmixing matrix, $\mathbf{W} \in \Re^{M \times q}$
W	Sequence of grid points
w_j	jth grid point
$\mathbf{w}_k$	Plant disturbances in state space model
$\mathbf{x}(t)$	Measurement vector at time t
$\mathbf{X}^{\mathrm{E}}$	Lagged trajectory matrix
x_k	kth state variable
$\mathbf{x}_k$	kth state vector
y	Variable
$\mathbf{y}_{f,k}$	kth future process measurement
$\mathbf{Y}_i$	ith transformed data set $\mathbf{X}_i$
y_k	kth measured variable
$\mathbf{y}_k$	kth measurement vector
$\mathbf{y}_{\mathrm{p},k}$	kth past process measurement vector
$\mathbf{z}$	Whitened measurement vector, i.e. with no cross-correlation between components
$\mathbf{z}_k$	kth vector of canonical variates

Chapter 3
Artificial Neural Networks

3.1 Generalized Framework for Data-Driven Fault Diagnosis by the Use of Artificial Neural Networks

The two basic models, viz. the derivation of forward mapping and reverse mapping between process or system variables and features, can be handled in a variety of different ways by the use of neural networks, since this machine learning paradigm is exceptionally diverse, as will be highlighted in this chapter. Some models, like autoassociative neural networks, would enable the development of both forward and reverse mapping models simultaneously, similar to principal component models, while other approaches would require the use of different models for each of the operations.

3.2 Introduction

As a machine learning paradigm, artificial neural networks have originated from studies of the human brain. Although the brain is a complex organ that is not completely understood, it is clear that it operates in a massively parallel mode based on highly interconnected processing units or neurons. The typical human brain contains between 10^{10} and 10^{11} neurons, each of which can be connected to as many as 10^4 other neurons. As a result, the overall performance of the brain is highly advanced, despite the switching time of neurons being approximately a million times slower than the 10^{-9} s or so of the processors in modern computers.

The origins of the first artificial neural networks can be traced back to the McCulloch–Pitts model in the 1940s (McCulloch and Pitts 1943). The model contained all the necessary elements to perform logic operations, but implementation was not feasible with the bulky vacuum tubes prevalent at the time. This was followed by a book *The Organization of Behaviour* (Hebb 1949), as well as

C. Aldrich and L. Auret, *Unsupervised Process Monitoring and Fault Diagnosis with Machine Learning Methods*, Advances in Computer Vision and Pattern Recognition, DOI 10.1007/978-1-4471-5185-2_3, © Springer-Verlag London 2013

Hebb's proposed learning scheme for updating a neuron's connections. Although not technically significant, these ideas laid the foundation for developments in the 1950s, when the first neurocomputers that could adapt their connections automatically were built and tested (Minsky 1954). The *Snark* was a device that consisted of 300 vacuum tubes and 40 variable resistors (weights of the network) and could be trained to run a maze.

The first practical neurocomputer was invented by Frank Rosenblatt and co-workers in 1957 at the Cornell Aeronautics Laboratory. Called the perceptron, it was used for the recognition of characters mounted on an illuminated board. A 20×20 array of cadmium sulphide photosensors provided the input to the neural network, while an 8×8 array of servomotor driven potentiometers constituted the adjustable weights of the neural network.

This was followed by other artificial neural networks in the 1960s (Widrow and Hoff 1960) and related developments in the 1970s, and although some success was achieved in this early period, the machine learning theorems were too limited at the time to support application to more complicated problems. It was only in the 1980s that neural network research and development really gained momentum, with developments such as those of Amari (1977, 1990), Fukushima (1980), Kohonen (1977, 1982, 1984, 1988, 1990, 1995), Anderson et al. (1977), Grossberg and Carpenter (Grossberg 1976, 1982; Carpenter and Grossberg 1990), Hopfield (1982, 1984), Rumelhart and McClelland (1986) and Werbos (1974).

In general terms neural networks are therefore simply computers or computational structures consisting of large numbers of primitive process units connected on a massively parallel scale. These units, nodes or artificial neurons are relatively simple devices by themselves, and it is only through the collective behaviour of these nodes that neural networks can realize their powerful ability to form generalized representations of complex relationships and data structures. Technically speaking, artificial neural networks are electronic devices, but software simulations of these devices are also referred to as artificial neural networks or simply neural networks, where confusion with their biological equivalent is unlikely, and in this book the same conventions will be followed. Neural networks is a collective term to describe a diverse variety of computational structures, some of which will be described in more detail below, owing to their technical or industrial importance. These include multilayer perceptrons, radial basis function neural networks, Kohonen or self-organizing maps as well as a few variations of these.

3.3 Multilayer Perceptrons

Despite the fact that neural networks really denote a class of diverse methods or algorithms, multilayer perceptron neural networks are by far the most popular (and simple) neural networks in use in the process engineering industries, and they are therefore considered in more detail below.

In multilayer perceptron neural networks the processing nodes are usually divided into disjoint subsets or layers, as indicated in Fig. 3.1. A distinction is

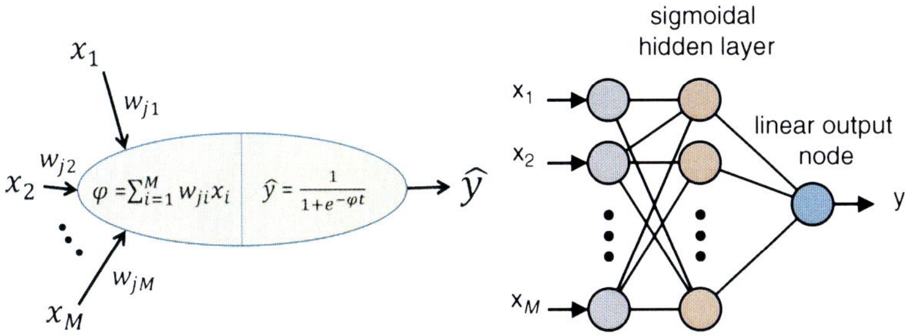

Fig. 3.1 Model of a single neuron (*left*) and structure of a typical multilayer perceptron neural network (*right*)

made between input, hidden and output layers, depending on their relation to the information environment of the network. The nodes in a particular layer are linked to other nodes in successive layers by means of artificial synapses or weighted connections (adjustable numeric values), as indicated by the bold lines connecting nodes in Fig. 3.1. These weights form the crux of the model, in that they define a distributed internal relationship between the input and output activations of the neural network. The development of neural network models thus consists of first determining the overall structure of the neural network (number of layers, number of nodes per layer, types of nodes, etc.). Once the structure of the network is fixed, the parameters (weights) of the network have to be determined. Unlike the case with a single node, a network of nodes requires that the output error of the network be apportioned to each node in the network.

3.3.1 Models of Single Neurons

Each node in the model consists of a processing element with a set of input connections, as well as a single output connection, as illustrated in Fig. 3.1. Each of these connections is characterized by a numerical value or weight, which is an indication of the strength of the connection. The flow of information through the node is unidirectional, as indicated by the arrows in this figure.

The output of the neuron can be expressed as follows:

$$z = f\left(\sum_{i=1}^{M} \mathbf{w}_i \mathbf{x}_i\right) = f\left(\mathbf{w}^{\mathrm{T}}\mathbf{x}\right) \tag{3.1}$$

where $\mathbf{w}$ is the weight vector of the neural node, defined as $\mathbf{w} = [w_1, w_2, w_3, \ldots, w_M]^{\mathrm{T}}$ and $\mathbf{x}$ is the input vector, defined as $\mathbf{x} = [\mathbf{x}_1, \mathbf{x}_2, \mathbf{x}_3, \ldots, \mathbf{x}_M]^{\mathrm{T}}$. Like all other vectors in the text, these vectors are column vectors, of which the superscript T denotes the transposition.

The function $f(\mathbf{w}^T\mathbf{x})$ is referred to as the activation function of the node, defined on the set of activation values, which are the scalar product of the weight and input vectors of the node. The argument of the activation function is sometimes referred to as the potential of the node, in analogy to the membrane potentials of biological neurons.

An additional input can be defined for some neurons, i.e. x_0, with associated weight w_0. This input is referred to as a bias and has a fixed value of -1. Like the other weights w_1, w_2, w_3, ..., w_M, the bias weight is also adaptable. The use of a bias input value is sometimes necessary to enable neural networks to form accurate representations of process trends, by offsetting the output of the neural network. Although the above model is used commonly in the application of neural networks, some classes of neural networks can have different definitions of potential ($\neq \mathbf{w}^T\mathbf{x}$). Also, in some neural networks, nodes can perform different functions during different stages of the application of the network.

Sigmoidal activation functions are used widely in neural network applications, that is, bipolar sigmoidal activations functions (with $\lambda > 0$):

$$f\left(\mathbf{w}^T\mathbf{x}\right) = \frac{2}{1 + e^{-\lambda\phi}} - 1 \tag{3.2}$$

and their hard-limiting equivalent (bipolar sign functions, or bipolar binary functions):

$$f(\mathbf{w}^T\mathbf{x}) = \text{sgn}(\phi) = +1, \quad \text{if } \phi > 0 \tag{3.3}$$

$$f(\mathbf{w}^T\mathbf{x}) = \text{sgn}(\phi) = -1, \quad \text{if } \phi < 0 \tag{3.4}$$

and unipolar sigmoidal activation functions:

$$f\left(\mathbf{w}^T\mathbf{x}\right) = \frac{1}{1 + e^{-\lambda\phi}} \tag{3.5}$$

with their hard-limiting version (unipolar sign functions or unipolar binary functions) the same as for the bipolar activations functions. The parameter λ is proportional to the gain of the neuron and determines the steepness of the continuous activation function. These functions are depicted graphically in Fig. 3.2. For obvious reasons, the sign function is also called a bipolar binary function.

3.3.2 *Training of Multilayer Perceptrons*

Training means fitting the weights of the neural network to a training data set. This is typically accomplished by first initializing the neural network (assigning small random values to the weights of the network, typically ranging from approximately

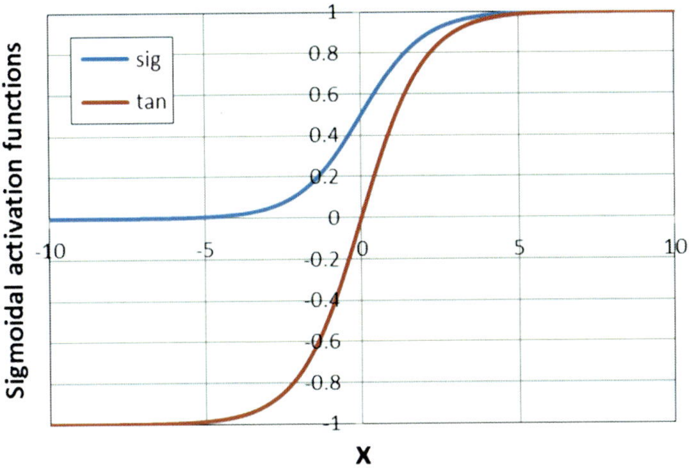

Fig. 3.2 Sigmoidal activation functions, (*blue*) unipolar, $sig = \frac{1}{1+e^{-x}}$ and bipolar (*red*), $\tan = \frac{1-e^{-x}}{1+e^{-x}}$

−0.1 to 0.1).[1] In the following step, some of the data records (exemplars) are presented to the initialized neural network, and the error generated by the difference between the output of the network and the actual output is subsequently used to iteratively update the weights of the network.

For example, given a multilayer perceptron with $n = 1, 2 \ldots, N$ weights, $p = 1, 2, \ldots, P$ training samples, and $m = 1, 2, \ldots, M$ outputs, trained with a steepest gradient descent algorithm, the gradient g or first-order error derivative of the total error function is used, i.e.

$$\mathbf{g}_k = \frac{\partial E(\mathbf{x}, \mathbf{w})}{\partial \mathbf{w}} = \begin{bmatrix} \frac{\partial E(\mathbf{x}, \mathbf{w})}{\partial \mathbf{w}_1} \\ \frac{\partial E(\mathbf{x}, \mathbf{w})}{\partial \mathbf{w}_2} \\ \vdots \end{bmatrix} \tag{3.6}$$

where

$$E(\mathbf{x}, \mathbf{w}) = \sum_{n=1}^{N} \sum_{m=1}^{N} e_{p,m}^2 \tag{3.7}$$

[1]It can be shown that regardless of the complexity of the network that small weight values amount to initially fitting a (hyper)plane through the data. With training (larger weight values), the plane gradually develops curvature to fit the data. Starting with zero-valued weights defeats the gradient descent algorithms typically used during training, as it usually means gradients cannot be calculated.

and $e_{p,m} = d_{p,m} - o_{p,m}$ is the training error at output m when sample p is presented to the neural network resulting from the difference between the desired output $d_{p,m}$ and the actual output $o_{p,m}$ of the network This gives a weight update rule of

$$\mathbf{w}_{k+1} = \mathbf{w}_k - \alpha \mathbf{g}_k \tag{3.8}$$

α is a learning coefficient or step size. More advanced algorithms can similarly be derived. This includes the Newton algorithm, with

$$\mathbf{w}_{k+1} = \mathbf{w}_k - \mathbf{H}_k^{-1} \mathbf{g}_k \tag{3.9}$$

where $\mathbf{H}_k$ is the Hessian matrix at the kth iteration, defined as

$$\mathbf{H}_k = \begin{bmatrix} \frac{\partial^2 E(\mathbf{x},\mathbf{w})}{\partial w_1^2} & \frac{\partial^2 E(\mathbf{x},\mathbf{w})}{\partial w_1 \partial w_2} & \cdots & \frac{\partial^2 E(\mathbf{x},\mathbf{w})}{\partial w_1 \partial w_N} \\ \frac{\partial^2 E(\mathbf{x},\mathbf{w})}{\partial w_2 \partial w_2} & \frac{\partial^2 E(\mathbf{x},\mathbf{w})}{\partial w_2^2} & \cdots & \frac{\partial^2 E(\mathbf{x},\mathbf{w})}{\partial w_2 \partial w_N} \\ \vdots & \vdots & \vdots & \vdots \\ \frac{\partial^2 E(\mathbf{x},\mathbf{w})}{\partial w_N \partial w_1} & \frac{\partial^2 E(\mathbf{x},\mathbf{w})}{\partial w_N \partial w_2} & \cdots & \frac{\partial^2 E(\mathbf{x},\mathbf{w})}{\partial w_N^2} \end{bmatrix} \tag{3.10}$$

the Gauss–Newton algorithm

$$\mathbf{w}_{k+1} = \mathbf{w}_k - \left(\mathbf{J}_k^{\mathrm{T}} \mathbf{J}_k\right)^{-1} \mathbf{J}_k \mathbf{e}_k \tag{3.11}$$

where $\mathbf{J}_k$ is the Jacobian matrix at the kth iteration, defined as

$$\mathbf{J}_k = \begin{bmatrix} \frac{\partial \mathbf{e}_{1,1}}{\partial w_1} & \frac{\partial \mathbf{e}_{1,1}}{\partial w_2} & & \frac{\partial \mathbf{e}_{1,2}}{\partial w_N} \\ \frac{\partial \mathbf{e}_{1,2}}{\partial w_1} & \frac{\partial \mathbf{e}_{1,2}}{\partial w_2} & \cdots & \frac{\partial \mathbf{e}_{1,2}}{\partial w_N} \\ \vdots & \vdots & \vdots & \vdots \\ \frac{\partial \mathbf{e}_{1,P}}{\partial w_1} & \frac{\partial \mathbf{e}_{1,P}}{\partial w_1} & & \frac{\partial \mathbf{e}_{1,P}}{\partial w_N} \\ \cdots & \cdots & & \cdots \\ \frac{\partial \mathbf{e}_{P,1}}{\partial w_1} & \frac{\partial \mathbf{e}_{P,1}}{\partial w_2} & & \frac{\partial \mathbf{e}_{P,1}}{\partial w_N} \\ \frac{\partial \mathbf{e}_{P,2}}{\partial w_1} & \frac{\partial \mathbf{e}_{P,2}}{\partial w_2} & \cdots & \frac{\partial \mathbf{e}_{P,2}}{\partial w_N} \\ \vdots & \vdots & & \vdots \\ \frac{\partial \mathbf{e}_{P,M}}{\partial w_1} & \frac{\partial \mathbf{e}_{P,M}}{\partial w_2} & & \frac{\partial \mathbf{e}_{P,M}}{\delta w_N} \end{bmatrix} \tag{3.12}$$

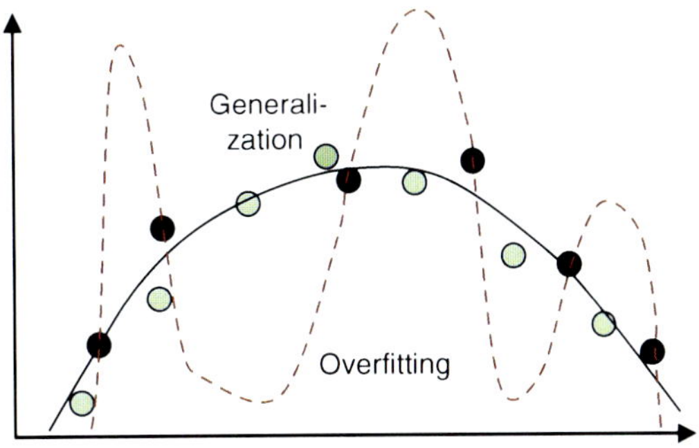

Fig. 3.3 Overfitting of data (*broken line*), compared with generalization (*solid line*) by a neural network. The *solid* and *empty circles* indicate training and test data, respectively

and $\mathbf{e}_k$ is an error vector of the form

$$\mathbf{e} = \begin{bmatrix} e_{1,1} \\ e_{1,2} \\ \vdots \\ e_{1,M} \\ \vdots \\ e_{p,1} \\ e_{p,2} \\ \vdots \\ e_{p,P} \end{bmatrix}. \tag{3.13}$$

Likewise, for the popular Levenberg–Marquardt algorithm,

$$\mathbf{w}_{k+1} = \mathbf{w}_k - \left(\mathbf{J}_k^{\mathrm{T}}\mathbf{J}_k + \lambda\mathbf{I}\right)^{-1}\mathbf{J}_k\mathbf{e}_k \tag{3.14}$$

where λ is a coefficient, very large values of which amount to standard gradient descent, while very small values amount to the Newton method.

$\mathbf{I}$ is simply a unity matrix, i.e. $\mathbf{I} = \begin{bmatrix} 1\ 0 \ldots \\ 0\ 1 \ldots \\ \vdots\ \vdots \ddots \end{bmatrix}$.

Training or learning by the neural network is terminated when the network has learnt to generalize the underlying trends or relationships exemplified by the data. Generalization implies that the neural network can interpolate sensibly at points not contained in its training set, as indicated in Fig. 3.3. The ability of the neural network to do so is typically assessed by means of cross-validation, where the performance of the network is evaluated against a separate set of test data, not used during training.

Neural networks, such as the multilayer perceptron, are known as universal approximators, that is, they can fit any continuous curve or response surface to an arbitrary degree of accuracy, given sufficient data. However, great care should be taken not to use these models to extrapolate, as they are not designed for that purpose (typical of nonlinear models). This is a real problem in process engineering, as process data tend to change with time and it is something that the user should be aware of.

3.4 Neural Networks and Statistical Models

Despite their different origins, neural networks are closely related to or equivalent to statistical models, especially multivariate methods such as generalized linear models, regression methods, discriminant analysis, principal component analysis and cluster analysis.

For example, consider a simple multilayer perceptron with no hidden layer and an output layer with linear activation functions, as shown in Fig. 3.4a, (for m input variables and a single output y). This network is equivalent to a multiple linear regression model, while the almost identical network shown in Fig. 3.4b, with a sigmoidal output node, is equivalent to a logistic regression model. The activation functions in multilayer perceptrons are analogous to the inverse of the link functions used in generalized linear models (McCullagh and Nelder 1989). Activation functions are usually bounded, while inverse link functions, such as identity, reciprocal or exponential functions, tend not to be.

If the activation function of the network is changed to a threshold function, $f(x) = 0$, if $x < 0$, or $f(x) = 1$ otherwise, then the network is equivalent to a linear discriminant model, suitable for the classification of data into two classes, as indicated in Fig. 3.4c. With only one output, these networks are also referred to as ADALINE networks, while MADALINE networks can accommodate more than one output. Note that instead of threshold functions, logistic functions can also be used to estimate the conditional probabilities of each class.

Figure 3.4d shows a functional link neural network, with a hidden layer that expands its single output into polynomials of orders 0 to p (the bias node is not shown). This neural network is clearly equivalent to a polynomial regression model. In general, functional links can be any transformation not requiring additional parameters, i.e. only one layer of weights need to be determined. Finally, in Fig. 3.4e, a multilayer perceptron with a sigmoidal hidden layer is shown.

This network is equivalent to multiple nonlinear regression. If the input nodes are also connected direct to the output nodes (sometimes referred to as a fully connected neural network), these connections can be seen as equivalent to what are referred to as main effects in the statistical literature. Since multilayer perceptrons with nonlinear activation functions are nonlinear models in the true sense of the word (i.e. nonlinear in their parameters), they tend to take more computer time to fit than polynomials or splines. On the other hand, they may also be more stable numerically, than high-order polynomials.

Moreover, multilayer perceptrons do not require the specification of knot positions and also have different extrapolation properties than their statistical

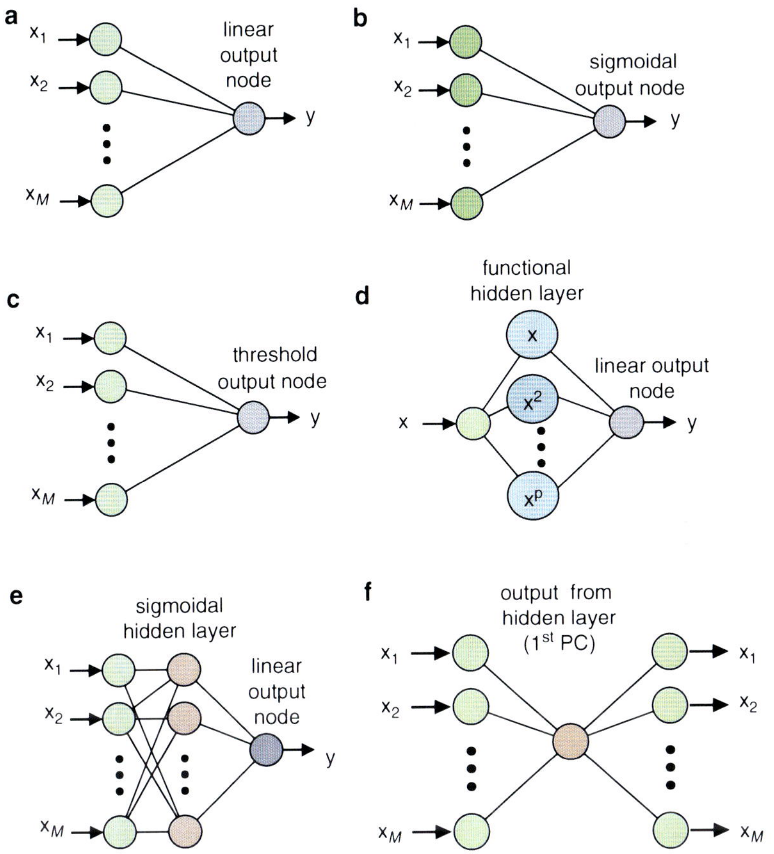

Fig. 3.4 Regression and classification with neural networks, i.e. the neural network equivalent of (**a**) linear regression, (**b**) logistic regression, (**c**) linear discriminant analysis, (**d**) polynomial regression, (**e**) nonlinear (non-parametric) regression and (**f**) principal component analysis

counterparts. Polynomials extrapolate to infinity, whereas multilayer perceptrons tend to flatten out. Nonetheless, this does not somehow make them more suitable for extrapolation than polynomial models.

Also important is the fact that nonlinear statistical models are supported by extensive statistical theory, allowing the estimation of various diagnostics, confidence intervals, prediction intervals, etc. These options are generally not readily available when multilayer perceptron models are developed.

The same parallels can be drawn with principal component analysis. The extraction of features from data by the use of neural networks is typically accomplished via unsupervised learning. Unsupervised learning entails the formulation of a suitable criterion, which has to be optimized during the training of the neural network. For example, the extraction of features from the data should be such that the original observations can be reconstructed from the features.

In the simple neural network structure shown in Fig. 3.4f, a set of variables $x_1, x_2, \ldots, x_m$ is presented to a multilayer perceptron with linear activation nodes. The multilayer perceptron has a hidden layer with a single node, and the target variables are identical to the input variables. The neural network is therefore forced to reconstruct the original variables from a single feature. As will be shown in more detail later on, this feature is equivalent to the first principal component of the data. Although this is not a particularly efficient way of extracting principal components, the addition of hidden layers before and after the hidden layer shown in Fig. 3.4f allows the extraction of nonlinear principal components from the data.

These examples can be extended to cluster analysis, where competitive neural networks, such the self-organizing map of Kohonen (1995) and adaptive resonance theory neural networks, can be shown to be closely related to k-means cluster analysis and the leader algorithm, respectively. The same goes for kernel-based methods, such as embodied in the probabilistic neural network and the general regression neural network (Specht 1991), that are direct derivatives of their statistical counterparts.

3.5 Illustrative Examples of Neural Network Models

3.5.1 Example 1

In the following example, two input variables x_1 and x_2 should be used to classify the data generated artificially as shown in Fig. 3.5. The data belong to two uniformly distributed classes distributed to resemble part of a chessboard, and the idea is to

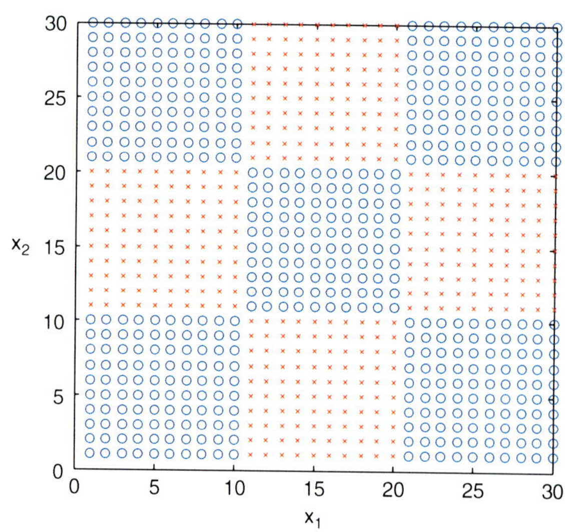

Fig. 3.5 Two classes ("x" and "o") as a function of two variables x_1 and x_2

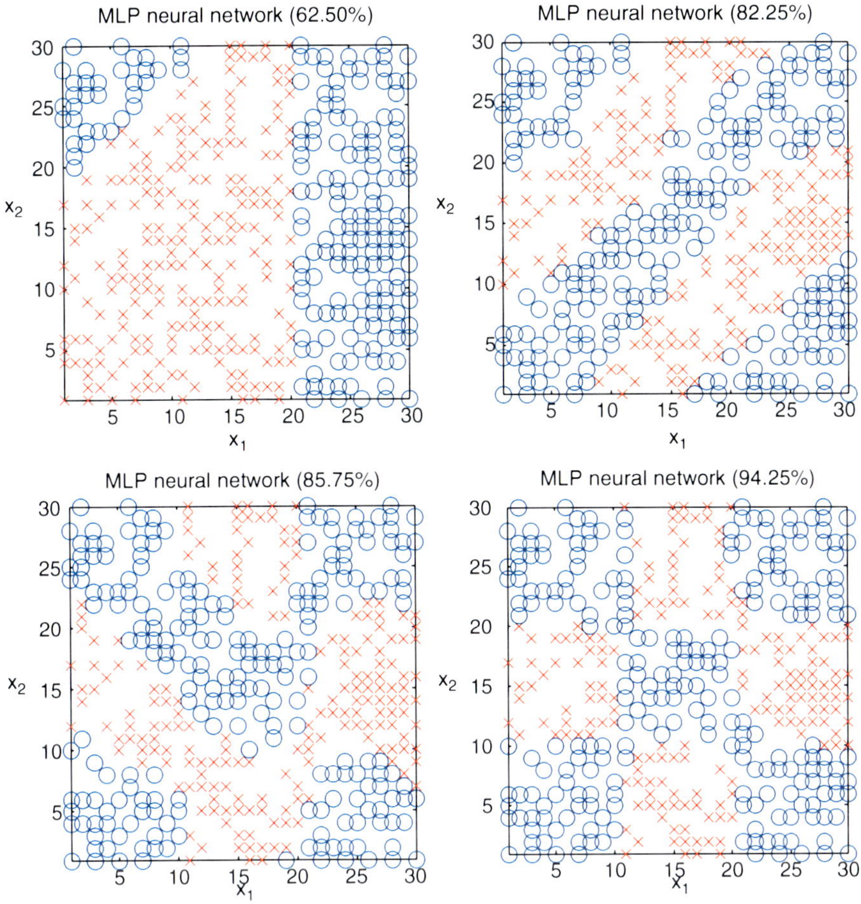

Fig. 3.6 Classification of data in Fig. 3.5 by means of a multilayer perceptron with a single hidden layer containing 2, 4, 6 and 12 tansigmoidal nodes (clockwise from *top left*) and a single node linear output layer

develop a model capable of classifying any point (x_1, x_2) correctly as either class "x" or class "o". The problem is relatively complicated, owing to the nonlinear boundaries separating the two classes.

Figure 3.6 shows the averaged results obtained with multilayer perceptrons containing from 2 to 12 hidden nodes. The networks with relatively few hidden nodes were unable to capture the boundaries between the two classes with a high degree of accuracy. In each case the neural network was trained with the Levenberg–Marquardt rule to 200 epochs.

3.5.2 Example 2

In the second example, the response surface $z = 1.21(1 + 2x_1 + 2x_1^2)(1 + 2x_2 + 2x_2^2)$ $\exp(-\frac{1}{2}x_1^2 - \frac{1}{2}x_2^2)$ shown in Fig. 3.7a has to be approximated by a neural network. The multilayer perceptron neural network has two input nodes (for x_1 and x_2) and a single output node for the response surface. The response surface generated by neural networks with one (b–e) or two (f) hidden layers is shown in Fig. 3.7. In all case, the hidden nodes had tansigmoidal activation functions, and the training was done with the Levenberg–Marquardt algorithm. The R^2 values indicated that the quality of the fits is also indicated in these figures.

3.5.3 Example 3

As was mentioned previously, neural networks (and other nonlinear models) can be used to interpolate or generalize trends exemplified by data with a high degree of accuracy, but are not good at extrapolating. In example 3, this issue is explored further.

In all cases, multilayer perceptron neural networks were used to approximate the relationship between x_1 and x_2, that is, $x_2 = -x_1^2$. The training data consisted of samples $x_1 \in [-3,-1] \cup [1,3]$, and all the networks had tansigmoidal hidden nodes and linear output nodes.

As can be seen from these figures, the more complex the network, the poorer its ability to extrapolate in the unshaded regions. In contrast, a simple neural network with only two hidden nodes was able to extrapolate reasonably well (Fig. 3.8d).

3.5.4 Interpretation of Multilayer Perceptron Models

A major drawback of neural network models is that information is encoded implicitly in the weights of the network and is therefore not directly available for interpretation (Gevrey et al. 2003). Nonetheless, several approaches have recently been proposed to assess the influence of input variables on response variables contained in neural network models. These methods include general influence measures, where the weights of the neural networks are analysed (Howes and Crook 1999; Nord and Jacobsson 1998). Other techniques concentrate on the effects observed via variation of input variables. Some of these methods are discussed in more detail below.

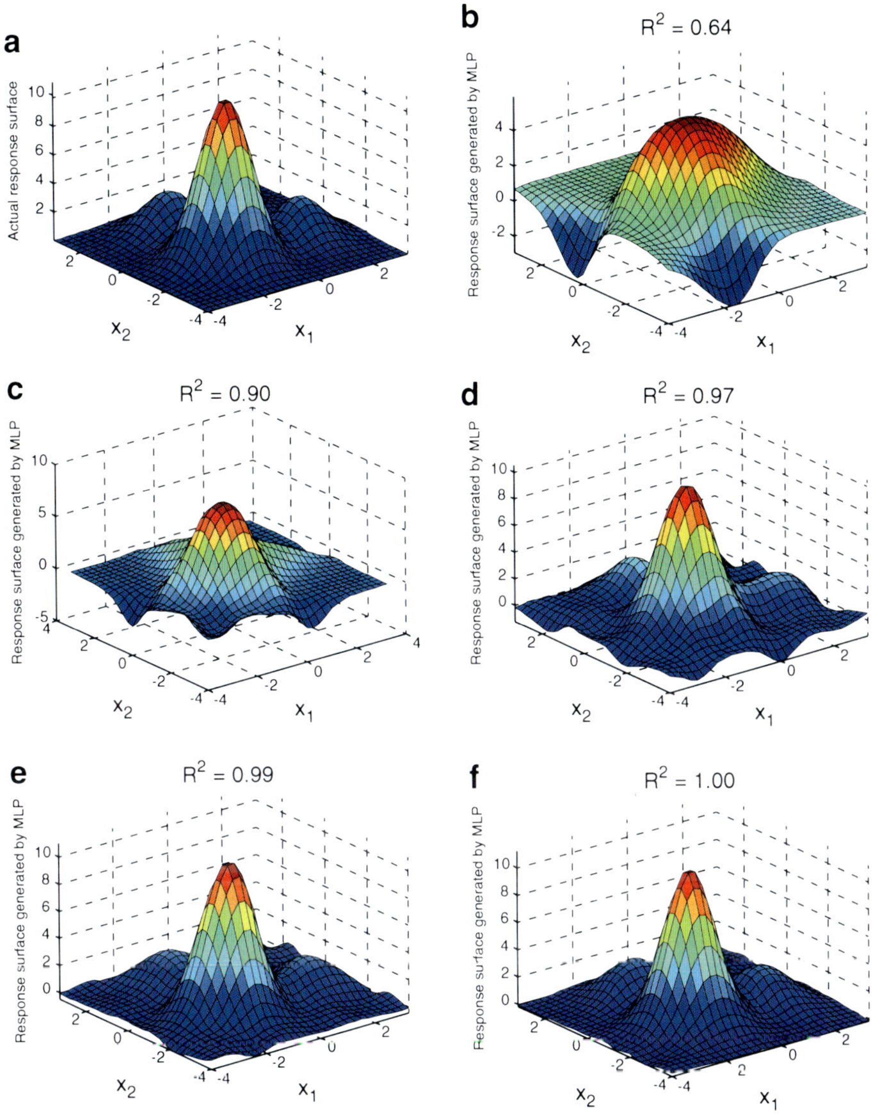

Fig. 3.7 Approximation of response surface $z = 1.21(1 + 2x_1 + 2x_1^2)(1 + 2x_2 + 2x_2^2)\exp(-x_1^2/2-x_2^2/2)$ in (**a**), with a tansigmoidal multilayer perceptron with (**b**) 3, (**c**) 6, (**d**) 12, (**e**) 15 and (**f**) 6 × 3 hidden nodes (two hidden layers)

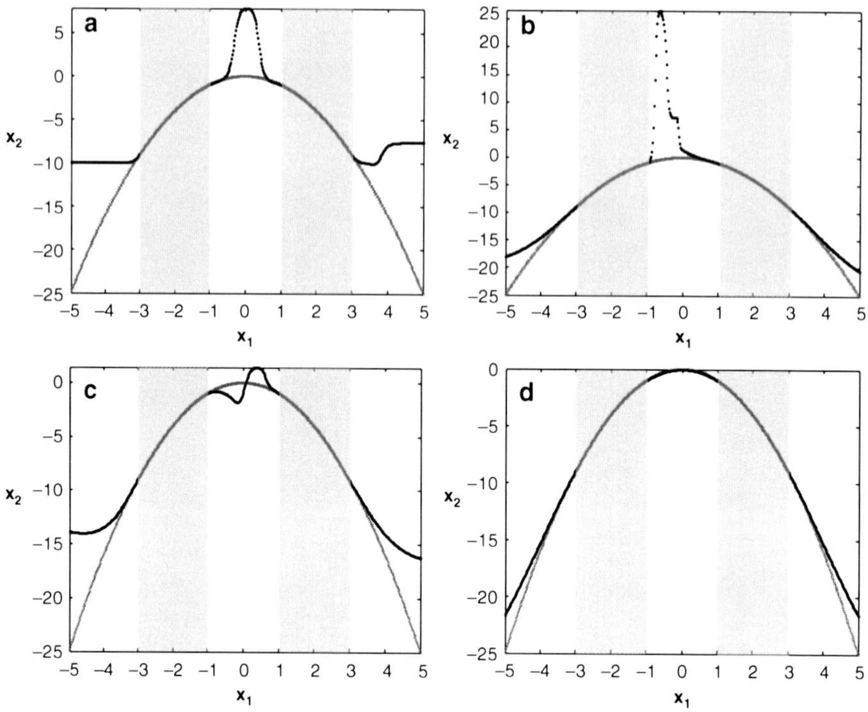

Fig. 3.8 Results of interpolation and extrapolation with a multilayer perceptron neural network with (**a**) two hidden layers with 12 and 6 tansigmoidal nodes, (**b**) two hidden layers with 6 and 3 tansigmoidal nodes, (**c**) one hidden layer with 6 tansigmoidal nodes and (**d**) one hidden layer with 2 tansigmoidal nodes. All the networks had a single linear output node and were trained with the Levenberg–Marquardt algorithm to fit the function $x_2 = -x_1^2$. The training data are indicated by the shaded regions, i.e. $x_1 \in [-3,-1] \cup [1,3]$

3.5.5 *General Influence Measures*

Howes and Crook (1999) have proposed assessment of the influence of input variables via analysis of the weights. In a fully connected neural network with m input nodes, a single hidden layer with k nodes and one output node, the partial influence, the partial influence (I_{ji}) of the ith input variable through the jth hidden node is given by

$$I_{ji} = \frac{w_{ji} w_j^h}{\left|w_{j0}\right| + \sum_{j=1}^{k} \left|w_{ji}^h\right|} \tag{3.15}$$

where w_{ji} connects to the jth hidden neuron, w_{j0} is the bias of the hidden neuron and w_j^h is the weight connecting the jth hidden neuron to the output node. The general influence of the ith input node can then be calculated as follows (Papadokonstantakis et al. 2006):

$$G_i = \frac{\sum_{j=1}^{k} |I_{ji}|}{|w_{o0}| + \sum_{j=1}^{k} \left| w_{ji}^h \right|} \tag{3.16}$$

3.5.6 *Sequential Zeroing of Weights*

The sequential zeroing of weights approach (Nord and Jacobsson 1998) is based on setting all the weights of the ith input node to the hidden layer to zero to exclude the contribution of this variable to the network response. The influence of the variable (V_i) is subsequently calculated as follows:

$$V_i = e_{\text{RMSE},i} - e_{\text{RMSE}} \tag{3.17}$$

where $e_{\text{RMSE},i}$ is the root-mean-squared error of the trained network with exclusion of the input of the ith variable and e_{RMSE} is the root-mean-squared error of the trained network without exclusion of any variables. The greater V_i, the more important the variable.

3.5.7 *Perturbation Analysis of Neural Networks*

Perturbation analysis is one of the earliest approaches to isolating the effects of individual variables in neural network models, such as described by Van der Walt et al. (1993). With perturbation methods, the inputs of the neural network are subjected to small changes, one input at a time, while keeping all the other inputs at constant values (Yao et al. 1998; Scardi and Harding 1999). The input variable whose perturbation affects the response the most is deemed to be the most important variable. Typical variable perturbations range from $\delta - 10$ to 50 %.

3.5.8 *Partial Derivatives of Neural Networks*

With this approach, the partial derivatives of the output of a neural network with respect to its input are calculated (Dimopoulos et al. 1995, 1999). For a network with M inputs, a hidden layer with k neurons and one output node, the partial derivatives of the output (y) with respect to the ith input, x_i, ($i = 1, 2, \ldots, N$ observations), are

$$d_{ij} = S_j \sum_{j=1}^{k} w_{jo} I_{ji} \left(1 - I_{ji}\right) w_{ij} \tag{3.18}$$

assuming that the activation functions of the network are logistic. S_j is the derivative of the output neuron with respect to its input, I_{ji} is the response of the jth hidden neuron while w_{jo} and w_{ij} are the weights connecting the output neuron and the jth hidden neuron and connecting the ith input neuron and the jth hidden neuron.

The influence of each input variable can be assessed by plotting the partial derivatives of the variables. A negative value of the partial derivative at a specific observation means that the response will decrease with an increase in the input variable and vice versa.

Variable contributions or variable importance analyses can also be done with partial derivatives. In this case, calculations are based on sum of square derivatives ($S_{\mathrm{SD},i}$) for each variable:

$$S_{\mathrm{SD},i} = \sum_{i=1}^{N} d_{ji}^2. \tag{3.19}$$

The higher $S_{\mathrm{SD},i}$ the more important the variable.

3.6 Unsupervised Feature Extraction with Multilayer Perceptrons

Autoassociative neural networks or autoencoders are multilayer perceptron neural networks that are trained with identical input and output data. As a consequence, the number of nodes in the input and output layers (M) is the same (and equal to the number of variables to be compressed). In these networks, a hidden layer with fewer nodes (p) than the input layer (M) acts as a bottleneck that forces the network to represent the inputs as best as possible with fewer features than M. Unless the mapped variables have the same intrinsic dimension as the number of nodes in the bottleneck layer of the neural network, the identity mapping from inputs to outputs is generally not attainable.

Instead of using single hidden layer autoassociative neural networks, the use of such networks with multiple hidden layers has been proposed by Kramer (1991, 1992). These networks are similar to single hidden layer autoassociative neural networks, except that they usually have three hidden layers.

3.6.1 Standard Autoassociative Neural Networks

The topology of the standard autoassociative neural network typically comprises five layers, viz. an input layer, three hidden layers and an output layer. The second hidden layer contains a few nodes only and forms a bottleneck in the network. Figure 3.9 shows the generic structure of a standard autoassociative neural network with a single node in its bottleneck layer.

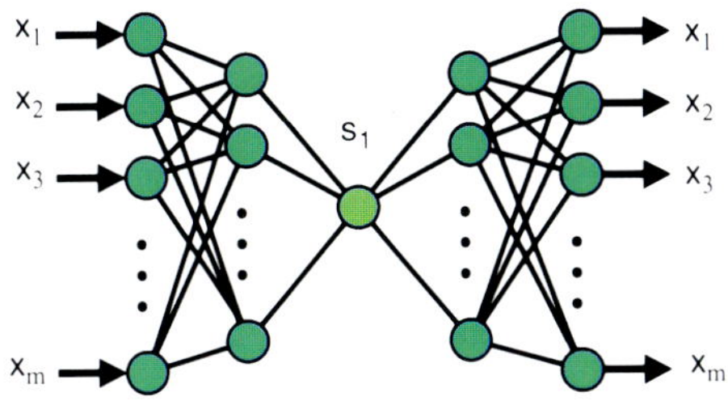

Fig. 3.9 An autoassociative neural network with three hidden layers

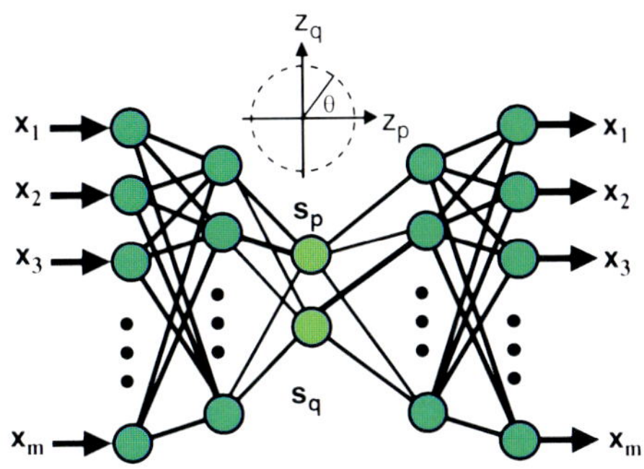

Fig. 3.10 An autoassociative neural network with three hidden layers

3.6.2 *Circular Autoassociative Neural Networks*

The use of a circular unit in the bottleneck layer of the autoencoder allows circular or highly curved data structures to be represented by closed curves (Kirby and Miranda 1996). As indicated in Fig. 3.10, a circular consists of a pair of nodes p and q whose output values s_p and s_q are constrained to lie on a unit circle, i.e.

$$s_p^2 + s_q^2 = 1. \tag{3.20}$$

This, the two units collectively represent a single variable θ, so that

$$s_p = \sin(\theta) \text{ and } s_q = \cos(\theta). \tag{3.21}$$

Training of the network is the same as for the standard multilayer perceptron, except that the arguments of the activation functions of the two nodes are first corrected before the outputs of the bottleneck nodes are calculated, i.e.

$$a_p = \frac{\sum_{i=1}^{m} w_{p,i} z_i}{\sqrt{\left(\sum_{i=1}^{m} w_{p,i} z_i\right)^2 + \left(\sum_{i=1}^{m} w_{q,i} z_i\right)^2}} \tag{3.22}$$

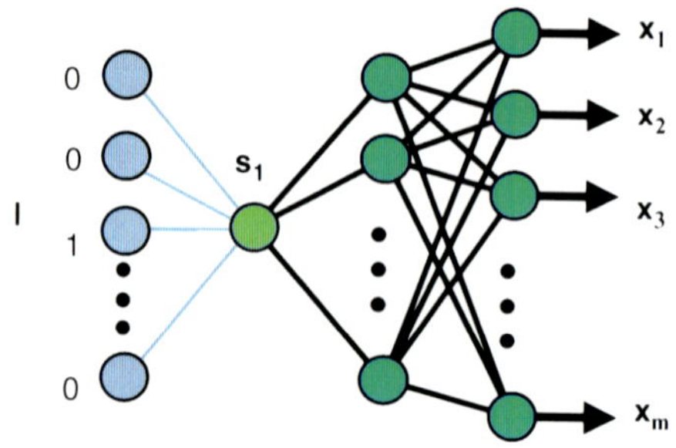

Fig. 3.11 An inverse autoassociative neural network, consisting of the 2nd reconstructivepart of the standard model only

and

$$a_q = \frac{\sum_{i=1}^{m} w_{q,i} z_i}{\sqrt{\left(\sum_{i=1}^{m} w_{p,i} z_i\right)^2 + \left(\sum_{i=1}^{m} w_{q,i} z_i\right)^2}} \tag{3.23}$$

The training algorithm is described in more detail elsewhere (Scholz 2007).

3.6.3 Inverse Autoassociative Neural Networks

The feature extraction problem can also be defined as an inverse problem, where the inputs matching a given output are estimated, instead of estimating the outputs from a given set of inputs. This is referred to as a *blind inverse problem*, since the model or data generating process is unknown. With the inverse approach, a single error function is used to simultaneously estimate both the input data and the parameters of the model, the details of which are described among other by Scholz et al. (2008).

Figure 3.11 shows the structure of an inverse autoassociative neural network, which consists of the constructive part of the standard model only. The component values of z are unknown inputs that are estimated by backpropagation of the partial errors to the input layer z indicated in light blue. The weights of the layer represent the component values of the input layer, while the inputs are then a sample-by-sample identity matrix I. For the 3rd sample, all inputs are zero, except the 3rd, which is one, as indicated in the Fig. 3.11.

Inverse models have the advantage that in principle they can deal with intersecting curves, i.e. where input data for two or more different features are identical. In addition, the approach can be extended to accommodate incomplete data sets, since the data are only used to determine the error of the model output.

3.6.4 Hierarchical Autoassociative Neural Networks

With hierarchical feature extraction, the features are not equivalent, but are extracted in a constrained sequence, so that the addition of features progressively complies

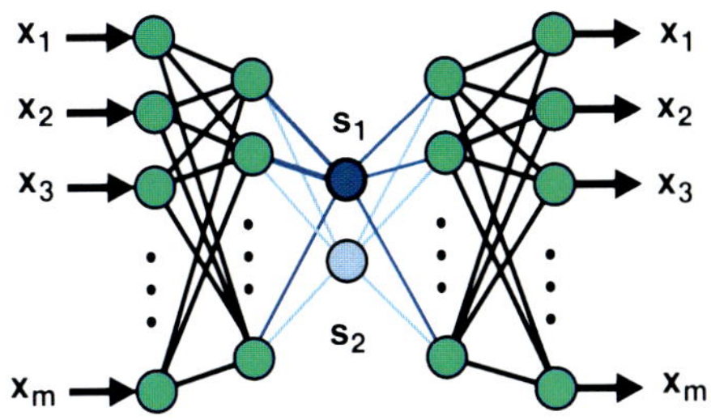

Fig. 3.12 Extraction of two components with a hierarchical autoassociative neural network where the subnetworks associated with each component are considered explicitly in the error function

with an optimization criterion. A hierarchical autoassociative neural network can be seen as a nonlinear equivalent of principal component analysis, where the extracted features explain the variance of the data, with the first feature explaining as much as possible of the variance, followed by the 2nd feature capturing as much of the variance as possible from the residuals of the 1st feature, the 3rd feature as much of the variance as possible from the residuals of the 2nd feature, etc. Learning is therefore asymmetric, unlike the standard approach.

This can be accomplished in different ways. The most obvious is by extracting features sequentially with single node bottleneck layers. After extraction of a feature, the data are deflated and the next component is then extracted from the deflated data. The process is repeated until all components have been extracted.

A different approach, proposed by Scholz and Vigário (2002) and Scholz et al. (2008), extends the standard model hierarchically by explicitly considering the subnetworks in the model, as indicated in Fig. 3.12. In this figure, the extraction of two hierarchical components is illustrated. The first component (s_1) is represented by the top unit that is the bottleneck layer, while the second component (s_2) is represented by the bottom unit. The total error of the network is expressed as a function of the errors of the subnetworks, $E = E_1 + E_{1,2}$, where E_1 is the error of the subnetwork associated with the first component (s_1) and $E_{1,2}$ is the collective error of the two subnetworks.

This approach is more effective than the sequential deflationary approach in which the components are successively extracted from the residuals of the previous components, since it is difficult to interpret the variance remaining after the extraction of each successive component.

In the following examples, the data are located in a subspace of the original variable space and can be represented by factors that can be extracted by nonlinear methods, such as autoassociative neural networks.

3.6.5 *Example 1: Nonlinear Principal Component Analysis (NLPCA) with Autoassociative Neural Networks*

Networks with a single hidden node in the bottleneck layer are similar to principal component methods, and both give the principal component solution when s_f and g

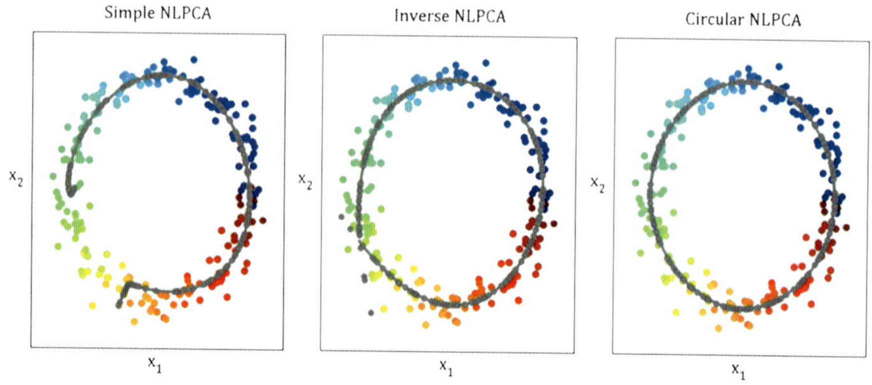

Fig. 3.13 An autoassociative neural network with three hidden layers

are linear. Moreover, the neural network defines a function $g:\Re^1 \rightarrow \Re^M$, which is a curve in $\Re^M$. The network also defines a function s_f: $\Re^M \rightarrow \Re^1$. The functions g and s_f are both continuous in the case of the autoassociative neural network.

In the following example, 200 data points were generated on two variables, x_1, and x_2, constrained to lie on circumference of circle with unit radius one. Random Gaussian noise with a standard deviation of 0.1 was added to both variables. Standard autoassociative artificial neural network, inverse autoassociative artificial neural network and circular autoassociative artificial neural network models were built using Scholz's NLPCA toolbox (Scholz et al. 2008; Scholz 2011), to extract a single component with maximum of 1,000 iterations.

Linear activations functions were used in the input, output and combination of feature layers, while the remaining layers consisted of nodes with hyperbolic tangent activation functions. Prescaling of the data was applied as follows:

Scaling factor for simple NLPCA:	0.2/max(std(x))
Scaling factor for inverse NLPCA:	0.1/max(std(x))
Scaling factor for circular NLPCA:	0.1/max(std(x))

The input spaces with reconstructions are shown in Fig. 3.13 below. Exploratory work suggested that non-inverse NLPCA (i.e. bottleneck versions, such as simple NLPCA and hierarchical NLPCA) requires larger scaling factors to allow nonlinear manifolds. In contrast, the inverse models were prone to overfitting when the scaling factors were too large.

In this figure, samples are coloured according to the rank of the generating angle $\theta \in [0,2\pi]$; $x_1 = \cos(\theta)$; $x_2 = \sin(\theta)$. The grey line in each of the panels indicates the 1D component in the input space, while grey markers indicate the positions of reconstructed points. The mean sum of the squared residuals (MSSR) for each case is summarized in Table 3.1. The standard encoder used an unscaled input variable range.

Table 3.1 Mean sum of the squared residuals of the different one-dimensional encoder models

Encoder	Standard	Inverse	Circular
MSSR	0.0529	0.0081	0.0086

3.6.6 Example 2: Nonlinear Principal Component Analysis (NLPCA) with Autoassociative Neural Networks

500 data points were generated on three variables x_1, x_2 and x_3. The first two variables were constrained to lie on circumference of circle with a radius of 6 units, and random noise was added to all variables x_1 and x_2 as before. The third variable x_3 consisted of random uniform data in the range [0,8] to yield an overall 3D distribution of data distributed on the face of a cylinder.

As before, standard autoassociative artificial neural network, inverse autoassociative artificial neural network and circular autoassociative artificial neural network models were built using Scholz's NLPCA toolbox (Scholz et al. 2008; Scholz 2011). In addition, a hierarchical autoassociative neural network was built. These networks were designed to extract two components, trained to a maximum of 2,000 iterations. The error function of the hierarchical models was calculated as $E_{\text{total}} = E_1 + E_{12} + 0.1{*}E_2$, where E_1 is the error for the hierarchical layer including only the first component, E_{12} is the error for the hierarchical layer with both components and E_2 is the hierarchical layer including only the second component.

Linear activation functions were again used in the input, output and combination of feature layers, while the remaining layers used the hyperbolic tangent activation function. The data were prescaled as follows:

Scaling factor for standard NLPCA:	0.2/max(std(x_i)) for $i = 1,2,3$
Scaling factor for hierarchical NLPCA:	0.2/max(std(x_i)) for $i = 1,2,3$
Scaling factor for inverse NLPCA:	0.1/max(std(x_i)) for $i = 1,2,3$
Scaling factor for circular NLPCA:	0.1/max(std(x_i)) for $i = 1,2,3$

Figure 3.14 shows the reconstructed input spaces as a mesh grid in the data space for each of the different autoencoder configurations. The data are coloured according to the rank of the generating angle $\theta \in [0,2\pi]$; $x_1 = 6{*}\cos(\theta)$; $x_2 = 6{*}\sin(\theta)$. The grey surfaces indicate the two-dimensional manifolds in input space, while the grey markers indicate the positions of the reconstructed data. The quality of the reconstructed data is summarized in Table 3.2 via the values of the mean sum of the squared residuals of the manifolds. For the standard encoder in Table 3.2, the ranges of the variables in the 3D data set differed, but a constant scaling factor was retained to emphasis the circularity of the data for fitting.

The feature spaces of the models are shown in Fig. 3.15, with samples coloured according to the rank of the generating angle $\theta \in [0,2\pi]$; $x_1 = 6{*}\cos(\theta)$; $x_2 = 6{*}\sin(\theta)$. The numbers in parentheses indicate the variance of that specific

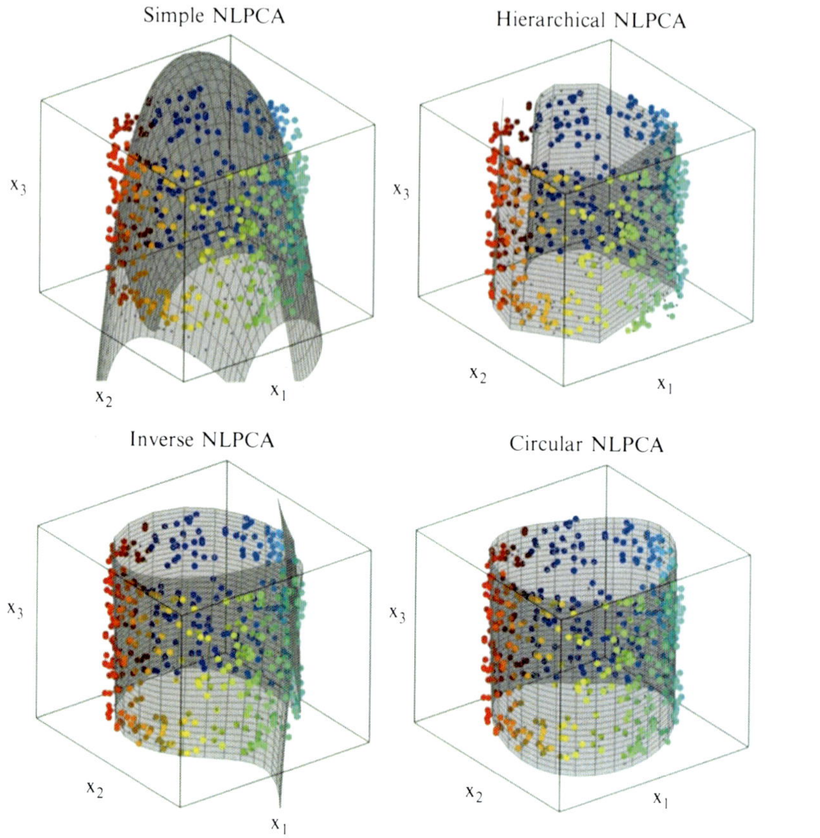

Fig. 3.14 Manifolds fitted to data lying on the exterior of a cylinder by the use of a standard autoassociative neural network (*top, left*), a hierarchical autoassociative neural network (*top, right*), an inverse autoassociative neural network (*bottom, left*) and a circular autoassociative neural network (*bottom, right*)

Table 3.2 Mean sum of the squared residuals of the different two-dimensional encoder models

Encoder	Standard	Hierarchical	Inverse	Circular
MSSR	0.631	1.219	0.0914	0.1085

feature. Note that these results do not show larger variance of the second simple NLPCA component compared to the first.

As can be seen from the top left figure, the standard autoassociative neural network has attempted to bend an initially flat 2D surface to fir the data, while the other configurations of autoassociative neural networks could identify the manifolds more easily.

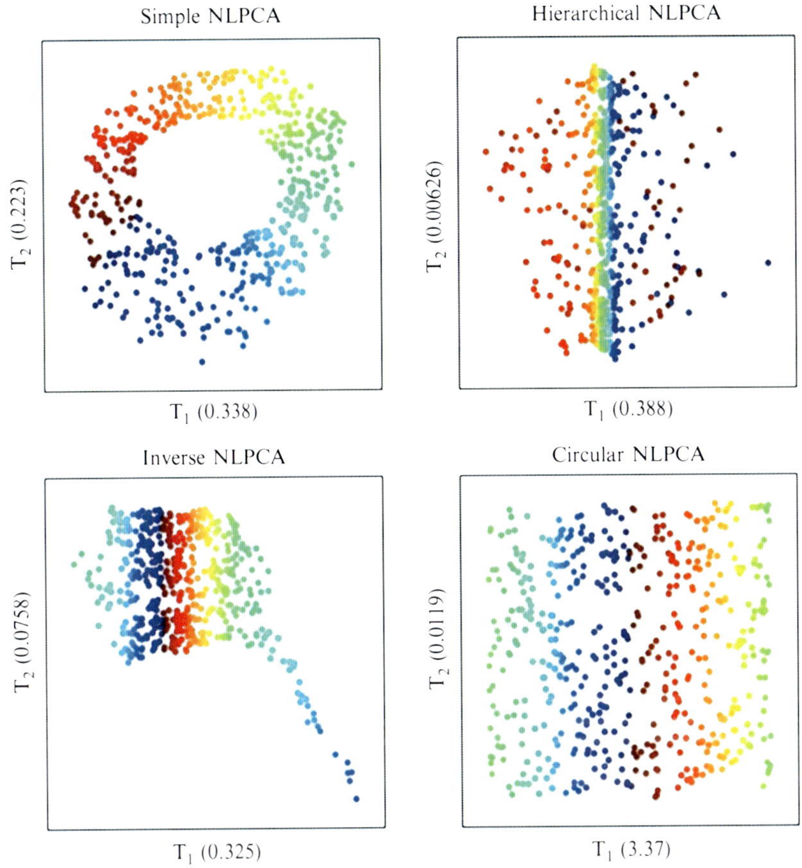

Fig. 3.15 Feature space generated by the use of a standard autoassociative neural network (*top, left*), a hierarchical autoassociative neural network (*top, right*), an inverse autoassociative neural network (*bottom, left*) and a circular autoassociative neural network (*bottom, right*)

3.7 Radial Basis Function Neural Networks

When solving problems concerning nonlinearly separable patterns, there is practical benefit to be gained in mapping the input space into a new space of sufficiently high dimension. This nonlinear mapping in effect turns a nonlinearly separable problem into a linearly separable one. The idea is illustrated in Fig. 3.16, where two interlocked two-dimensional patterns are easily separated by mapping them to three dimensions where they can be separated by a flat plane. In the same way it is possible to turn a difficult nonlinear approximation problem into an easier linear approximation problem.

Consider therefore without loss of generality a feedforward neural network with an input layer with p input nodes, a single hidden layer and an output layer with

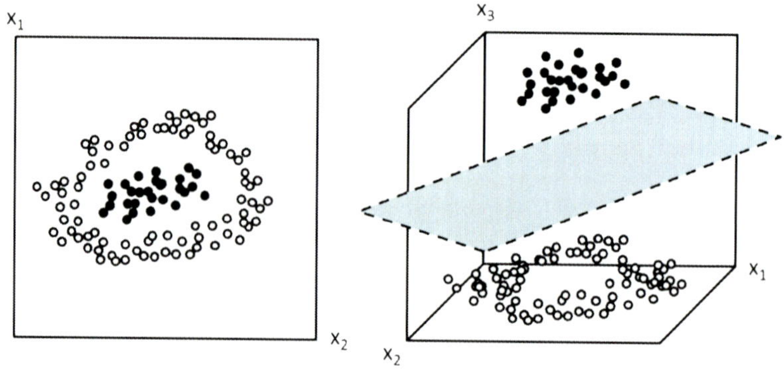

Fig. 3.16 Linear separation of two nonlinearly separable classes indicated by black and white markers (*left*), after mapping to a higher dimension (*right*) and fitting a decision plane

one node. This network is designed to perform a nonlinear mapping from the input space to the hidden space and a linear mapping from the hidden space to the output space.

Overall the network represents a mapping from the M-dimensional input space to the one-dimensional output space, as follows, s: $\Re^M \to \Re^1$ and the map s can be thought of as a hypersurface $\Gamma \subset \Re^{M+1}$ in the same way as we think of the elementary map s: $\Re^1 \to \Re^1$, where $s(x) = x^2$, as a parabola drawn in $\Re^2$-space. The curve Γ is a multidimensional plot of the output as a function of the input. In practice the surface Γ is unknown but exemplified by a set of training data (input–output pairs).

As a consequence, training constitutes a fitting procedure for the hypersurface Γ, based on the input–output examples presented to the neural network. This is followed by a generalization phase, which is equivalent to multivariable interpolation between the data points, with interpolation performed along the estimated constrained hypersurface (Powell 1985).

In a strict sense the interpolation problem can be formulated as follows: Given a set of N different observations on m variables $\{\mathbf{x}_i \in \Re^M \mid i = 1, 2, \ldots, N\}$ and a corresponding set of N real numbers $\{z_i \in \Re^1 \mid i = 1, 2, \ldots, N\}$, find a function F: $\Re^M \to \Re^1$ that complies with the interpolation condition $F(\mathbf{x}_i) = z_i$, for $i = 1, 2, \ldots, N$. Note that in the strict sense specified, the interpolation surface is forced to pass through all the training data points.

Techniques based on radial basis functions are based on the selection of a function F of the following form:

$$F(\mathbf{x}) = \sum_{i=1}^{n} w_i \varphi(\|\mathbf{x} - \mathbf{x}_i\|) \tag{3.24}$$

where $\{\varphi(\|\mathbf{x}-\mathbf{x}_i\|) \mid i = 1, 2, \ldots, N\}$ is a set of n arbitrary functions, known as radial basis functions. $\|\cdot\|$ denotes a norm that is usually Euclidean. The known data points $\boldsymbol{x}_i$ typically form the centres of the radial basis functions. Examples of such functions are multiquadrics, $\varphi(r) = (r^2 + c^2)^{1/2}$, inverse multiquadrics, $\varphi(r) = (r^2 + c^2)^{-1/2}$, Gaussian functions $\varphi(r) = \exp\{-r^2/(2\sigma^2)\}$ and thin-plate splines $\varphi(r) = (r/\sigma)\log(r/\sigma)$, where c and σ are positive constants and $r \in \Re$.

By the use of the interpolation condition $F(\mathbf{x}_i) = z_i$ and Eq. (3.25), a set of simultaneous linear equations for the unknown coefficients or weights (w_i) of the expansion can be obtained:

$$\begin{bmatrix} \varphi_{11} & \varphi_{12} & \cdots & \varphi_{1N} \\ \varphi_{21} & \varphi_{22} & \cdots & \varphi_{2N} \\ \vdots & \vdots & \vdots & \vdots \\ \varphi_{N1} & \varphi_{N2} & \cdots & \varphi_{NN} \end{bmatrix} \begin{bmatrix} w_1 \\ w_2 \\ \vdots \\ w_N \end{bmatrix} = \begin{bmatrix} z_1 \\ z_2 \\ \vdots \\ z_N \end{bmatrix} \tag{3.25}$$

where $\varphi_{ij} = \varphi(||\mathbf{x}_i - \mathbf{x}_j||)$, for $i, j = 1, 2, \ldots, N$. Moreover, the $N \times 1$ vectors $\mathbf{w} = [w_1, w_2, \ldots, w_N]^{\mathrm{T}}$ and $\mathbf{z} = [z_1, z_2, \ldots, z_N]^{\mathrm{T}}$ represent the linear weight vector and target or desired response vector, respectively. With $\boldsymbol{\Phi} = \{\varphi_{ij} \mid i, j = 1,2, \ldots, N\}$ the $N \times N$ interpolation matrix, $\boldsymbol{\Phi}\mathbf{w} = \mathbf{z}$, represents a more compact form of the set of simultaneous linear equations.

For a certain class of radial basis functions, such as inverse multiquadrics (Eq. 3.25) and Gaussian functions (Eq. 3.26), the $N \times N$ matrix $\boldsymbol{\Phi}$ is positive definite:

$$\varphi(r) = \frac{1}{\sqrt{r^2 + c^2}} \tag{3.26}$$

for $c > 0$ and $r \geq 0$, and

$$\varphi(r) = e^{-\frac{r^2}{2\sigma^2}} \tag{3.27}$$

for $\sigma > 0$ and $r \geq 0$

If all the data points are distinct and the matrix Φ is positive definite, then the weight vector can be obtained from $\mathbf{w} = \Phi^{-1}\mathbf{z}$. If the matrix is arbitrarily close to singular, perturbation of the matrix can help to solve for $\mathbf{w}$. These radial basis functions are used for interpolation, where the number of basis functions is equal to the number of data points.

Although the theory of radial basis function neural networks is intimately linked with that of radial basis functions themselves (a main field of study in numerical analysis), there are some differences. For example, with radial basis function neural networks, the number of basis functions need not be equal to the number of data points and is typically much less. Moreover, the centres of the radial basis functions need not coincide with the data themselves, and the widths of the basis functions also do not need to be the same. The determination of suitable centres and widths for the basis functions is usually part of the training process of the network. Finally,

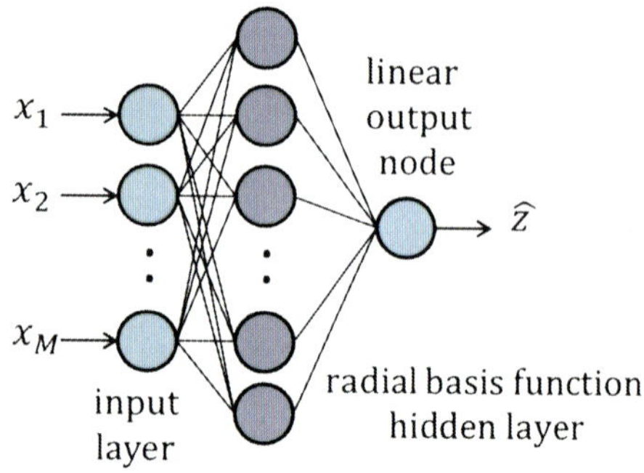

Fig. 3.17 Structure of a radial basis function neural network

bias values are typically included in the linear sum associated with the output layer to compensate for the difference between the average value of the targets and the average value of the basis functions over the data set (Bishop 2007).

In its most basic form, the construction of a radial basis function neural network involves three different types of layers. These networks typically consist of input layers, hidden (pattern) layers as well as output layers, as shown in Fig. 3.17. The input nodes (one for each input variable) merely distribute the input values to the hidden nodes (one for each exemplar in the training set) and are not weighted. In the case of multivariate Gaussian functions, the hidden node activation functions can be described by

$$z_{ij}\left(x_j, \alpha_i, \beta_i\right) e^{-\frac{\left\|\alpha_i - x_j\right\|^2}{\beta_i^2}} \tag{3.28}$$

where $\mathbf{x}_j = \{x_1, x_2, \ldots, x_M\}_j$ is the jth input vector of dimension m presented to the network and $z_{ij}\left(x_j, \alpha_i, \beta_i\right)$ is the activation of the ith node in the hidden layer in response to the jth input vector $\mathbf{x}_j$. $M + 1$ parameters are associated with each node, viz. $\boldsymbol{\alpha}_i = \{\alpha_1, \alpha_2, \ldots, \alpha_M\}_i$, as well as $\boldsymbol{\beta}_i$, a distance scaling parameter which determines the distance in the input space over which the node will have a significant influence.

The parameters $\boldsymbol{\alpha}_i$ and $\boldsymbol{\beta}_i$ function in much the same way as the mean and standard deviation in a normal distribution. The closer the input vector to the pattern of a hidden unit (i.e. the smaller the distance between these vectors, the stronger the activity of the unit. The hidden layer can thus be considered to be a density function for the input space and can be used to derive a measure of the probability that a new input vector is part of the same distribution as the training vectors. Note that the training of the hidden units is unsupervised, i.e. the pattern layer representation is constructed solely by self-organization.

Whereas the $\boldsymbol{\alpha}_i$ vectors are typically found by vector quantization, the $\boldsymbol{\beta}_i$ parameters are usually determined in an ad hoc manner, such as the mean distance to the first k nearest $\boldsymbol{\alpha}_i$ centres. Once the self-organizing phase of training is complete, the output layer can be trained using standard least mean square error techniques.

Each hidden unit of a radial basis function network can be seen as having its own receptive field, which is used to cover the input space. The output weights leading from the hidden units to the output nodes subsequently allow a smooth fit to the desired function. Radial basis function neural networks can be used for classification, pattern recognition and process modelling and can model local data more accurately than multilayer perceptron neural networks. They perform less well as far as representation of the global properties of the data is concerned.

The classical approach to training of radial basis function neural networks consists of unsupervised training of the hidden layer, followed by supervised training of the output layer, as explained in more detail below.

3.7.1 Estimation of Clusters Centres in Hidden Layer

The distribution of the hidden nodes in the input space (unsupervised training of the hidden layer) is similar to the k-means clustering procedure in that the data are partitioned into a set number of clusters or segments. It can be summarized as follows:

- Start with a random set of cluster k centres $\mathbf{c} = \{c_1, c_2, \ldots, c_k\}$.
- Read rth input vector $\mathbf{x}_r$.
- Modify closest cluster centre (the learning coefficient η is usually reduced with time):

$$c_k^{\text{new}} = c_k^{\text{old}} + \eta \left(x_r - c_k^{\text{old}} \right). \tag{3.29}$$

- Terminate after a fixed number of iterations, or when $\eta = 0$.

Once the hidden nodes have been distributed in the input space, their regions of influence can be adjusted by setting the widths of the basis functions.

3.7.2 Estimation of Width of Activation Functions

- The width of the transfer functions of each of the Gaussian kernels or receptive fields is based on a P nearest neighbour heuristics (Moody and Darken 1989):

$$\sigma_k = \frac{1}{P} \sqrt{\sum_{p=1}^{P} \left\| c_k - c_{kp} \right\|^2} \tag{3.30}$$

where c_{kp} represents the pth nearest neighbour of the kth cluster c_k.

3.7.3 Training of the Output Layer

The output layer is trained by minimization of a least-squares criterion and is equivalent of parameter estimation in linear regression, i.e. it does not involve a lengthy process, since there is only one linear (output) layer.

In summary, when compared with multilayer perceptrons:

- Radial basis function neural networks have single hidden layers, whereas multilayer perceptrons can have more than one hidden layer. It can be shown that radial basis function neural networks require only one hidden layer to fit an arbitrary function (as opposed to the maximum of two required by multilayer perceptrons). This means that training is considerably faster in radial basis function networks.
- In contrast with radial basis function neural networks, a common neuron model can be used for all the nodes in a multilayer perceptron. In radial basis function networks, the hidden layer neurons differ markedly from those in multilayer perceptrons.
- In radial basis function networks the hidden nodes are nonlinear and the output nodes linear, while in multilayer perceptrons the hidden and output nodes can be linear or nonlinear.

One of the drawbacks of kernel-based approximations (such as radial basis function neural networks) is that they suffer from the so-called curse of dimensionality, which is associated with the exponential increase in the required number of hidden nodes with an increase in the dimensionality of the input space. These problems are particularly acute in large-scale problems, such as those concerned with image analysis and speech recognition.

Other approaches to the training of radial basis function neural networks include one-phase learning, where only the output weights are adjusted through some kind of supervised optimization. The centres are subsampled from the input data, and the widths of the Gaussians are all equal and predefined. Support vector learning is a special example of one-phase training. Variants of two-phase learning differ mainly in the way that the radial basis functions are determined. For example, (Kubat 1998) has proposed a method to transform the disjoint hyperrectangular regions represented by the leaves of a decision tree into a set of centres and scaling parameters in order to initialize radial basis function neural networks. Similarly, Kohonen's learning vector quantization can also be used to determine prototypes for basis functions. Finally, three-phase learning (Schwenker et al. 2001) entails separate training of the hidden and output layers of the neural network, followed by a third phase of optimization of the entire network architecture. An example of the output of a radial basis function neural network on a classification problem is shown in Fig. 3.18.

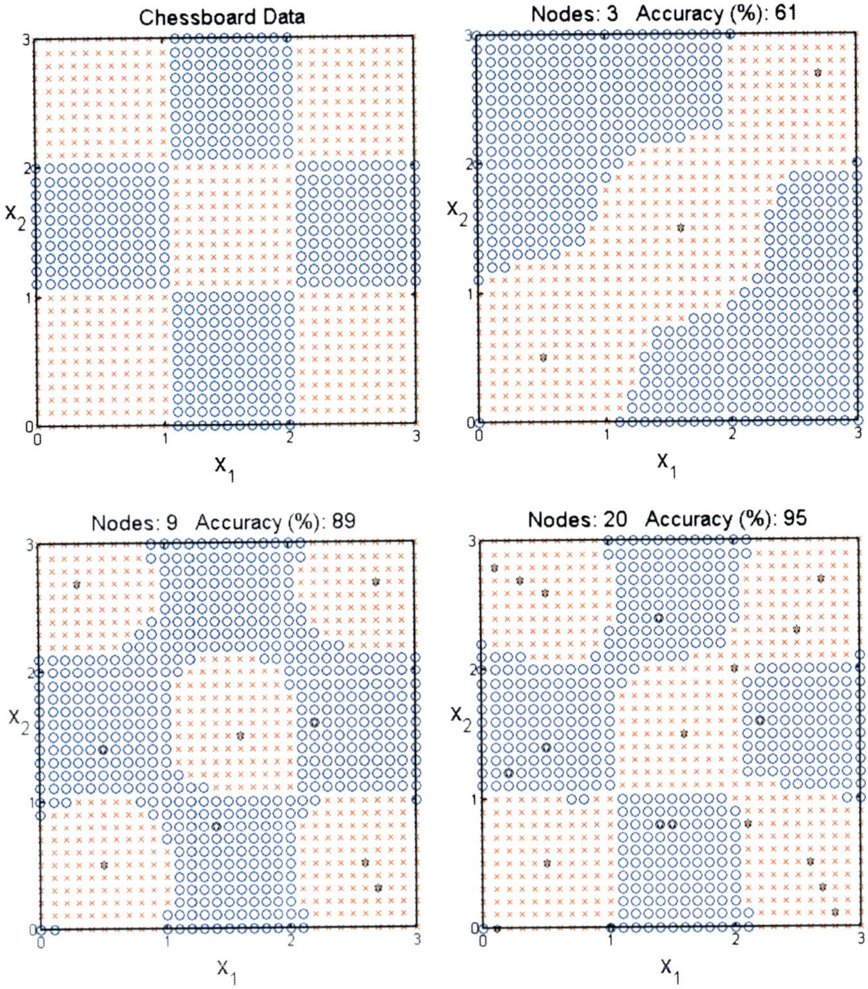

Fig. 3.18 Classification of data (*top right*) by a Gaussian radial basis function neural network. The positions of the nodes in the input space are indicated by *black stars*

3.8 Kohonen Self-Organizing Maps

Self-organizing neural networks or self-organizing (feature) maps (SOM) are systems that typically create two- or three-dimensional feature maps of input data in such a way that order is preserved. These networks do not require output data, i.e. training is unsupervised. This characteristic makes them useful for cluster analysis and the visualization of topologies and hierarchical structures of higher-dimensional input spaces.

Self-organizing systems are based on competitive learning, i.e. the outputs of the network compete among themselves to be activated or fired, so that only one node can win at any given time. These nodes are known as winner-takes-all nodes.

Each node in the Kohonen layer measures the (Euclidean) distance of its weights to the input values (exemplars) fed to the layer. For example, if the input data consist of M-dimensional vectors of the form $\mathbf{x} = \{x_1, x_2, \ldots, x_M\}$, then each Kohonen node will have m weight values, which can be denoted by $\mathbf{w}_i = \{w_{i1}, w_{i2}, \ldots, w_{iM}\}$. The Euclidean distances $D_i = ||\mathbf{x} - \mathbf{w}_i||$ between the input vectors and the weights of the network are then computed for each of the Kohonen nodes, and the winner is determined by the minimum Euclidean distance.

The weights of the winning node, as well as its neighbouring nodes, which constitute the adaptation zone associated with the winning node are subsequently adjusted in order to move the weights closer to the input vector. The adjustment of the weights of the nodes in the immediate vicinity of the winning node is instrumental in the preservation of the order of the input space and amounts to an order preserving projection of the input space onto the (typically) two-dimensional Kohonen layer. As a result similar inputs are mapped to similar regions in the output space, while dissimilar inputs are mapped to different regions in the output space.

More formally, as non-parametric latent variable models that are topologically constrained, the forward and reverse mappings of the model are defined as follows during training:

$$\Im_{t+1}(y) = E\left\{x|\aleph(x) \in N\left(y_j t\right)\right\}, \quad \text{for } \forall y \in \left\{y_j\right\}_{j=1}^{K} \tag{3.31}$$

$$\aleph(x) = \text{argmin}_{\mu \in \left\{y_j\right\}_{j=1}^{K}} \|x - \Im_t(\mu)\|. \tag{3.32}$$

In the above, $N(y,t)$ denotes the set of K nodes located within a shrinking neighbourhood of the scores $y \in \Re^q$ at iteration t. The neighbourhood is defined with respect to a chosen latent variable topology in $\Re^q$. Such topologies include lines ($q = 1$) and square or hexagonal grids ($q = 2$). Figure 3.19 shows the basic structure of a self-organizing map with a two-dimensional hidden layer.

With continued training, the neighbourhood of $\mathbf{y}$ shrinks, until at convergence of the network, it contains $\mathbf{y}$ only, i.e. $\lim_{t \to \infty} N(\mathbf{y},t) = \mathbf{y}$. At this point, the topological constraints disappear and the forward and reverse mappings can then be expressed as

$$\Im(y) = E\{x|\aleph(x) = y\}, \quad \text{for } \forall y \in \{y_k\}_{k=1}^{K} \tag{3.33}$$

$$\aleph(x) = \text{argmin}_{\mu \in \{y_k\}_{k=1}^{K}} \|x - \Im(\mu)\|. \tag{3.34}$$

Once the self-organized map has converged, it is similar to a discretized version of principal surfaces. Moreover, it is computationally efficient, since only $K \ll n$ nodes are used to represent the manifold. Owing to the spherical distance measure

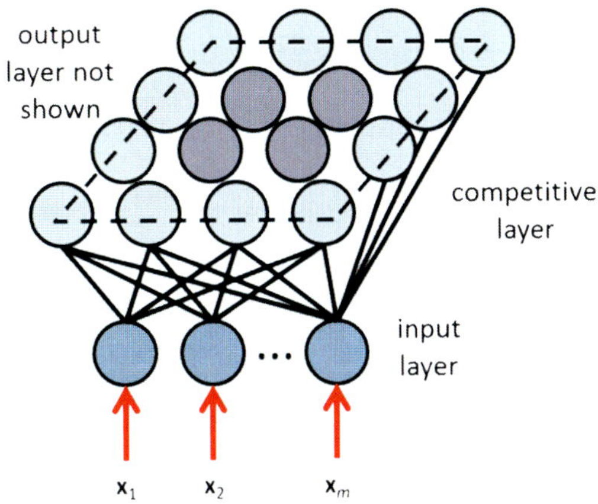

Fig. 3.19 A self-organizing mapping neural network (Kohonen)

used by the network, it is also unbiased. Variations in the distribution of the input are reflected in the feature map, in that regions in the input space X from which samples are drawn with a higher probability than other regions are mapped with better resolution and onto larger domains of the output space Y than samples drawn with a lower probability. However, in self-organizing maps, reverse mappings compute the average of the point-to-node distances, instead of the more general point to manifold projected distances. As a consequence, it fails the self-consistency requirement of principal surfaces.

From a more practical point of view, one of the problems that have to be dealt with as far as the training of self-organizing neural networks is concerned is the non-participation of neurons in the training process. This problem can be alleviated by modulation of the selection of (a) winning nodes or (b) learning rates through frequency sensitivity. Frequency sensitivity entails a history-sensitive threshold in which the level of activation of the node is proportional to the amount by which the activation exceeds the threshold. This threshold is constantly adjusted, so that the thresholds of losing neurons are decreased and those of winning neurons are increased. In this way output nodes which do not win sufficiently frequently become increasingly sensitive. Conversely, if nodes win too often, they become increasingly insensitive. Eventually this enables all neurons to be involved in the learning process.

3.8.1 Example: Using Self-Organizing Maps to Generate Principal Curves

Self-organizing maps can be used to generate principal curves from data (Kumar et al. 2004). In the following example, the use of self-organizing maps to construct principal curves from data is demonstrated. This is accomplished by fitting self-organizing maps with one-dimensional output grids to the data.

During training, the negative Euclidean distances between each neuron's weight vector and the input vector were calculated to get the weighted inputs to the network. These inputs competed, so that only the neuron with the most positive net input outputs a 1.

Training

The SOM maps were trained in batch mode. First, the network identifies the winning neuron for each input vector. Each weight vector is then moved to the average position of all of the input vectors for which it was a winner or for which it was in the neighbourhood of a winner. The distance that defined the size of the neighbourhood was altered during training through an ordering phase and a tuning phase.

Ordering Phase

During the ordering phase, the neighbourhood distance was decreased to the tuning neighbourhood distance. During this phase, the neurons of the network ordered themselves in the input space with the same topology in which they were ordered physically.

Tuning Phase

Training was concluded by a tuning phase, where the small neighbourhood fine-tuned the network while retaining the ordering learnt in the previous phase. To summarize, the weight vectors of the neurons are initially moved in large steps towards the region in the input space populated by the data. As the neighbourhood size decreases to unity, the map tends to order itself topologically over the data, but then training continues to give the neurons time to spread out evenly across the input vectors.

As with other competitive layers, the neurons of a self-organizing map will order themselves with approximately equal distances between them if the data appear with an even probability throughout a section of the input space. If input vectors occur with varying frequency throughout the input space, the feature map layer tends to allocate neurons to an area in proportion to the data density in the region.

Figure 3.20 shows the identification of curves in two-dimensional space with batch trained self-organizing maps. The number of nodes and neighbourhood sizes in the 1D layers of the maps are indicated in the figure. In each case, the weights of the networks were initialized in the centre of the input space, and these weights were updated with a distance weighted average of active input functions for node and its current neighbours. The neighbourhood sizes were linearly reduced to unity over 100 iterations during the ordering phase, and the nearest neighbour linking was used as a distance function.

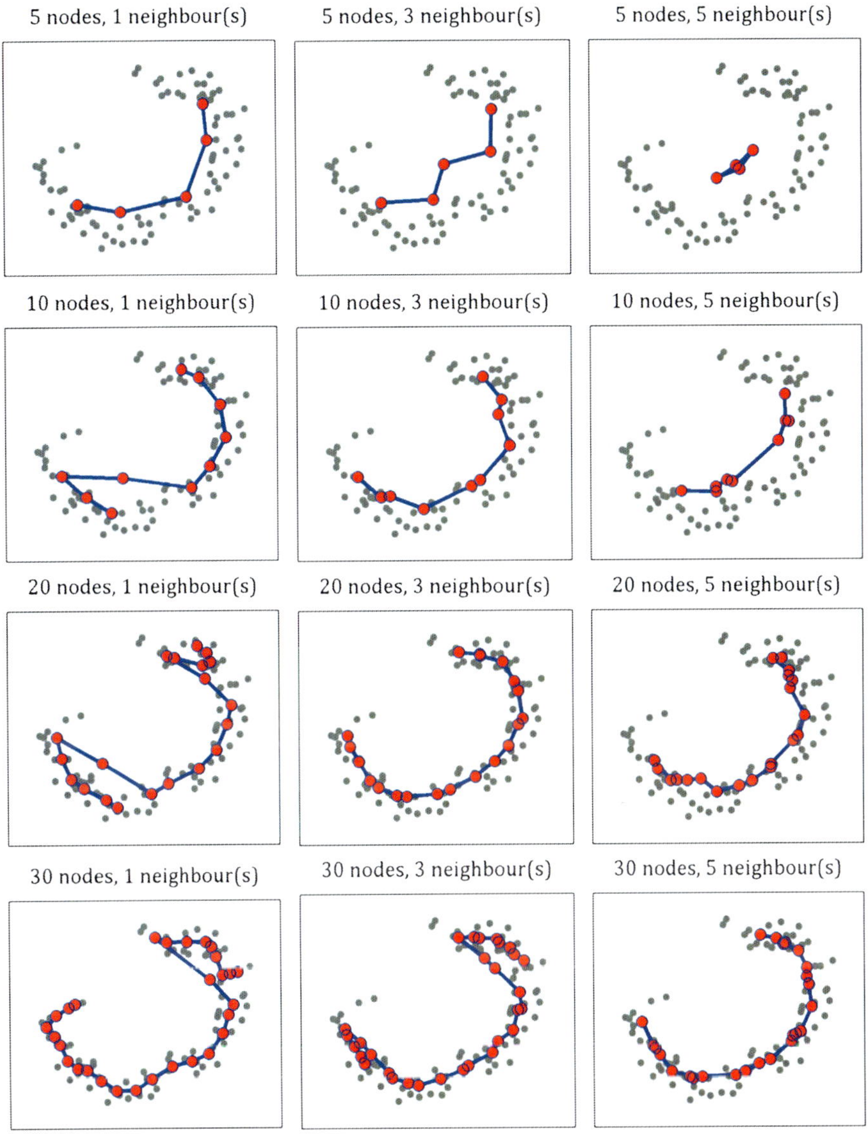

Fig. 3.20 Manifold identification with self-organizing maps

3.9 Deep Learning Neural Networks

One of the problems associated with the construction of data-based models is the so-called curse of dimensionality (Bellman 1957), since learning complexity grows exponentially with a linear increase in the dimensionality (number of variables) of the data. The classical approach to dealing with this problem is to make use of variable selection or to reduce the dimensionality of the data by the use of feature extraction. Variable selection algorithms find it difficult to deal with highly correlated data, while feature extraction can likewise be a difficult problem in its own right, and incomplete or erroneous reduction of the dimensionality of the data can limit the model upfront.

Approaches based on deep learning theory (Bengio 2007, 2009; Larochelle et al. 2009) do not attempt to reduce the dimensionality of the data, but rely on models of the mammalian neocortex that allows sensory signals to be filtered by a complex hierarchy of models. These models learn to represent observations over time, based on the regularities that these observations exhibit, and capturing spatiotemporal dependencies in data is therefore seen as a fundamental goal of deep learning systems. Deep learning models include convolutional neural networks, deep belief networks, stacked autoencoders, hierarchical temporal memory as well as deep spatiotemporal inference networks, as discussed in more detail below.

3.9.1 Deep Belief Networks

Deep belief neural networks are probabilistic generative models that can provide joint probability distributions over observable data and labels. That is, they can be used to estimate both the probability of an observation being realized, given a specific label $p(observation|label)$, as well as the probability of a label realizing, given a specific observation $p(label|observation)$, while discriminative models are limited to the latter.

Deep belief networks consist of multiple layers of stochastic units, the two top layers of which have undirected, symmetric connections between them and form an associative memory. The lower layers receive top-down, directed connections from the layer above. The states of the units in the lowest layer, or the visible units, represent an input data vector. They are trained one layer at a time, which can be followed by other learning methods to fine-tune the weights of the network for enhanced performance.

Deep belief networks do not suffer from the problems encountered when backpropogation training is applied to neural networks with multiple hidden layers, i.e. they do not require large data sets, they converge comparatively rapidly and are less prone to get stuck in local minima during training. On a slightly higher level, these networks can be viewed as composed of simple learning modules, particularly restricted Boltzmann machines and autoencoders, which are discussed in more detail below.

3.9.2 Restricted Boltzmann Machines

Restricted Boltzmann machines (RBMs) are special types of Markovian random fields consisting of two connected layers, as indicated in Fig. 3.21. One layer

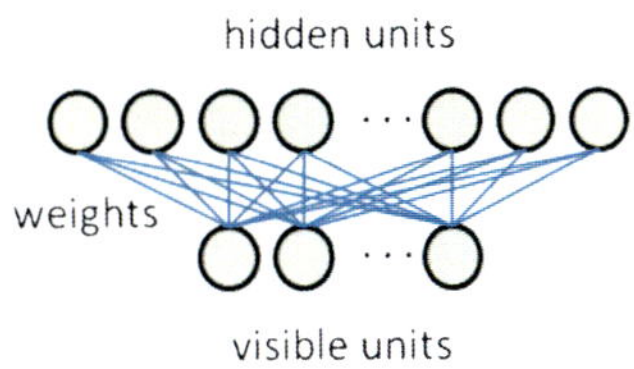

Fig. 3.21 Structure of a restricted Boltzmann machine

comprises (typically Gaussian or Bernoulli) stochastic visible or input units and the other (typically Bernoulli) stochastic hidden units. They are restricted in the sense that there is a single visible and single hidden layer only, and only units between layers are connected, i.e. units within a layer are not connected. In RBMs, the joint distribution $p(\mathbf{v},\mathbf{h},\theta)$ over the visible units **v** and the hidden units **h**, given the model parameters or network weights θ, is

$$p\,(v,h,\theta) = \frac{1}{Z}e^{-E(v,h,\theta)} \tag{3.35}$$

where Z is a normalization factor or partition function defined as

$$Z = \sum_{i=1}^{V}\sum_{i=1}^{H} e^{-E(v,h,\theta)} \tag{3.36}$$

E is an energy function, and for Bernoulli visible and hidden units, this function is defined as

$$E\,(v,h,\theta) = -\sum_{i=1}^{V}\sum_{i=1}^{H} w_{ij}v_ih_j - \sum_{i=1}^{H} b_iv_i - \sum_{j=1}^{H} a_jh_j \tag{3.37}$$

where w_{ij} is the weight of the connection between the ith visible unit and the jth hidden unit in the network consisting of V visible and H hidden units, while a_j and b_i are bias terms. The network is trained by updating the weights as follows:

$$\Delta w_{ij} - E\left(v_i,h_j\right) - E^*\left(v_i,h_j\right) \tag{3.38}$$

where $E(v_i, h_j)$ is the expectation observed in the training data set and $E^*(v_i, h_j)$ is the same expectation under the distribution defined by the model. Since the computation of $E^*(v_i, h_j)$ is intractable, it is replaced running a Gibbs sampler initialized at the data for one full step.

Deep belief networks composed of RBMs have the following attractive properties: (i) Unlabeled data are used effectively, (ii) the values of the hidden variables in the deepest layers are computed effectively, (iii) they can be interpreted as Bayesian probabilistic generative models and (iv) overfitting often observed in models with large numbers of parameters and underfitting of such networks are addressed effectively by the generative pretraining step (Yu and Deng 2011).

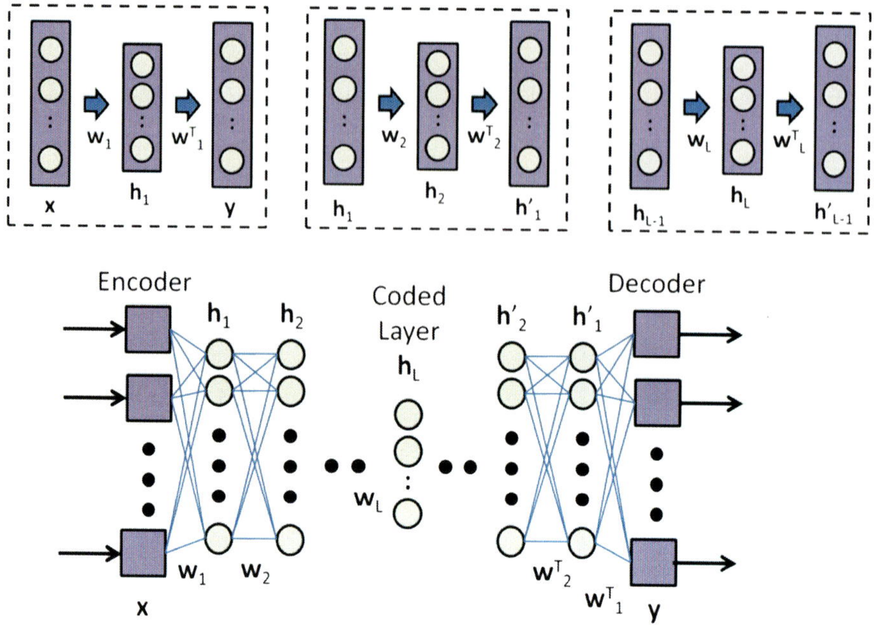

Fig. 3.22 Pretraining of RBMs (*top*) and stacking after training to form a deep-layered autoencoder (*bottom*)

3.9.3 Training of Deep Neural Networks Composed of Restricted Boltzmann Machines

Initial training or pretraining is done layer by layer in an unsupervised mode. This is accomplished by a Gibbs sampling process, where an input vector v is presented to the visible units in the first layer of the network. These units pass the values of the vector to the hidden layer, which then attempts to reconstruct the input vector. The activations of the hidden units then serve as the input to the next layer, until all the layers in the network have been trained, as indicated in Fig. 3.22.

A similar training scheme can be followed when the deep belief network is used in supervised training schemes. In this instance, the last layer would be trained in supervised mode. Training can also be extended to partial supervision and semisupervised learning problems, as discussed by Bengio (2007).

3.9.4 Stacked Autoencoders

As mentioned before, deep belief neural networks can also be composed of other learning modules, such as autoassociative neural networks or autoencoders. These modules are trained by the same greedy layer-by-layer methods in which each layer attempts to reproduce the data vector from the feature activations that it elicits (Bengio et al. 2007). More specifically, training consists of the following steps:

(i) Train the 1st layer of the autoencoder to minimize some reconstruction error on the raw input data.
(ii) The hidden units' outputs are subsequently used as input for another hidden unit, also trained in a purely unsupervised mode, as in (i).
(iii) Iterate (ii) until the desired number of hidden layers has been trained.
    (a) Use the output of the last hidden layer as input to a supervised layer and train to reproduce the labels as best as possible according to some criterion while keeping the parameters (weights) of the rest of the network fixed.
(iv) Fine-tune all network parameters (weights) with respect to the supervised criterion. Alternatively, if it is an unsupervised problem, the global reconstruction error of the network can be fine-tuned, as discussed, for example, by Hinton and Salakhutdinov (2006).

Larochelle et al. (2007) have found that since the variational bound associated with restricted Boltzmann machines no longer applies, autoencoder modules are less robust to random noise in the training data.

Deep architecture learning systems are not confined to systems based on restricted Boltzmann machines or stacked autoencoders, and convolutional neural networks, hierarchical temporal memories and deep spatiotemporal inference networks have all been applied successfully in certain niche areas.

3.10 Extreme Learning Machines

Extreme learning machines (ELMs) are single hidden layer feedforward neural networks that differ from conventional feedforward neural networks in that the hidden layer is not trained and can consist of random computational nodes not related to the training data (Huang et al. 2011). In multilayer perceptron variants, the hidden layers would typically be considerably larger than in standard multilayer perceptrons.

More formally, given n arbitrary training samples over m variables and d outputs, $\{(\mathbf{x}_i, \mathbf{y}_i)\}_{i=1}^{N} \in \Re^{Mxd}$, the model representing the extreme learning machine with L hidden nodes and activation functions $g(\cdot)$ can be defined as follows, for $i = 1, 2 \ldots, N$:

$$y_i = \sum_{j=1}^{L} \boldsymbol{\beta}_j g\left(\mathbf{w}_j, \mathbf{x}_i, b_j\right). \tag{3.39}$$

In this model, $\mathbf{w}_j$ is the input weight vector connecting the jth hidden node, and the input nodes, $\boldsymbol{\beta}_j$ is the weight vector connecting the jth hidden node and the output nodes, and b_j is the bias of the jth hidden node. The activation functions are general and can represent arbitrary activation functions, e.g. $g(\mathbf{w}_j\mathbf{x}_j + b_j)$ for multilayer perceptrons. This model is equivalent to $\mathbf{H\beta} = \mathbf{Y}$, where

$$\mathbf{H}\left(\mathbf{w}_1, \mathbf{w}_2, \ldots \mathbf{w}_L, b_1, b_2, \ldots b_L, \mathbf{x}_1, \mathbf{x}_2, \ldots \mathbf{x}_N\right) = \begin{bmatrix} g\left(\mathbf{w}_1, b_1, \mathbf{x}_1\right) & \cdots & g\left(\mathbf{w}_1, b_1, \mathbf{x}_1\right) \\ \vdots & \cdots & \vdots \\ g\left(\mathbf{w}_1, b_1, \mathbf{x}_1\right) & \cdots & g\left(\mathbf{w}_1, b_1, \mathbf{x}_1\right) \end{bmatrix}_{N \times L} \tag{3.40}$$

$$\boldsymbol{\beta} = \begin{bmatrix} \boldsymbol{\beta}_1^{\mathrm{T}} \\ \vdots \\ \boldsymbol{\beta}_L^{\mathrm{T}} \end{bmatrix}_{L \times M} \tag{3.41}$$

$$\mathbf{Y} = \begin{bmatrix} \mathbf{y}_1^{\mathrm{T}} \\ \vdots \\ \mathbf{y}_N^{\mathrm{T}} \end{bmatrix}_{N \times M} . \tag{3.42}$$

In the hidden layer output matrix of the network ($\mathbf{H}$), the jth column of $\mathbf{H}$ is the output of the jth hidden node with respect to inputs $\mathbf{x}_1, \mathbf{x}_2, \ldots \mathbf{x}_n$.

Training is accomplished as follows:

1. Assignment of random input weight vectors $\mathbf{w}_j$ and hidden node biases b_j, for $j = 1, 2, \ldots L$
2. Calculation of the hidden layer output matrix $\mathbf{H}$
3. Calculation of the output weights, $\boldsymbol{\beta} = \mathbf{H}^{+}\mathbf{Y}$, where $\mathbf{H}^{+}$ is the Moore–Penrose generalized inverse of matrix $\mathbf{H}$

Training of extreme learning machines is rapid and simple, and unlike traditional gradient-based methods, they can accommodate activation functions that are not necessarily differentiable. They have found wide use, since their inception less than a decade ago in biomedicine (Song et al. 2012), power generation (Li et al. 2012), the process industries (Zhang and Zhang 2011), economics (Yao et al. 1998), and others. Ensembles of extreme learning machines can be constructed by means of bagging and boosting (Hansen and Salamon 1990), as discussed in Chap. 5, and these have also been applied in various contexts (Lan et al. 2009; Wang and Alhamdoosh 2013; Butcher et al. 2013).

An example of the application of extreme learning machines to fit a two-dimensional surface is shown in Fig. 3.23. In this figure, the surface generated by $z = 1.21(1 + 2x_1 + 2x_1^2)(1 + 2x_2 + 2x_2^2)\exp(-x_1^2/2 - x_2^2/2)$ was fitted with extreme learning machines containing tansigmoidal hidden nodes. Training was rapid and with between 100 and 200 nodes, the response surface is approximated accurately (the variance explained by each model is shown in parentheses in each figure).

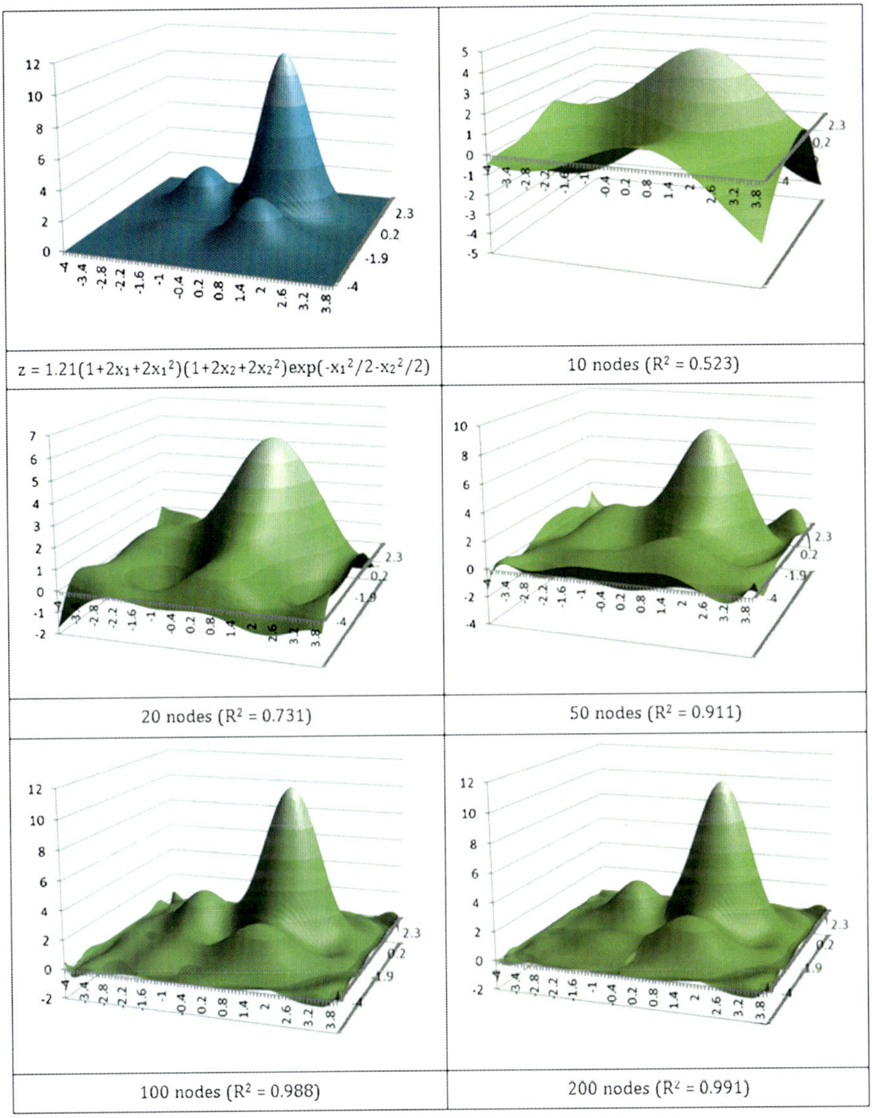

Fig. 3.23 Fitting of a two-dimensional surface (*top left*) with extreme learning machines with 10 nodes (*top right*), 20 hidden nodes (*middle left*), 50 nodes (*middle right*), 100 nodes (*bottom left*) and 200 nodes (*bottom right*)

3.11 Fault Diagnosis with Artificial Neural Networks

As alluded to in Sect. 3.1., neural networks can be used in a large variety of ways to set up fault diagnostic models. Some of these approaches have been referred to in Chap. 2, e.g. the application of autoassociative neural networks, where both

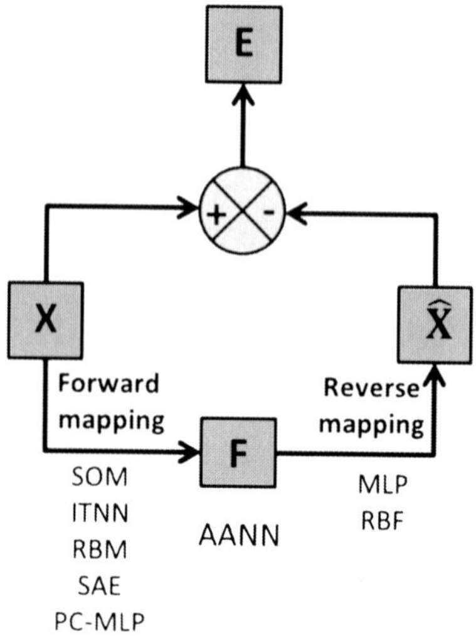

Fig. 3.24 Fault diagnosis with neural network models, e.g. self-organized feature maps (*SOM*), input training neural networks (*ITNN*), restricted Boltzmann machines (*RBM*), stacked autoencoders (*SAE*), principal curves combined with multilayer perceptrons (*PC–MLP*), autoassociative neural networks (*AANN*), multilayer perceptrons (*MLP*) and radial basis function neural networks (*RBFNN*)

the forward and reverse mapping models are constructed at the same time by the preceding and succeeding hidden layers around a middle bottleneck layer that serves as output and input to the respective models. Input training neural networks (Jia et al. 1998) have been used in a similar fashion (Reddy and Mavrovouniotis 1996).

Other approaches are also possible. For example, any of a number of two separate unsupervised and supervised learning models could be used for forward and reverse mapping. This could be a self-organized map and a multilayer perceptron, a Sammon map in conjunction with a multilayer perceptron for feature extraction as well as an extreme learning machine or radial basis function neural networks or a multilayer perceptron for the reverse mapping, etc. One could also envisage an optimization scheme incorporating a search for the best combination of such models, similar to schemes found in some data mining packages, such as StatSoft's *Statistica* or *NeuralWorks Professional*, where model optimization involves searching through a combination of different neural network structures, training algorithms and optimization criteria. Of course, any of the forward and reverse models need not even be neural network models, but could also be other machine learning models, such as regression trees or kernel-based methods, as considered in subsequent chapters of the book.

Figure 3.24 gives a conceptual summary of the generalized framework for constructing fault diagnostic systems with artificial neural networks. As indicated, some neural networks, such as autoassociative neural networks and stacked autoencoders, allow simultaneous construction of both forward and reverse mapping models. In addition to neural networks, such as self-organizing maps that are unsupervised, supervised neural networks can also be used in conjunction with other feature extraction algorithms, particularly those that do not generate models when features are extracted, such as principal curves (Dong and McAvoy 1992) and Sammon maps (Wang 2008).

References

Amari, S. (1977). Neural theory of association and concept-formation. *Biological Cybernetics, 26*(3), 175–185.

Amari, S. (1990). Mathematical foundations of neurocomputing. *Proceedings of the IEEE, 78*(9), 1443–1463.

Anderson, J. A., Silverstein, J. W., Ritz, S. A., & Jones, R. S. (1977). Distinctive features, categorical perception, and probability learning: Some applications of a neural model. *Psychological Review, 84*(5), 413–451.

Bellman, R. E. (1957). *Dynamic programming*. Princeton: Princeton University Press. ISBN 978-0-691-07951-6.

Bengio, Y. (2007). On the challenge of learning complex functions. *Progress in Brain Research, 165*, 521–534.

Bengio, Y. (2009). Learning deep architectures for AI. *Foundations and Trends in Machine Learning, 2*, 1–127.

Bengio, Y., Lamblin, P., Popovici, D., & Larochelle, H. (2007). *Greedy layer-wise training of deep networks* (NIPS 19, pp. 153–160). Cambridge: MIT Press.

Bishop, C. (2007). *Natural networks for pattern recognition* (Repr.). Oxford: Oxford University Press.

Butcher, J. B., Verstraeten, D., Schrauwen, B., Day, C. R., & Haycock, P. W. (2013). Reservoir computing and extreme learning machines for non-linear time-series data analysis. *Neural Networks, 38*, 76–89.

Carpenter, G., & Grossberg, S. (1990). ART 3: Hierarchical search using chemical transmitters in self-organizing pattern recognition architectures. *Neural Networks, 3*(2), 129–152.

Dimopoulos, Y., Bourret, P., & Lek, S. (1995). Use of some sensitivity criteria for choosing networks with good generalization ability. *Neural Processing Letters, 2*(6), 1–4.

Dimopoulos, Y., Chronopoulos, J., Chronopoulou-Sereli, A., & Lek, S. (1999). Neural network models to study relationships between lead concentration in grasses and permanent urban descriptors in Athens city (Greece). *Ecological Modelling, 120*, 157–165.

Dong, D., & McAvoy, T. J. (1992). Nonlinear principal component analysis – Based on principal curves and neural networks. *Computers and Chemical Engineering, 16*, 313–328.

Fukushima, K. (1980). Neocognitron: A self organizing neural network model for a mechanism of pattern recognition unaffected by shift in position. *Biological Cybernetics, 36*(4), 193–202.

Gevrey, M., Dimitropoulos, I., & Lek, S. (2003). Review and comparison of methods to study the contribution of variables in artificial neural network models. *Ecological Modelling, 160*, 249–264.

Grossberg, S. (1976). Adaptive pattern classification and universal recoding: I. Parallel development and coding of neural feature detectors. *Biological Cybernetics, 23*(3), 121–134.

Grossberg, S. (1982). *Studies of mind and brain*. Boston: Reidel.

Hansen, L. K., & Salamon, P. (1990). Neural network ensemble. *IEEE Transactions on Pattern Analysis and Machine Intelligence, 12*(10), 993–1001.

Hebb, D. O. (1949). *The organization of behavior*. New York: Wiley.

Hinton, G. E., & Salakhutdinov, R. R. (2006). Reducing the dimensionality of data with neural networks. *Science, 313*(5786), 504–507.

Hopfield, J. J. (1982). Neural networks and physical systems with emergent collective computational abilities. *Proceedings of the National Academy of Sciences, 79*(8), 2554–2558.

Hopfield, J. J. (1984). Neurons with graded response have collective computational properties like those of two-state neurons. *Proceedings of the National Academy of Sciences, 81*(10), 3088–3092.

Howes, P., & Crook, N. (1999). Using input parameter influences to support the decisions of feedforward neural networks. *Neurocomputing, 24*, 191–206.

Huang, G.-B., Wang, D. H., & Lan, Y. (2011). Extreme learning machines: A survey. *International Journal of Machine Learning and Cybernetics, 2*, 107–122. doi:10.1007/513042-011-0019-y.

Jia, F., Martin, E. B., & Morris, A. J. (1998). Non-linear principal components analysis for process fault detection. *Computers and Chemical Engineering, 22*, S851–S854.
Kirby, M. J., & Miranda, R. (1996). Circular nodes in neural networks. *Neural Computation, 8*(2), 390–402.
Kohonen, T. (1977). *Associative memory: A system-theoretical approach*. Berlin: Springer.
Kohonen, T. (1982). Self-organized formation of topologically correct feature maps. *Biological Cybernetics, 43*(1), 59–69.
Kohonen, T. (1984). *Self-organization and associative memory*. Berlin: Springer.
Kohonen, T. (1988). The "neural" phonetic typewriter. *Computer, 21*(3), 11–22.
Kohonen, T. (1990). The self-organizing map. *Proceedings of the IEEE, 78*(9), 1464–1480.
Kohonen, T. (1995). *Self-organizing maps*. Berlin: Springer.
Kramer, M. A. (1991). Nonlinear principal component analysis using autoassociative neural networks. *AICHE Journal, 37*(2), 233–243.
Kramer, M. A. (1992). Autoassociative neural networks. *Computers and Chemical Engineering, 16*(4), 313–328.
Kubat, M. (1998). Decision trees can initialize radial-basis function networks. *IEEE Transactions on Neural Networks, 9*(5), 813–821.
Kumar, G., Kaldra, G. A., & Dandhe, S. G. (2004). Curve and surface reconstruction from points: An approach based on self-organizing maps. *Applied Soft Computing, 5*(1), 55–66.
Lan, Y., Soh, Y. C., & Huang, G.-B. (2009). Ensemble of online sequential extreme learning machine. *Neurocomputing, 72*(13–15), 3391–3395.
Larochelle, H., Erhan, D., Courville, A., Bergstra, J., & Bengio, Y. (2007). An empirical evaluation of deep architectures on problems with many factors of variation. In Z. Ghahramani (Ed.), *24th International Conference on Machine Learning (ICML 2007)* (pp. 473–480). Corvallis: Omnipress. URL: http://www.machinelearning.org/proceedings/icml2007/papers/331.pdf
Larochelle, H., Bengio, Y., Louradour, J., & Lamblin, P. (2009). Exploring strategies for training deep neural networks. *Journal of Machine Learning Research, 1*, 1–40.
Li, G., Niu, P., Liu, C., & Zhang, W. (2012). Enhanced combination modeling method for combustion efficiency in coal-fired boilers. *Applied Soft Computing, 12*(10), 3132–3140.
McCullagh, P., & Nelder, J. A. (1989). *Generalized linear models* (2nd ed.). London: Chapman & Hall.
McCulloch, W. S., & Pitts, W. H. (1943). A logical calculus of the ideas imminent in nervous activity. *Bulletin of Mathematical Biophysics, 5*, 115–133.
Minsky, M. (1954). *Neural nets and the brain*. Princeton: Princeton University Press.
Moody, J., & Darken, C. J. (1989). Fast learning in networks of locally tuned processing units. *Neural Computation, 1*, 281–294.
Nord, L., & Jacobsson, S. P. (1998). A novel method for examination of the variable contribution to computational neural network models. *Chemometrics and Intelligent Laboratory Systems, 44*(1–2), 153–160.
Papadokonstantakis, S., Lygeros, A., & Jacobsson, S. (2006). Comparison of recent methods for inference of variable influence in neural networks. *Neural Networks, 19*(4), 500–513.
Powell, M. J. D. (1985). Radial basis functions for multivariate interpolation: A review. In *IMA conference on algorithms for the approximation of functions and data* (pp. 143–167). Oxford: Oxford University Press.
Reddy, V.N., & Mavrovouniotis, M. (1996, April 9–11). Plant monitoring and diagnosis using input training neural networks. *Proceedings of the 58th American Power Conference*. Part 1 (of 2), Chicago, IL, USA.
Rumelhart, D. E., & McClelland, J. L. (1986). *Parallel distribution processing: Exploration in the microstructure of cognition*. Cambridge, MA: MIT Press.
Scardi, M., & Harding, L. W. (1999). Developing an empirical model of phytoplankton primary production: A neural network case study. *Ecological Modelling, 120*(2–3), 213–223.
Scholz, M. (2007). Analysing periodic phenomena by circular PCA. In S. Hochreiter & R. Wagner (Eds.), *Bioinformatics research and development* (pp. 38–47). Berlin/Heidelberg: Springer. Available at: http://www.springerlink.com/index/10.1007/978-3-540-71233-6_4. Accessed 23 June 2011.

Scholz, M. (2011). Nonlinear PCA toolbox for Matlab - Matthias Scholz. *Nonlinear PCA*. Available at: http://www.nlpca.de/matlab.html. Accessed 22 June 2011.
Scholz, M., & Vigário, R. (2002). Nonlinear PCA: A new hierarchical approach. In *ESANN 2002 proceedings. European Symposium on Artificial Neural Networks* (pp. 439–444). Bruges: d-side publi.
Scholz, M., Fraunholz, M., & Selbig, J. (2008). Nonlinear principal component analysis: Neural network models and applications. In A. N. Gorban et al. (Eds.), *Principal manifolds for data visualization and dimension reduction* (pp. 44–67). Berlin/Heidelberg: Springer. Available at: http://www.springerlink.com/index/10.1007/978-3-540-73750-6_2. Accessed 22 June 2011.
Schwenker, F., Kestler, H. A., & Palm, G. (2001). Three learning phases for radial-basis-function networks. *Neural Networks: The Official Journal of the International Neural Network Society, 14*(4–5), 439–458.
Song, Y., Crowcroft, J., & Zhang, J. (2012). Automatic epileptic seizure detection in EEGs based on optimized sample entropy and extreme learning machine. *Journal of Neuroscience Methods, 210*(2), 132–146.
Specht, D. F. (1991). A general regression neural network. *IEEE Transactions on Neural Networks, 2*(6), 568–576.
Van der Walt, T. J., Van Deventer, J. S. J., & Barnard, E. (1993). Neural nets for the simulation of mineral processing operations: Part II. Applications. *Minerals Engineering, 6*(11), 1135–1153.
Wang, Q. (2008). *Use of topographic methods to monitor process systems*. M.Sc. Eng. thesis. University of Stellenbosch, South Africa.
Wang, D., & Alhamdoosh, M. (2013). Evolutionary extreme learning machine ensembles with size control. *Neurocomputing, 102*, 98–110.
Werbos, P. J. (1974). *Beyond regression: New tools for prediction and analysis in the behavioural sciences*. Cambridge, MA: Harvard.
Widrow, B.,& Hoff, M. E. (1960). Adaptive switching circuits. In *IRE WESCON convention record* (pp. 96–104). New York: IRE.
Yao, J., Teng, N., Poh, H.-L., & Tan, C. L. (1998). Forecasting and analysis of marketing data using neural networks. *Journal of Information Science and Engineering, 14*, 843–862.
Yu, D., & Deng, L. (2011). Deep learning and its applications to signal and information processing. *IEEE Signal Processing Magazine, 1*, 145–154. doi:10.1109/MSP.2010.939038.
Zhang, Y., & Zhang, P. (2011). Optimization of nonlinear process based on sequential extreme learning machine. *Chemical Engineering Science, 66*, 4702–4710.

Nomenclature

Symbol	Description
λ	Scaling parameter in activation function of a neural network
ϕ	Argument of an activation function in neural network
α	Learning coefficient
λ	Learning coefficient
δ	Perturbation of a variable
θ	Scalar variable
Γ	Hypersurface
σ	Width of radial basis function
$\boldsymbol{\Phi}$	Radial basis function matrix
η	Learning coefficient
$\boldsymbol{\theta}$	Parameters (weights) of a restricted Boltzmann machine
$\aleph(\cdot)$	Reverse mapping
$\boldsymbol{\beta}_j$	Weight vector connecting the jth hidden node with the output nodes of a neural network

(continued)

(continued)

Symbol	Description
$\Im_{t+1}(\cdot)$	Forward mapping a time $t+1$
$\mathbf{H}^+$	Moore–Penrose generalized inverse of matrix **H**
$\mathbf{e}_k$	Error vector of neural network at kth iteration of training
$\mathbf{g}_k$	Gradient vector of neural network error function with respect to weight vector
$\mathbf{w}_j$	Weight vector of jth hidden node in a neural network
$\mathbf{w}_k$	Weight vector of neural network at kth training iteration
G_i	General influence measure of ith input variable to a neural network model
I_{ji}	Partial influence of the ith input variable through the jth hidden node of a neural network
$S_{SD,i}$	Sum of squares derivative of ith variable
a_j	Bias term in a restricted Boltzmann machine associated with jth hidden unit
a_p	Activation of pth hidden node in a neural network
a_q	Activation of qth hidden node in a neural network
b_i	Bias term in a restricted Boltzmann machine associated with ith visible unit
b_j	Bias of jth hidden node in a neural network
c_k^{new}	kth updated cluster centre component during iterative clustering process
c_k^{old}	kth cluster centre component before updating during iterative clustering process
$e_{RMSE,i}$	Root-mean-squared error of the network with exclusion of the input of the ith variable
e_{RMSE}	Root-mean-squared error of the network without exclusion of any variables
$e_{p,m}$	Training error of mth output upon presentation of pth sample to a neural network
s_p	Output value of neural network node p
s_q	Output value of neural network node q
w_{j0}	Bias *of* jth hidden node in a neural network
w_j^h	Weight connecting jth hidden node to the output node in a neural network with a single output node
w_{ji}	Weight connecting ith input to jth hidden node in a neural network
w_{ji}^h	Weight connecting the ith input variable to the jth node in the hidden layer of a (single hidden layer) neural network
w_{o0}	Weight connecting bias of hidden layer to the output node of a (single-output) neural network
$w_{p,i}$	Weight of pth hidden neural network node associated with the ith input variable
$w_{q,i}$	Weight of qth hidden neural network node associated with the ith input variable
z_i	Output of the ith node in a neural network
z_j	Activation of jth radial basis function node
α_i, β_i	Parameters of ith hidden node in a Gaussian radial basis function neural network
φ_{ij}	Radial basis function
Δw_{ij}	Change in weight connecting ith visible unit with jth hidden unit during training of a restricted Boltzmann machine
c	Radial basis function parameter
$\mathbf{c}$	Cluster centre in k-dimensional space
c_j	jth component of a cluster centre in k-dimensional space
$\mathbf{c_{kp}}$	pth nearest neighbour of the kth cluster c_k
D_i	Euclidean distance between an input vector and the ith weight vector of a Kohonen neural network

(continued)

(continued)

Symbol	Description
d_{ij}	Partial derivatives of the output of a single-output neural network with respect to the *i*th input, x_i
$d_{p,m}$	Desired value of *m*th output upon presentation of *p*th sample to a neural network
E	Energy function of a restricted Boltzmann machine
$E(\mathbf{x},\mathbf{w})$	Error function of a neural network depending on input variables and weights
E_1	Training error of the first subnetwork in a hierarchical autoassociative neural network
$E_{1,2}$	Collective training error of two subnetworks in a hierarchical autoassociative neural network
g	Continuous function defined by an autoassociative neural network, reconstructing M variables from a lower-dimensional subspace, $g : \Re^1 \rightarrow \Re^M$
$\mathbf{h}$	Vector of hidden units in a restricted Boltzmann machine
H	Number of hidden units in a restricted Boltzmann machine
$\mathbf{H}$	Output matrix of hidden layer of an extreme learning machine, $\mathbf{H} \in \Re^{n \times L}$
$\mathbf{H}_k$	Hessian matrix of error function of neural network at *k*th iteration of training
I_{ji}	Response of the *j*th hidden neuron on *i*th input
$\mathbf{J}_k$	Jacobian matrix of error function of neural network at *k*th iteration of training
$o_{p,m}$	Actual value of *m*th output upon presentation of *p*th sample to a neural network
r	Scalar variable
s_f	Continuous function defines by an autoassociative neural network, M mapping inputs to a lower-dimensional subspace, $s_f : \Re^{\mathrm{M}} \rightarrow \Re^l$
S_j	Derivative of the output neuron with respect to its input
$\mathbf{v}$	Vector of visible units in a restricted Boltzmann machine
V	Number of visible units in a restricted Boltzmann machine
V_{i}	Influence of *i*th variable the output of a (neural network) model
$\mathbf{w}$	Weight vector of a node in a neural network
w_j	Weight of *j*th node in a neural network
w_{ji}	Weight connecting the *i*th input neuron and the jth hidden neuron in a neural network
w_{jo}	Weight connecting the output neuron and the jth hidden neuron of a neural network
z	Output of a neuron
Z	Partition function
$\mathrm{F}(\mathbf{x})$	Function of a vector of variables, $\mathbf{x}$
$\varphi(\cdot)$	Radial basis function

Chapter 4
Statistical Learning Theory and Kernel-Based Methods

4.1 Generalized Framework for Data-Driven Fault Diagnosis by Use of Kernel Methods

Like artificial neural networks, kernel methods constitute a rich class of analytical methods that have seen rapid assimilation in all spheres of data analysis, owing to their attractive characteristics, including the facility to derive models in high-dimensional spaces that can be identified by the use of convex optimization, as well as the ability to generate primal and dual model representations.

Kernel methods allow the development of a range of different forward and reverse mapping models within the generalized framework for the development of diagnostic models, as discussed in Chap. 2. More so, they can also be used to derive confidence limits as determined by the distribution of the data. For these reasons, kernel methods feature strongly in the literature, as was outlined in Chap. 2. The following sections in the chapter give a basic introduction to these powerful methods.

4.2 Statistical Learning Theory

Statistical learning theory studies the mathematical properties of learning machines. Through the development of statistical learning theory, insights into capacity control of learning machines have shaped the development of machine learning algorithms. An example of a machine learning framework with rigorous foundations in statistical learning theory is support vector machines. Basic concepts of statistical learning theory are introduced here, which will illuminate the development and advantageous properties of support vector machines and other kernel-based learning methods.

C. Aldrich and L. Auret, *Unsupervised Process Monitoring and Fault Diagnosis with Machine Learning Methods*, Advances in Computer Vision and Pattern Recognition,
DOI 10.1007/978-1-4471-5185-2_4,

4.2.1 *The Goals of Statistical Learning*

The broad aim of statistical learning is to discover insightful relations between variables in data generated by some specific process. This broad aim can manifest in four types of statistical learning approaches (Berk 2008).

The first approach, causal modelling, aims to recover the exact data generating process, a process assumed to consist of independent inputs that map unidirectionally to dependent outputs. This approach requires an assumed model, for which parameters are to be estimated. The second approach is that of conditional distribution modelling: No causal directions between inputs and outputs are assumed, with the main interest being the conditional distribution of the output given certain inputs. The third approach focuses on summarizing the data. The generating process is no longer of primary interest; rather, an accurate and accessible summary of the data is required. The fourth approach is that of forecasting: The main objective of a derived model is accuracy on some future, unknown data, irrespective of causal insights or conditional distributions.

4.2.2 *Learning from Data*

A supervised learning machine attempts to inductively discover dependencies between input and output data, often with the primary goal of generalizing accurately to unseen data. The data used to train the learning machine is known as training data, while unseen data on which generalization performance is to be determined is known as test data.

The inputs are presented as M-dimensional input vectors $\mathbf{x} \in \mathbb{R}^M$ with each input vector associated with an output, or response, y. Where the response is a class label, e.g. $y \in \{-1, +1\}$, the learning problem is known as classification. Where y is a continuous response ($y \in \mathbb{R}$), the learning problem is known as regression.

The learnt relations between the inputs and outputs are encapsulated in the function f:

$$y = f(\mathbf{x}) \tag{4.1}$$

The functional form of f can be specified a priori based on subject knowledge (parametric learning) or learnt, along with its parameters, from the training data (non-parametric learning).

Loss Functions

The generalization performance of a learning machine (how well the said learning machine performs on unseen data) can be expressed in terms of a loss function, $L(y, f(\mathbf{x}))$. The loss function is often used during training as well, as an explicit

Table 4.1 Examples of loss functions for classification and regression

Classification[a]	Loss function $L(y, f(\mathrm{x}))$	Regression	Loss function $L(y, f(\mathrm{x}))$
Zero-one	$\frac{1}{2}\lvert y - f(\mathrm{x})\rvert$	Squared	$\frac{1}{2}(y - f(\mathrm{x}))^2$
Linear soft margin	$\max(0, 1 - yf(\mathrm{x}))$	Absolute	$\lvert y - f(\mathrm{x})\rvert$
Quadratic soft margin	$\max(0, 1 - yf(\mathrm{x}))^2$	Huber's robust	$\begin{cases} \frac{1}{2\sigma}(y - f(\mathrm{x}))^2 \; if \; \lvert y - f(\mathrm{x})\rvert \leq \sigma \\ \lvert y - f(\mathrm{x})\rvert - \frac{\sigma}{2} \; if \; \lvert y - f(\mathrm{x})\rvert > \sigma \end{cases}$
Logistic regression	$\ln(1 + exp(-yf(\mathrm{x})))$	ε-insensitive	$\max(0, \lvert y - f(\mathrm{x})\rvert - \varepsilon)$

[a]Where response labelling is of the form $y \in \{-1, +1\}$

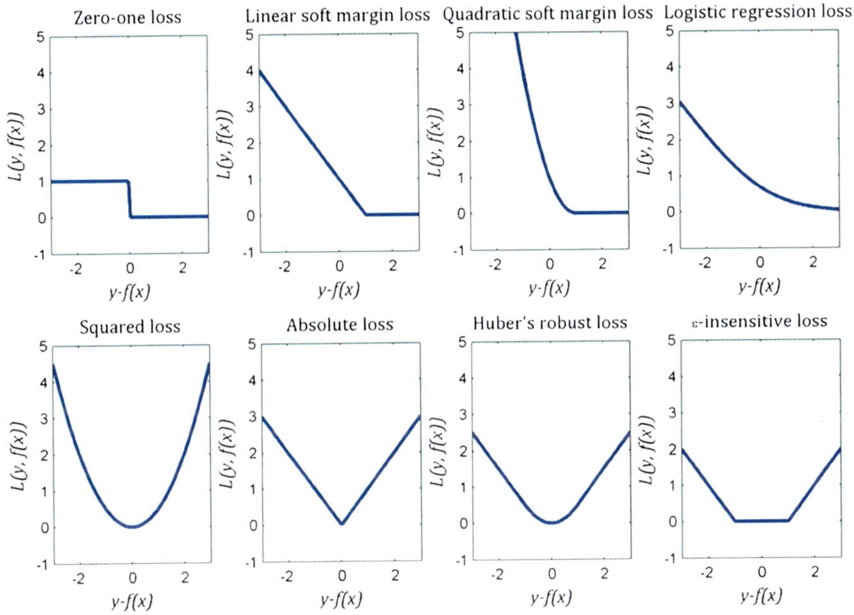

Fig. 4.1 Examples of loss functions for classification (*top row*) and regression (*bottom row*)

optimization goal, e.g. least-squares regression explicitly minimizes the average squared loss function over the training data set. Examples of loss functions for classification are the zero-one loss, the linear and quadratic soft margin losses and the logistic regression loss. Regression loss functions include squared loss, absolute loss, Huber's robust loss and ε-insensitive loss.

For the classification task, the zero-one loss function treats all misclassifications the same, while the soft margin and logistic regression loss functions give increased weight to large error margins (e.g. where $f(\mathbf{x})$ represents a probability or margin). For the regression task, the oft-used squared loss may erroneously place too much emphasis on outliers. To circumvent this, the absolute loss scales linear with error margin, while Huber's robust loss combines squared loss for small error margins and linear loss for large error margins. The ε-insensitive loss function does not penalize errors within a certain range $[-\varepsilon, +\varepsilon]$ at all, and linearly penalizes errors outside this range. (See Table 4.1 and Fig. 4.1).

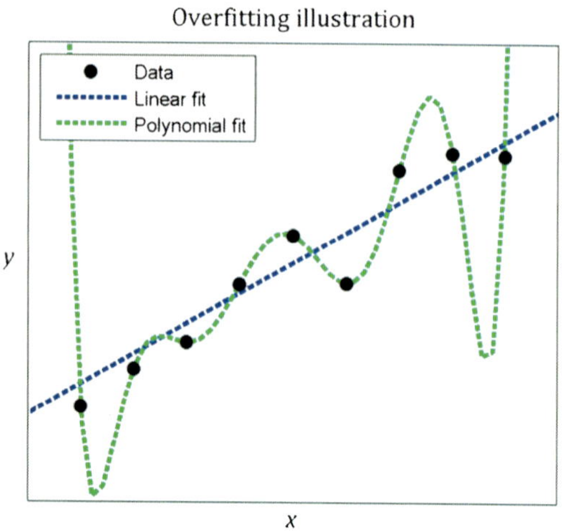

Fig. 4.2 Example of overfitting of data with a polynomial regression function

Requirements for Inductive Learning

A fundamental assumption of statistical learning is that the inputs and outputs of the training and test data are generated independently from some unknown, but fixed, probability distribution $P(\mathbf{x}, y)$. In other words, if a learning machine can adequately capture the true dependencies in the training data, these dependencies should be applicable for the test data as well. The great challenge in statistical learning lies in the uncertainty of the true probability distribution, exasperated by finite, often insufficient sample sizes. This uncertainty is related to a major caveat of inductive reasoning: Although inductive logical arguments suggest true relations with some probability, they do not guarantee it.

4.2.3 Overfitting and Riskminimization

An important consideration in statistical learning relates to the unknowability of $P(\mathbf{x}, y)$, specifically on how function fitting to noisy training data may give an erroneous representation of $P(\mathbf{x}, y)$. This consideration is known as the bias-variance dilemma or the underfitting versus overfitting trade-off.

Fitting Simple and Complex Functions

An example is shown in Fig. 4.2. Parametric regression functions are to be learnt based on a set of nine examples. Two functional forms are considered: a linear fit and a polynomial function of the eighth order. From least-squares regression, the eighth-order polynomial will fit the training data exactly, while the linear fit results

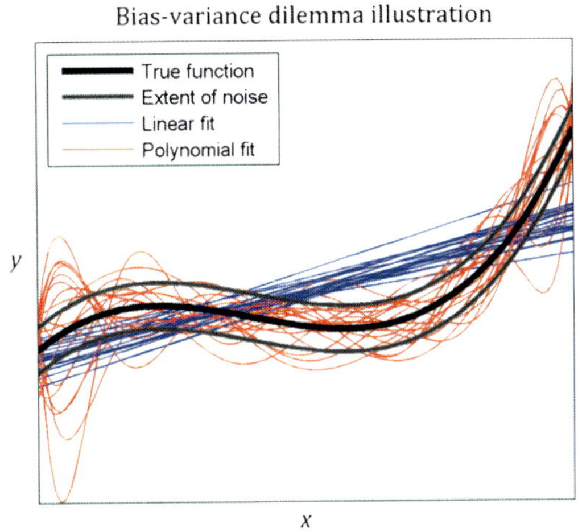

Fig. 4.3 Example of bias-variance dilemma

in non-zero residuals for all training samples. Based solely on empirical measures of fit, the eighth-order polynomial function performs much better than the linear function.

However, if the non-zero residuals resulting from the linear fit represent measurement errors, the linear fit could be seen as being a simpler, more robust summary of the relation between x and y. In the case of measurement errors, the higher-order polynomial fit may overfit the data, attempting to incorporate spurious data in its parameters. A higher-order polynomial fit also suffers from a more complex physical interpretation than its linear counterpart.

Bias-Variance Dilemma

This danger of overfitting is also known as the bias-variance dilemma. Preselecting a simple functional form (such as a linear function) imposes a large bias on the fit; whereas selecting a complex functional form (such as higher-order polynomial functions or neural networks) may be subject to large variance: large fluctuations in the fitted function depending on the specific training data set used.

An illustration of the bias-variance trade-off is shown in Fig. 4.3. The true underlying function is a third-degree polynomial function, with added Gaussian noise. For various manifestations of training data sampled from this underlying distribution, linear and eighth-order polynomial fits were calculated. The linear fitted functions cannot capture the underlying relationship (showing large bias), but is more stable than the fluctuating higher-order polynomial fitted functions (showing large variance).

Knowledge of the nature of the data, the underlying process and the intended goal of the analysis can guide the choice of functional form to be parameterized. Apart from such knowledge, much work has been done in statistical learning theory to gauge the complexity of functional forms and from this, to quantify the risk of overfitting.

Prediction Risk

In statistical learning, the risk of overfitting for a function f is described in terms of the prediction risk function R:

$$R(f) = \int L\,(y, f(\mathbf{x}))\,\mathrm{d}P(\mathbf{x}, y). \tag{4.2}$$

In (4.2), $L(y, f(\mathbf{x}))$ is a loss function, and $P(\mathbf{x}, y)$ is the unknown, fixed probability distribution of the inputs and outputs. Since the probability distribution $P(\mathbf{x}, y)$ is unknowable, the risk function in (4.2) cannot be calculated. However, an approximation of the risk function over the N samples of the training data can be calculated. This is known as the empirical risk, $R_e(f)$:

$$R_e(f) = \frac{1}{N}\sum_{i=1}^{N} L\,(y_i, f(\mathbf{x}_i)). \tag{4.3}$$

Empirical Risk Minimization

A learning machine can then be trained to minimize the empirical risk (4.3); this is called empirical risk minimization. The danger of empirical risk minimization is overfitting, as mentioned before. Given a very large selection of possible functions ($f \in \mathcal{F}$), a learning machine may find a particular f that minimizes R_e but still shows a large prediction risk R (performs poorly on test data). For example, a function may map each training data point exactly to its response but carry no information about all other possible input values. Such a function would minimize the empirical risk but would have a high overall prediction risk function (as gauged by prediction error on test data).

Bounding Risk with a Capacity Term

To counteract overfitting, statistical learning theory suggests restricting the set of possible functions $\mathcal{F}$ over which empirical risk minimization is conducted. The restriction is based on the complexity, or capacity, of the functions in $\mathcal{F}$, expressed as some capacity parameter h. A risk bound on the prediction risk function can then be expressed in terms of the empirical risk and a capacity term $c(h, N, \delta)$. This bounded risk can serve as a guide to restricting the possible function set for a learning machine in order to minimize the risk of overfitting.

For a certain probability $1 - \delta$, a function f trained on N samples, with a functional capacity of h, has an upper bound on its prediction risk function of:

$$R(f) \leq R_e(f) + c(h, N, \delta). \tag{4.4}$$

An example of a capacity parameter h is the Vapnik-Chervonenkis (VC) dimension. The binary classification capacity term for the VC dimension(where $h < N$) is:

$$c(h, N, \delta) = \sqrt{\frac{1}{N}\left(h\left(\ln\frac{2N}{h}+1\right)+\ln\frac{4}{\delta}\right)}. \tag{4.5}$$

This capacity term (4.5) increases monotonically with h, implying that more complex functional forms will result in higher risk bounds (4.4), given the same training error.

Vapnik-Chevornenkis Dimension and Shattering

The VC dimension is a capacity parameter defined in terms of the ability of a classification function to shatter a specific number of data points. For any set of N data points, there are 2^N possible labelling combinations, if each sample can be labelled as -1 or $+1$. A function that can achieve all 2^N possible labelling combinations is said to have a large capacity with respect to N. This function form shatters all N points. The VC dimension h for a specific function form is defined as the largest N such that the function form can shatter all N points. If no limit on the number of points that can be shattered exists, the VC dimension is ∞.

As an example, consider the VC dimension of a linear classifier in a two-dimensional input space (see Fig. 4.4). A linear classifier can shatter any two points, as well as any three points. However, a single linear classifier cannot separate certain label permutations of four points. The largest N that a linear classifier can shatter is then three, its VC dimension. In general, linear classifiers are able to shatter $M + 1$ number of points, where M is the dimensionality of the input space.[1]

Structural Risk Minimization

Structural risk minimization provides a framework for statistical learning where both the empirical risk (from training data) and risk due to capacity (from function complexity) are considered. The procedure is described here and illustrated in Fig. 4.5:

Structural Risk Minimization Procedure

- **Select a class of functions $\mathcal{F}$ (e.g. polynomials up to degree k and neural networks with k hidden layers) where this selection is based on some previous knowledge of the nature of the data.**

(continued)

[1] As long as the data points are not situated exactly on a linear hyperplane.

(continued)

- Subdivide the class of functions $\mathcal{F}$ into a hierarchy of sets of functions $\mathcal{F}_i$ of increasing capacity (e.g. polynomials with increasing degree, neural networks with increasing number of nodes): $\mathcal{F}_1 \subset \mathcal{F}_2 \subset \cdots \subset \mathcal{F}$.
- For each subset of functions $\mathcal{F}_i$ calculate the optimal function parameters by empirical risk minimization.
- For the series of nested function sets $\mathcal{F}_1 \subset \mathcal{F}_2 \subset \cdots \subset \mathcal{F}$, determine the bounded risk of each set.
- The function f^* with the lowest risk bound $R(f^*)$ is the best statistical learner.

Structural risk minimization is restricted by the validity of the capacity parameter. The capacity term in (4.4) is reliant on sensible representation of functional complexity, as well as the effect of this complexity on the risk of overfitting. Some functional forms have infinite VC dimensions (resulting in very large, even infinite

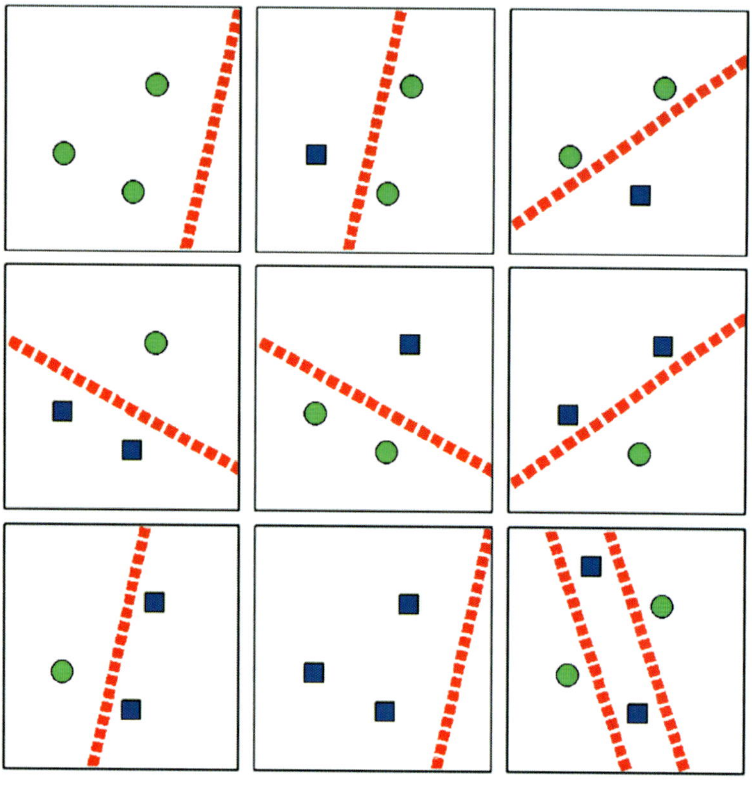

Fig. 4.4 Shattering of points in two dimensions with a linear classification function (*Lines* represent possible decision boundaries). A linear classifier can shatter three points, but not four

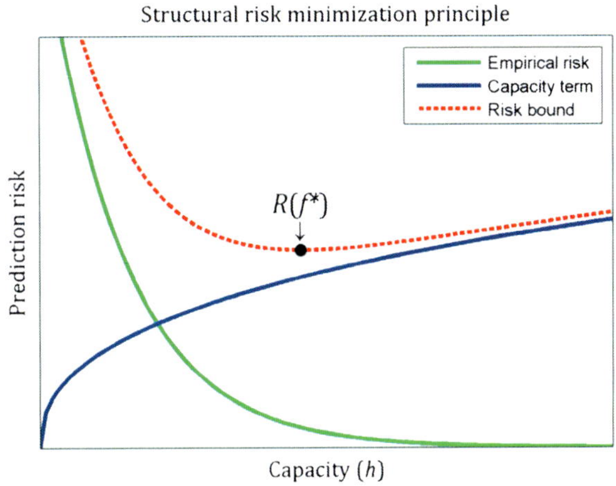

Fig. 4.5 Illustration of how structural risk minimization incorporates empirical risk (training error) and capacity penalization to find an optimal function f^* which minimizes the prediction risk function R

risk bounds) but still perform well on both training and test data. Functional forms with well-behaved capacity parameters are suitable for structural risk minimization. One such functional family is margin hyperplane classifiers.

4.3 Linear Margin Classifiers

The first learning algorithm for classification was suggested by Fisher, where a linear decision function is constructed based on the mean and variance differences between classes (Fisher 1936), based on assumptions of Gaussian distributions. An improvement to Fisher's discriminant analysis was framed as the generalized portrait algorithm by Vapnik and Lenner in 1963 (Schölkopf and Smola 2001). This algorithm (also known as margin classifiers) considers optimal separating hyperplanes, where the optimality of the hyperplane refers to the size of the margin between samples of opposing classes.

Optimal margin classifiers have certain advantageous characteristics: No distributional assumptions (bar linear separability) are made, and the capacity of optimal margin classifiers may be more preferable for structural risk minimization than Fisher's method.

4.3.1 Hard Margin Linear Classifiers

The construction of a linear hyperplane classifier for linearly separable data is now considered. Linearly separable data are data labelled $y \in \{-1, +1\}$ such that it can be correctly separated into two subsets by a linear hyperplane, with one subset labelled -1 and the other $+1$.

A hyperplane in the input space $\mathbb{R}^M$ can be expressed as the dot product of a weight vector $\mathbf{w}$ and input vectors $\mathbf{x}$, shifted by some bias b:

$$\mathbf{w} \cdot \mathbf{x} + b = \mathbf{w}^{\mathrm{T}}\mathbf{x} + b = 0. \tag{4.6}$$

Above, $\mathbf{w}{\cdot}\mathbf{x}$ is the dot product of $\mathbf{w}$ and $\mathbf{x}$. In (4.6), the hyperplane is defined by $\mathbf{w}$, a weight vector orthogonal to the hyperplane and b, a threshold.

The two-class classification problem considered consists of inputs presented as M-dimensional input vectors $\mathbf{x} \in \mathbb{R}^M$ with each input vector associated with an output, or response, $y \in \{-1, +1\}$. The learning task is to determine an optimal separating hyperplane (defined by $\mathbf{w}$ and b) that leads to a prediction function f:

$$y = f(\mathbf{x}) = \operatorname{sgn}\left(\mathbf{w}^{\mathrm{T}}\mathbf{x} + b\right). \tag{4.7}$$

The distance of any point $\mathbf{x}$ (with associated label y) from a separating hyperplane defined by $\mathbf{w}$ and b is known as the geometrical margin $m_G(\mathbf{x}, y)$:

$$m_G(\mathbf{x}, y) = \frac{y\left(\mathbf{w}^{\mathrm{T}}\mathbf{x} + b\right)}{\|\mathbf{w}\|}. \tag{4.8}$$

To remove a degree of freedom from the definition of the hyperplane (which will prove beneficial for optimization), the canonical form of the hyperplane is used. Given inputs $\mathbf{x}_1, \ldots, \mathbf{x}_N$, a canonical hyperplane such that the following is satisfied:

$$\min_{i=1,\ldots,N} \left|\mathbf{w}^{\mathrm{T}}\mathbf{x}_i + b\right| = 1. \tag{4.9}$$

The restriction in (4.9) ensures that the closest possible distance of any point ($\mathbf{x}_1, \ldots, \mathbf{x}_N$) to the hyperplane will be $1/\|\mathbf{w}\|$. The separating hyperplane ($\mathbf{w}^{\mathrm{T}}\mathbf{x} + b = 0$) is now effectively flanked by two hyperplanes: one that indicates the boundary of the -1 labelled region ($\mathbf{w}^{\mathrm{T}}\mathbf{x} + b = -1$) and another that indicates the boundary for $+1$ labelled region. No points occur within this margin, giving rise to the term "hard margin".

Points that lie exactly on the -1 or $+1$ hyperplanes are known as support vectors, as they "support" the boundaries. Since the closest distance from a point to the separating hyperplane is $1/\|\mathbf{w}\|$, the distance between the -1 and $+1$ hyperplanes is $2/\|\mathbf{w}\|$. This is known as the margin m of the separating hyperplane defined by $\mathbf{w}$ and b:

$$m(\mathbf{w}, b) = \frac{2}{\|\mathbf{w}\|}. \tag{4.10}$$

Figure 4.6 illustrates the separating -1 and $+1$ hyperplanes of a linearly separable classification problem.

An infinite number of separating hyperplanes can be defined that will separate linearly separable data. However, there exists a separating hyperplane that will

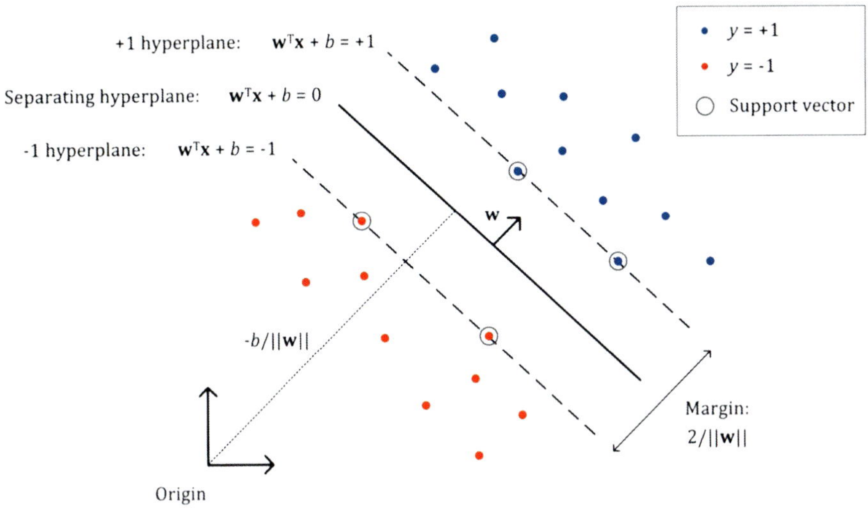

Fig. 4.6 Illustration of the separating (defined by **w** and b) −1 and +1 hyperplanes for a linearly separable classification problem

result in the maximum margin m. It will be shown later that such a maximal margin hyperplane has favourable characteristics, especially in terms of statistical learning and computational properties. The optimal separating hyperplane is then determined in an optimization framework:

$$\max_{\mathbf{w},b}\left(m = \frac{2}{\|\mathbf{w}\|}\right)$$
$$\text{subject to } y_i\left(\mathbf{w}^T\mathbf{x}_i + b\right) \geq 1; \quad i = 1, \ldots, N. \tag{4.11}$$

The constraints in (4.11) ensure that the hyperplane is in the canonical form and that the training data are classified correctly. Maximizing the margin is equivalent to minimizing $\|\mathbf{w}\|$, so (4.11) can be rewritten as:

$$\min_{\mathbf{w},b} \frac{1}{2}\|\mathbf{w}\|^2$$
$$\text{subject to } y_i\left(\mathbf{w}^T\mathbf{x}_i + b\right) \geq 1; \quad i = 1, \ldots, N. \tag{4.12}$$

The squared norm of **w**, rather than the norm of **w**, is minimized in (4.11) (which leads optimal hyperplane) in order to cast the optimization problem in the preferable form of a quadratic programming problem. Such a quadratic programming problem with inequality constraints can be solved with the aid of Lagrangian multipliers. An overview of optimization with Lagrangian multipliers is given below (Boyd and Vandenberghe 2004), as it neatly illustrates the advantageous computational properties of maximal margin classifiers. If the reader is not interested in the technicalities of constrained optimization, this subsection can be skipped.

Constrained Optimization with Lagrangian Multipliers

Let $f_0(\mathbf{x})$ be an objective function to be minimized, subject to a number of inequality constraints $f_i(\mathbf{x})$ ($i = 1, \ldots, n_c$). Here, $\mathbf{x}$ represents a variable in a real-numbered space $\mathbb{R}$ over which f_0 must be minimized, and not the input vectors as in the previous sections. The primal formulation of the optimization problem is then[2]:

$$\min_{\mathbf{x}} f_0(\mathbf{x})$$

$$\text{subject to } f_i(\mathbf{x}) \geq 0; \quad i = 1, \ldots, n_c. \tag{4.13}$$

The solution of the optimization will not necessarily occur where the derivative of the objective function $f_0'(\mathbf{x})$ is zero, as this may occur at a point $\mathbf{x}$ outside of the feasible region defined by the inequality constraints. Instead of searching for a local extremum of f_0, a suitable stationary point of a weighted combination of f_0 and f_i is the goal. The Lagrangian formulation of the optimization problem incorporates the inequality constraints into the objective function. The Karush-Kuhn-Tucker (KKT) conditions include (among others) stationarity considerations, providing sufficient criteria for convex optimization.

The Lagrangian function $L(\mathbf{x}, \boldsymbol{\alpha})$ augments the objective function f_0 with the inequality constraints, where each inequality constraint f_i is weighted by a Lagrangian multiplier $\alpha_i > 0$:

$$L(\mathbf{x}, \boldsymbol{\alpha}) = f_0(\mathbf{x}) - \sum_{i=1}^{n_c} \alpha_i f_i(\mathbf{x}). \tag{4.14}$$

The Lagrangian dual function $g(\boldsymbol{\alpha})$ is defined as the infimum of the Lagrangian function over all $\mathbf{x}$, for a given $\boldsymbol{\alpha}$ (the infimum is the greatest lower bound of a set):

$$g(\boldsymbol{\alpha}) = \inf_{\mathbf{x}} L(\mathbf{x}, \boldsymbol{\alpha}). \tag{4.15}$$

Above, the inf operator gives the infimum. The Lagrangian dual function $g(\boldsymbol{\alpha})$ gives a lower bound to the optimal value f_0^* of the objective function $f_0(\mathbf{x})$:

$$g(\boldsymbol{\alpha}) \leq f_0^*. \tag{4.16}$$

[2]The general formulation of constrained optimization problems states the inequality constraints as less than or equal to zero. For ease of visualization and generalization to SVM, inequality constraints are stated here as larger than or equal to zero, without loss of generality.

The existence of this lower bound in (4.16) is supported by considering any point $\tilde{\mathbf{x}}$ that satisfies the inequality constraints f_i. This point will have an objective function value $f_0(\tilde{\mathbf{x}})$ less than or equal to the optimal value f_0^*. In turn, the Lagrangian $L(\tilde{\mathbf{x}}, \boldsymbol{\alpha})$ at this point will be less than or equal to $f_0(\tilde{x})$, since only non-negative weighted inequality functions (which in themselves are non-negative) are subtracted from the objective function $f_0(\tilde{\mathbf{x}})$. The Lagrangian dual function $g(\boldsymbol{\alpha})$ will be less than or equal to Lagrangian $L(\tilde{\mathbf{x}}, \boldsymbol{\alpha})$ at this point, since it is the minimum of the Lagrangian over all possible $\mathbf{x}$. Since $L(\mathbf{x}, \boldsymbol{\alpha})$ for $\mathbf{x} \in \mathbb{R}$ is an infinite set, the lower bound of $g(\boldsymbol{\alpha})$ is $-\infty$ if the Lagrangian is unbounded from below. This exposition of the bound is shown in (4.17):

$$-\infty \leq g(\boldsymbol{\alpha}) = \inf_{\mathbf{x}} L(\mathbf{x}, \boldsymbol{\alpha}) \leq L(\tilde{\mathbf{x}}, \boldsymbol{\alpha}) \leq f_0(\tilde{\mathbf{x}}) \leq f_0^*. \tag{4.17}$$

The Lagrangian dual function $g(\boldsymbol{\alpha})$ is concave, irrespective of whether the optimization problem is convex or not. This concavity is well-suited for maximization. Maximizing $g(\boldsymbol{\alpha})$, the lower bound of $L(\mathbf{x}, \boldsymbol{\alpha})$, will produce the optimal Lagrangian multipliers $\boldsymbol{\alpha}$ for (4.14). This leads to the dual formulation of the optimization problem:

$$\begin{aligned} &\max_{\alpha} g(\boldsymbol{\alpha}) \\ &\text{subject to } \alpha_i \geq 0; \; i = 1, \ldots, n_c. \end{aligned} \tag{4.18}$$

If the original optimization problem is convex, the maximum of $g(\boldsymbol{\alpha})$ will be equal to the optimal value f_0^*. However, if the original problem is non-convex, the maximum of $g(\boldsymbol{\alpha})$ will be less than the optimal value f_0^*. This difference is known as the duality gap. Figure 4.7 shows simple non-convex and convex optimization problems with one inequality constraint each, where this duality gap is visible for the non-convex case.

The original optimization problem (4.13) where $f_0(\mathbf{x})$ is to be minimized can now be restated as maximization over $\boldsymbol{\alpha}$ (from (4.18)) given minimization over $\mathbf{x}$ of the Lagrangian function (4.14):

$$\begin{aligned} &\max_{\alpha} \left(\max_{\mathbf{x}} L(\mathbf{x}, \boldsymbol{\alpha}) \right) = \max_{\alpha} \left(\min_{\mathbf{x}} \left(f_0(\mathbf{x}) - \sum_{i=1}^{n_c} \alpha_i f_i(\mathbf{x}) \right) \right) \\ &\text{subject to } f_i(\mathbf{x}) \geq 0; \; i = 1, \ldots, n_c \\ &\text{and } \alpha_i \geq 0; \; i = 1, \ldots, n_c. \end{aligned} \tag{4.19}$$

For convex optimization problems, the KKT conditions provide sufficient conditions for optimal $\mathbf{x}$ and $\boldsymbol{\alpha}$ for continually differentiable f_0 and f_i.

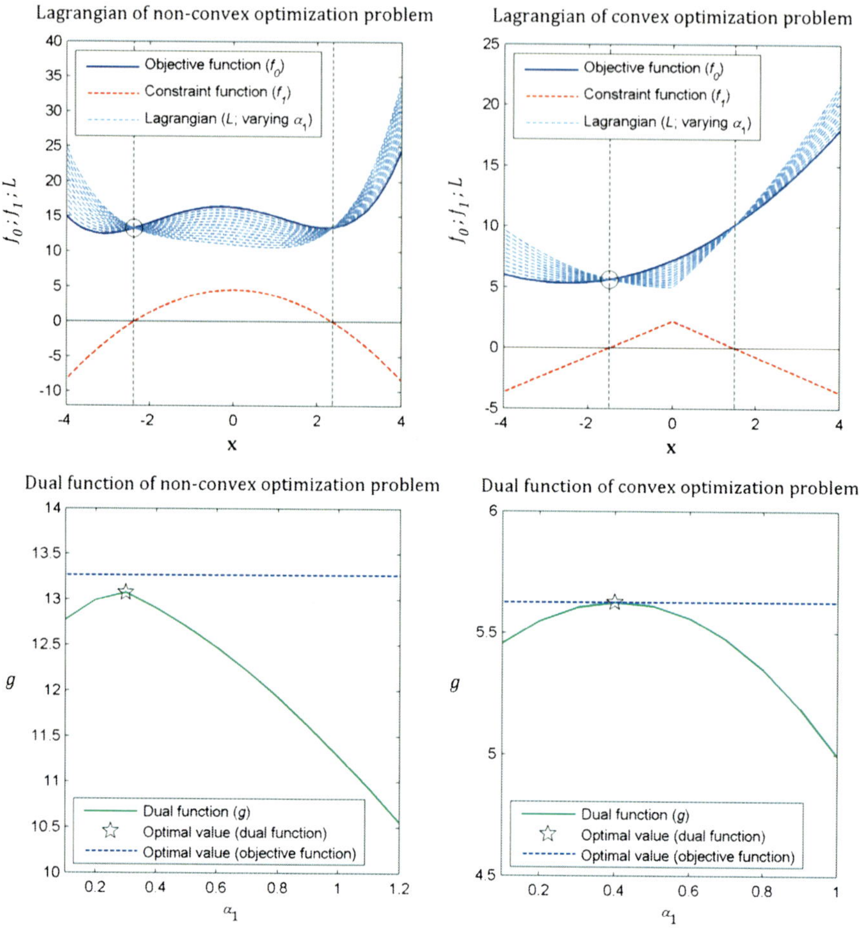

Fig. 4.7 Illustration of optimization with Lagrangian multipliers and the dual formulation for non-convex and convex optimization problems

The first KKT condition derives from the stationarity for $\mathbf{x}$ required for the minimization of $L(\mathbf{x}, \boldsymbol{\alpha})$:

$$\frac{\partial L(\mathbf{x}, \boldsymbol{\alpha})}{\partial \mathbf{x}} = f_0'(\mathbf{x}) - \sum_{i=1}^{n_c} \alpha_i f_i'(\mathbf{x}) = 0. \tag{4.20}$$

The second KKT condition ensures primal feasibility in terms of the inequalities of the primal formulation (from (4.13)):

$$f_i(\mathbf{x}) \geq 0; \quad i = 1, \ldots, n_c. \tag{4.21}$$

The third KKT condition ensures dual feasibility in terms of the inequalities of the dual formulation (from (4.18)):

$$\alpha_i \geq 0; \quad i = 1, \ldots, n_c. \tag{4.22}$$

The fourth KKT condition of complementary slackness (4.24) derives from the zero duality gap of convex optimization problems. At optimal $\mathbf{x}$ and $\boldsymbol{\alpha}$, $f_0(\mathbf{x})$, $L(\mathbf{x}, \boldsymbol{\alpha})$ and $g(\boldsymbol{\alpha})$ (as related in (4.17)) are equal:

$$g(\boldsymbol{\alpha}) = L(\mathbf{x}, \boldsymbol{\alpha}) = f_0(\mathbf{x})$$

$$\therefore\ f_0(\mathbf{x}) - \sum_{i=1}^{n_c} \alpha_i f_i(\mathbf{x}) = f_0(\mathbf{x})$$

$$\therefore\ \sum_{i=1}^{n_c} \alpha_i f_i(\mathbf{x}) = 0. \tag{4.23}$$

Since $f_i(\mathbf{x}) \geq 0$ for primal feasibility and $\alpha_i \geq 0$ for dual feasibility ($i = 1, \ldots, n_c$), the sum in (4.23) can only hold when the following is satisfied:

$$\alpha_i f_i(\mathbf{x}) = 0; \quad i = 1, \ldots, n_c. \tag{4.24}$$

This condition is known as complementary slackness. For each $i = 1, \ldots, n_c$, it follows that when the dual variable equality for α_i is tight (equal to zero), the primal inequality $f_i(\mathbf{x})$ can be slack (larger than zero). Similarly, it follows that when the primal inequality $f_i(\mathbf{x})$ is tight (equal to zero), the dual variable equality for α_i can be slack (larger than zero).

Optimal Margin Classifier

Now that the mechanics of quadratic optimization have been detailed, we return to optimization problem for a maximal margin classifier, given here in its primal formulation:

$$\min_{\mathbf{w},b} \frac{1}{2}\|\mathbf{w}\|^2$$

$$\text{subject to } y_i\left(\mathbf{w}^{\mathrm{T}}\mathbf{x}_i + b\right) - 1 \geq 0; \quad i = 1, \ldots, N. \tag{4.25}$$

Above, the primal variables are $\mathbf{w}$ and b, with $(\mathbf{x}_i, y_i; i = 1, \ldots, N)$ the training data inputs and responses. The Lagrangian for (4.25) is constructed by subtracting

α_i-weighted inequalities $y_i\left(\mathbf{w}^{\mathrm{T}}\mathbf{x}_i + b\right) - 1 \geq 0$ from the objective function, where there are N (the number of training data points) number of inequalities:

$$L(\mathbf{w}, b, \boldsymbol{\alpha}) = \frac{1}{2}\|\mathbf{w}\|^2 - \sum_{i=1}^{N} \alpha_i\left(y_i\left(\mathbf{w}^{\mathrm{T}}\mathbf{x}_i + b\right) - 1\right); \quad i = 1, \ldots, N. \tag{4.26}$$

The KKT stationarity condition (4.20) concerns the derivatives of $L(\mathbf{w}, b, \boldsymbol{\alpha})$ with respect to the primal variables. This condition leads to the following expressions:

$$\frac{\partial L(\mathbf{w}, b, \boldsymbol{\alpha})}{\partial b} = \sum_{i=1}^{N} \alpha_i y_i = 0. \tag{4.27}$$

$$\frac{\partial L(\mathbf{w}, b, \boldsymbol{\alpha})}{\partial \mathbf{w}} = \mathbf{w} - \sum_{i=1}^{N} \alpha_i y_i \mathbf{x}_i = 0$$

$$\therefore \mathbf{w} = \sum_{i=1}^{N} \alpha_i y_i x_i. \tag{4.28}$$

The KKT complementary slackness condition (4.24) gives rise to the following set of N equalities, in terms of the training data $(\mathbf{x}_i, y_i)$ and the Lagrangian multipliers α_i:

$$\alpha_i(y_i(\mathbf{w}^{\mathrm{T}}\mathbf{x}_i + b) - 1) = 0; \quad i = 1, \ldots, N. \tag{4.29}$$

Substituting (4.27) and 4.28 into the Lagrangian (4.26) and considering maximization over the Lagrangian multipliers in the manner of (4.19) results in the dual formulation of the optimization problem:

$$\max_{\alpha}\left(\sum_{i=1}^{N} \alpha_i - \frac{1}{2}\sum_{i,j=1}^{N} \alpha_i \alpha_j y_i y_j \mathbf{x}_i^{\mathrm{T}} \mathbf{x}_j\right)$$

$$\text{subject to } \alpha_i \geq 0; \quad i = 1, \ldots, N$$

$$\text{and } \sum_{i=1}^{N} \alpha_i y_i = 0. \tag{4.30}$$

The dual formulation of the margin optimization problem is simpler to solve than the primal formulation, and this form generalizes well to the case of soft margin, nonlinear and regression expansions. Another important characteristic of the dual formulation is the presence of a dot product relation between the input vectors, $\mathbf{x}^{\mathrm{T}}\mathbf{x}$. As will be seen later, this dot product can easily be replaced by a kernel function of $\mathbf{x}$, which expands the capacity of the hyperplane classifier.

If $\boldsymbol{\alpha}$ is the solution to the optimization problem above, $\alpha_j > 0$ indicates input training points $\mathbf{x}_j$ that lie exactly on the -1 or $+1$ hyperplanes, the support vectors (see Fig. 4.6). The optimal weight vector $\mathbf{w}$ can be calculated from (4.28). For all support vectors $\mathbf{x}_j$ ($j = 1, \ldots, N_{\mathrm{SV}}$), the following equality holds:

$$y_j \left(w^{\mathrm{T}} x_j + b \right) = 1 \tag{4.31}$$

Rearranging (4.31), substituting (4.28) for the weight vector $\mathbf{w}$ and remembering that $y_j = (y_j)^{-1}$ gives the expression for calculating the optimal bias b (where j is the index for the N_{SV} support vectors):

$$b = \frac{1}{N_{\mathrm{SV}}} \sum_{j=1}^{N_{\mathrm{SV}}} \left(y_j - \sum_{i=1}^{N} \alpha_i y_i x_i^{\mathrm{T}} x_j \right); \; j = 1, \ldots, N_{\mathrm{SV}}. \tag{4.32}$$

Although the bias can be calculated from any individual support vector input/response pair, the averaging calculation in (4.28) provides a stable estimate. The bias can also be adjusted post priori (in conjunction with a validation data set) to deliver a user-specified fraction of false positives or false negatives.

The prediction function $f(\mathbf{x})$ for the optimal margin classifier can be expressed in terms of the dual variable $\boldsymbol{\alpha}$, the bias b and the support vector input training points $\mathbf{x}_j$ ($j = 1, \ldots, N_{\mathrm{SV}}$):

$$y = f(\mathbf{x}) = \mathrm{sgn}(\mathbf{w}^{\mathrm{T}}\mathbf{x} + b) = \mathrm{sgn}\left(\sum_{j=1}^{N_{SV}} \alpha_j y_j \mathbf{x}^{\mathrm{T}} \mathbf{x}_j + b \right). \tag{4.33}$$

The decision function given in (4.33) is interesting in that it only depends on the support vectors. Given a large training data set, the optimal margin may be dependent on only a small number of support vectors. The decision function is expressed in terms of dot products of a new input $\mathbf{x}$ with the support vectors $\mathbf{x}_j$ ($j = 1, \ldots, N_{\mathrm{SV}}$). These dot products can easily be replaced by kernel functions, which allow the expansion of this linear space margin to a nonlinear space margin.

VC Dimension of Optimal Margin Classifier

Previously, it was shown that the VC dimension h of a hyperplane (linear) classifier is $M + 1$, where M is the dimensionality of the input vectors. A margin classifier restricts the set of possible hyperplanes, and this restricts the capacity of maximal margin classifiers. The VC dimension h of a maximal margin classifier is related to the size of the margin m and the extent of the input training data, represented by a diameter D_s of the smallest data enclosing sphere.

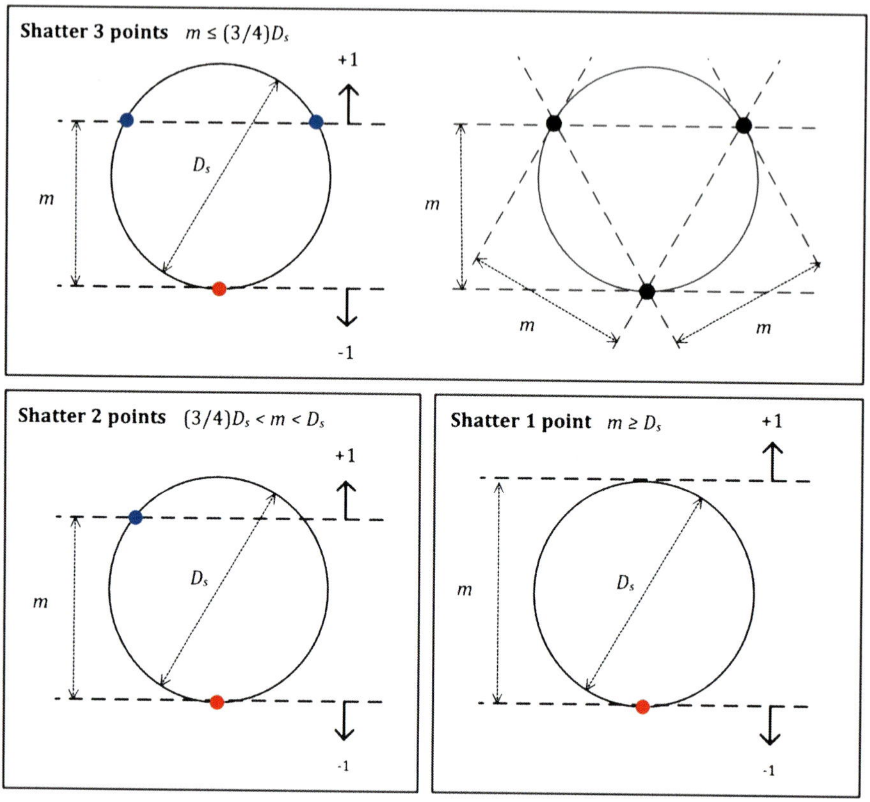

Fig. 4.8 Illustration of the shattering capability of a margin classifier in two dimensions (*m* is the size of the margin, and D_s is the diameter of a circle enclosing the data to be classified)

Figure 4.8 illustrates the reduced shattering capability of a margin classifier in two dimensions. Recall that a simple linear hyperplane classifier is able to shatter any three points in two dimensions. Where the margin m is less than $\frac{3}{4}D_s$, the margin classifier can also shatter any three points. However, when the margin increases ($\frac{3}{4}D_s < m < D$), only two points can be shattered; when the margin increases beyond D, only one point can be shattered.

The VC dimension of a maximal margin classifier is given by the following relation (Burges 1998):

$$h = \min\left(\frac{D_s^2}{m^2}, M\right) + 1. \tag{4.34}$$

A result from the reduced shattering capability is that the VC dimension of a margin classifier can be decreased by maximizing its margin. This property is attractive for the purpose of structural risk minimization.

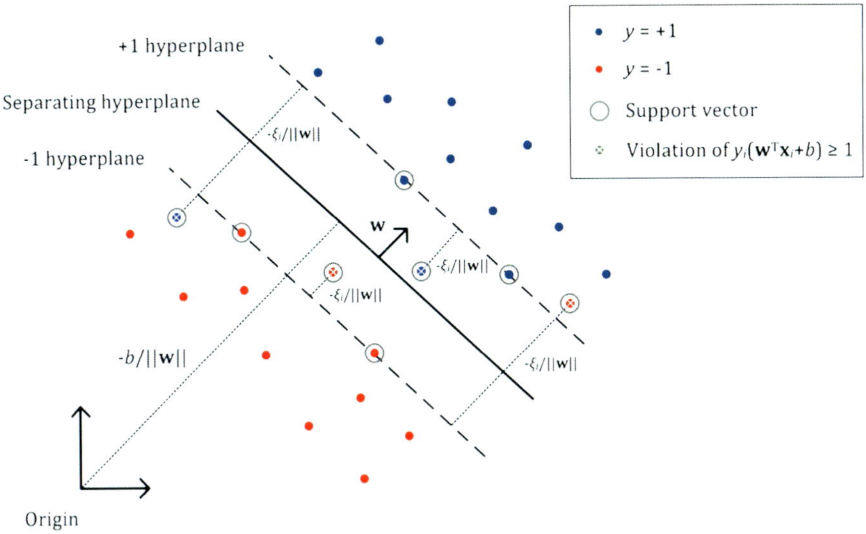

Fig. 4.9 Illustration of the separating (defined by **w** and b), -1 and $+1$ hyperplanes for a linearly inseparable classification problem, with slack variables ξ_i

4.3.2 *Soft Margin Linear Classifiers*

The hard margin classifier in the previous subsection can only exist when the training data are linearly separable. When training data are not linearly separable, the constraint $y_i\left(\mathbf{w}^{\mathrm{T}}\mathbf{x}+b\right) \geq 1$ in (4.12) is violated, resulting in no feasible region and no solution to the optimization problem. An adjustment can be made for linearly inseparable data to allow a certain number of violations in terms of $y_i\left(\mathbf{w}^{\mathrm{T}}\mathbf{x}+b\right) \geq 1$ to occur. A slack variable $\boldsymbol{\xi}$ (with elements $\xi_i, i = 1, \ldots, N$) is introduced to slacken the empty margin constraint:

$$y_i(\mathbf{w}^{\mathrm{T}}\mathbf{x}_i + b) \geq 1 - \xi_i; \quad i = 1, \ldots, N. \tag{4.35}$$

The slackened constraint in (4.21) allows points to reside in the margin between the $+1$ and -1 hyperplanes (margin errors) or even to cross over to a different classification zone (classification errors), as illustrated in Fig. 4.9.

The primal soft margin optimization problem is now:

$$\begin{aligned} &\min_{\mathbf{w},b,\boldsymbol{\xi}} \left(\frac{1}{2}\|\mathbf{w}\|^2 + C\sum_{i=1}^{N}\xi_i\right) \\ &\text{subject to } y_i(\mathbf{w}^{\mathrm{T}}\mathbf{x}_i + b) - 1 + \xi_i \geq 0;\ i = 1, \ldots, N \\ &\text{and } \xi_i \geq 0;\ i = 1, \ldots, N. \end{aligned} \tag{4.36}$$

Above, C is a user-defined positive parameter that controls the trade-off between minimizing margin errors and maximizing the margin. This approach to soft margin classification gives rise to C-SVM.

The dual formulation of (4.36) is given here:

$$\max_{\alpha}\left(\sum_{i=1}^{N}\alpha_i - \frac{1}{2}\sum_{i,j=1}^{N}\alpha_i\alpha_j y_i y_j \mathbf{x}_i^{\mathrm{T}}\mathbf{x}_j\right)$$

$$\text{subject to } 0 \le \alpha_i \le C; \quad i = 1,\ldots,N$$

$$\text{and } \sum_{i=1}^{N}\alpha_i y_i = 0 \tag{4.37}$$

Once the optimal $\boldsymbol{\alpha}$ has been determined from (4.37), the weight vector $\mathbf{w}$ and bias b can be calculated as for the hard margin case (according to (4.28) and (4.32), respectively). All points with $\alpha_i > 0$ are support vectors, while points satisfying $0 < \alpha_i < C$ are support vectors on the $+1$ and -1 hyperplanes. Points with $\alpha_i \ge C$ lie within the margin or in the zone of incorrect classification. A limitation of the above approach is the interpretability of C and its selection.

The soft margin optimization problem can be stated in terms of a more intuitive parameter, ν. The parameter ν enforces an upper bound on the fraction of margin errors and a lower bound on the fraction of support vectors, relative to the number of training data, N.

The primal formulation for this ν-parameter approach to the soft margin classifier is given here:

$$\min_{\mathbf{w},b,\rho,\xi}\left(\frac{1}{2}\|\mathbf{w}\|^2 - \nu\rho + \frac{1}{N}\sum_{i=1}^{N}\xi_i\right)$$

$$\text{subject to } y_i(\mathbf{w}^{\mathrm{T}}\mathbf{x}_i + b) - \rho + \xi_i \ge 0; \quad i = 1,\ldots,N$$

$$\text{and } \xi_i \ge 0; \quad i = 1,\ldots,N$$

$$\text{and } \rho \ge 0. \tag{4.38}$$

Above, ρ is an effective margin parameter to be optimized as well. In the dual formulation of (4.38), ρ (along with $\mathbf{w}$, b and $\boldsymbol{\xi}$) disappears:

$$\max_{\alpha}\left(-\frac{1}{2}\sum_{i,j=1}^{N}\alpha_i\alpha_j y_i y_j \mathbf{x}_i^{\mathrm{T}}\mathbf{x}_j\right)$$

$$\text{subject to } 0 \le \alpha_i \le \frac{1}{N}; \; i = 1,\ldots,N$$

$$\text{and} \sum_{i=1}^{N} \alpha_i y_i = 0$$

$$\text{and} \sum_{i=1}^{N} \alpha_i \geq \nu. \tag{4.39}$$

The weight vector $\mathbf{w}$ corresponding to the optimal $\boldsymbol{\alpha}$ from (4.39) is determined according to (4.28). To determine the bias b, two sets S_+ and S_- of equal size s representing support vectors for the $+1$ and -1 classes are constructed. The bias can now be calculated from the union of these sets, $S_\pm$:

$$b = -\frac{1}{2s} \sum_{\mathbf{x} \in S_\pm} \sum_{i=1}^{N} \alpha_i y_i \mathbf{x}_i^{\mathrm{T}} \mathbf{x}. \tag{4.40}$$

This approach to soft margin classification with the intuitive ν-parameter gives rise to ν-SVM.

4.3.3 *Primal and Dual Formulation of Problems*

As discussed in the previous sections, models can be cast as primal or dual problems to be solved. In practice, the choice of such formulations would depend on the nature of the problem. For example, with linear regression problems associated with N samples and M variables, the primal formulation would be given by

$$\hat{y} = \boldsymbol{w}^{\mathrm{T}} \boldsymbol{x} + \boldsymbol{b}. \tag{4.41}$$

In contrast, the dual formulation would amount to

$$\hat{y} = \sum_{i=1}^{M} \boldsymbol{\alpha}_i \boldsymbol{x}_i^{\mathrm{T}} \boldsymbol{x} + b. \tag{4.42}$$

The set of parameters in the primal formulation would be elements of the M-dimensional space, as follows, $\boldsymbol{w} \in \mathbb{R}^M$, while the set of parameters in the dual formulation would be elements of the N-dimensional space, as follows, $\alpha \in \mathbb{R}^N$ ($N \times N$ kernel matrix).

Therefore, the dual formulation would be more advantageous in problems associated with large numbers of variables and small numbers of samples (e.g. in multivariate image analysis, with each pixel in a vectorized image representing a variable and each image a sample). In contrast, the primal formulation would be better in problems associated with few variables (small M) and many samples (large N).

4.4 Kernels

A recurring feature of the margin classifiers above is the presence of the dot product $\mathbf{x}\cdot\mathbf{x}'$ (between sample vectors $\mathbf{x}$ and $\mathbf{x}'$) in the parameter estimation and decision function formulations. The dot product of two vectors is related to their lengths and angle θ in the vector space $\mathcal{X} \in \mathbb{R}^M$:

$$\mathbf{x}\cdot\mathbf{x}' = \sum_{i=1}^{M} \mathbf{x}_i \mathbf{x}'_i = \mathbf{x}^{\mathrm{T}}\mathbf{x}' = \|\mathbf{x}\| \left\|\mathbf{x}'\right\| \cos\theta. \tag{4.43}$$

This dot product can be considered a linear similarity measure between two vectors: As the angle θ increases between the vectors, the cosine of θ decreases, and the vectors are less similar. A function that estimates the similarity between two vectors is known as a kernel function k. The dot product above can be considered a linear kernel:

$$k(\mathbf{x},\mathbf{x}') = \mathbf{x}^{\mathrm{T}}\mathbf{x}'. \tag{4.44}$$

The margin classifiers discussed previously use this linear kernel in the calculation of an optimal separation of linearly separable data. Even with soft margin allowances, these margin classifiers may fail to find suitable solutions for linear inseparable data. To extend the application of margin classifiers to nonlinear data, more general kernel functions have been explored.

4.4.1 *Nonlinear Mapping and Kernel Functions*

Cover's theorem suggests that data that are linearly inseparable in an input space $\mathcal{X}$ will be more likely to be linearly separable in some higher-dimensional space $\mathcal{H}$, where the mapping function Φ from $\mathcal{X}$ to $\mathcal{H}$ is nonlinear (Cover 1965). An example of such a nonlinear mapping is given in Fig. 4.10.

Above, two classes (represented by red and blue markers) are linearly inseparable in the original input space $\mathcal{X} \in \mathbb{R}^2$. The following mapping function $\Phi(\mathbf{x})$ to $\mathcal{H} \in \mathbb{R}^3$ is considered:

$$\Phi(\mathbf{x}) = \Phi(x_1, x_2) = (z_1, z_2, z_3) = \left(x_1^2, \sqrt{2}x_1x_2, x_2^2\right). \tag{4.45}$$

The general nonlinear transformation of $\mathcal{X}$ to $\mathcal{H}$ can be seen from the change in shape of the square grid. As can be seen in Fig. 4.10, the mapped data in $\mathcal{H}$ becomes linearly separable. The utility of nonlinear mapping having been illustrated on this simple example, the general properties of kernel functions as representations of dot products in nonlinear spaces $\mathcal{H}$ will now be considered.

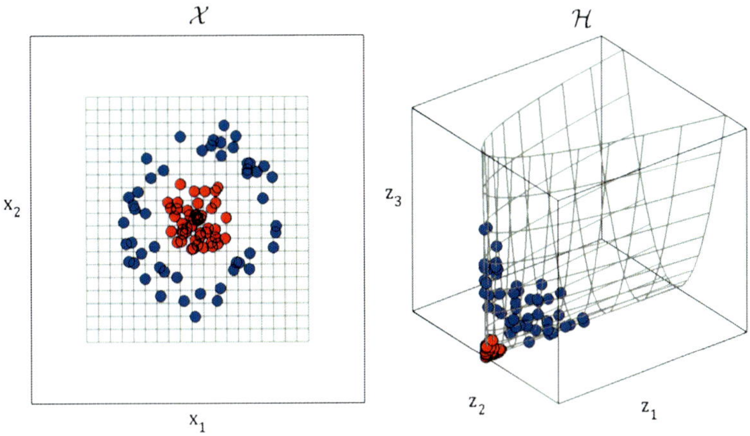

Fig. 4.10 Illustration of a mapping from a low-dimensional space $\mathcal{X}$ to a high-dimensional space $\mathcal{H}$ through a nonlinear map $\Phi(x_1, x_2) = (z_1, z_2, z_3) = \left(x_1^2,\ \sqrt{2}x_1x_2, x_2^2\right)$. The *grid* represents the same square in both low- and high-dimensional spaces

Let Φ be a mapping function from an input space $\mathcal{X}$ to some space $\mathcal{H}$ (known as a Hilbert space, where dot products are defined) such that $\Phi(\mathbf{x})$ is the mapping of a sample vector $\mathbf{x}$. A kernel function that corresponds to a dot product in $\mathcal{H}$ is then defined as:

$$k(\mathbf{x}, \mathbf{x}') = \Phi(\mathbf{x}) \cdot \Phi(\mathbf{x}'). \tag{4.46}$$

A benefit of a kernel function satisfying the above equation lies in the fact that only the input patterns $\mathbf{x}$ and $\mathbf{x}'$ are required to calculate a nonlinear similarity, the dot product in $\mathcal{H}$ mapped by Φ. The choice of a kernel function k fixes the choice of Φ and $\mathcal{H}$. From4.46, the question arises: Which functional forms of k represent a valid dot product in some space $\mathcal{H}$, defined by a mapping Φ?

From the mathematical field of functional analysis, a special case of Mercer's theorem states that 4.46 will hold if and only if the kernel function k is a symmetric positive semi-definite function. Let $\mathbf{K}$ be the Gram matrix generated by the kernel function k for N input patterns, with elements K_{ij}; $i, j = 1, \ldots, N$:

$$K_{ij} = k(\mathbf{x}_i, \mathbf{x}_j). \tag{4.47}$$

Then $\mathbf{K}$ will be symmetric if:

$$K_{ij} = K_{ji}. \tag{4.48}$$

And positive semi-definite if, for any vector $\mathbf{v} \in \mathbb{R}^N$:

$$\mathbf{v}^{\mathrm{T}}\mathbf{K}\mathbf{v} \geq 0. \tag{4.49}$$

A kernel function k is then said to be symmetric positive semi-definite if it gives rise to a symmetric positive semi-definite Gram matrix $\mathbf{K}$, satisfying the conditions above.

If it can be shown that a function is symmetric positive semi-definite, it can serve as a kernel function, expressing a general form of similarity that is guaranteed to be a dot product in some Hilbert space $\mathcal{H}$ defined by some mapping Φ. With the kernel function known, it is not necessary to have an explicit formulation of the mapping Φ. The Hilbert space $\mathcal{H}$ can be any viable dot product space, even of infinite dimension. By implicitly calculating dot products, the computational problem associated with possibly very large vectors is circumvented.

4.4.2 Examples of Kernel Functions

A number of kernel functions are well known and oft-used in the machine learning field, especially the polynomial, Gaussian and sigmoid kernels. The dth-order inhomogeneous polynomial kernel function is shown below:

$$k(\mathbf{x}, \mathbf{x}') = (\mathbf{x}^{\mathrm{T}}\mathbf{x}' + 1)^d. \tag{4.50}$$

The dth-order homogeneous polynomial kernel function omits an offset:

$$k(\mathbf{x}, \mathbf{x}') = (\mathbf{x}^{\mathrm{T}}\mathbf{x}')^d. \tag{4.51}$$

Returning to Fig. 4.10, a dot product between two vectors $\mathbf{z} = \Phi(\mathbf{x})$ and $\mathbf{z}' = \Phi(\mathbf{x}')$ in $\mathcal{H}$ would be:

$$\begin{aligned}\mathbf{z}^T\mathbf{z}' &= \sum_{i=1}^{3} \mathbf{z}_i\mathbf{z}_i' = x_1^2 x_1'^2 + 2x_1x_2x_1'x_2' + x_2^2x_2'^2 \\ &= (x_1x_1' + x_2x_2')^2 = (\mathbf{x}^T\mathbf{x}')^2.\end{aligned} \tag{4.52}$$

From above, a second order homogeneous polynomial function would be able to calculate dot products in the feature space $\mathcal{H}$ without computing the mapping function Φ.

Another popular kernel function is the Gaussian kernel function (with a kernel width parameter $\sigma > 0$):

$$k(\mathbf{x}, \mathbf{x}') = \exp\left(-\frac{\|\mathbf{x} - \mathbf{x}'\|^2}{2\sigma^2}\right). \tag{4.53}$$

The sigmoid kernel (with parameters $\kappa > 0$ and $\vartheta < 0$) makes use of the hyperbolic tangent function often used in multilayer perceptrons:

$$k(\mathbf{x}, \mathbf{x}') = \tanh(\kappa \mathbf{x}^{\mathrm{T}} \mathbf{x}' + \vartheta). \quad (4.54)$$

The sigmoid kernel function is interesting in that it does not necessarily satisfy Mercer's conditions, but has been used successfully with margin classifiers in practice.

The kernel functions above all have free parameters (d, σ, κ, ϑ). For a particular application of a kernel function in some algorithm, these parameters can be estimated based on some a priori knowledge, from heuristics based on properties of the training data or by finding optimal values for the particular application through cross-validation. Scaling of the kernel function arguments (the training data matrix $\mathbf{X}$) is sometimes required to ensure sensical and sensitive similarity measures.

Kernel functions can be created by performing certain operations on other kernel functions (Hsieh 2009). For example, given two kernel functions k_1 and k_2 that satisfy Mercer's conditions, the following are examples of operations that will result in kernel functions that also satisfy Mercer's conditions:

$$k(\mathbf{x}, \mathbf{x}') = c k_1(\mathbf{x}, \mathbf{x}'), c \geq 0. \quad (4.55)$$

$$k(\mathbf{x}, \mathbf{x}') = k_1(\mathbf{x}, \mathbf{x}') + k_2(\mathbf{x}, \mathbf{x}'). \quad (4.56)$$

$$k(\mathbf{x}, \mathbf{x}') = k_1(\mathbf{x}, \mathbf{x}') k_2(\mathbf{x}, \mathbf{x}'). \quad (4.57)$$

$$k(\mathbf{x}, \mathbf{x}') = (k_1(\mathbf{x}, \mathbf{x}'))^d. \quad (4.58)$$

4.4.3 Kernel Trick

Kernel functions, which implicitly calculate dot products in higher-dimensional, nonlinear dot product spaces, are powerful tools for expanding statistical learning algorithms. If a learning algorithm can be expressed in terms of training data dot products, these dot products can be replaced by a kernel function of the training data.

This is the so-called kernel trick, first applied by Cortes and Vapnik (1995): If an algorithm is formulated in terms of a valid kernel function k, an alternative algorithm can be constructed by substituting k with an alternate valid kernel function k'. This allows linear algorithms that can be stated in dot product form (such as the maximal margin classifier discussed previously) to be extended as nonlinear algorithms, with appropriate choice of nonlinear kernel functions.

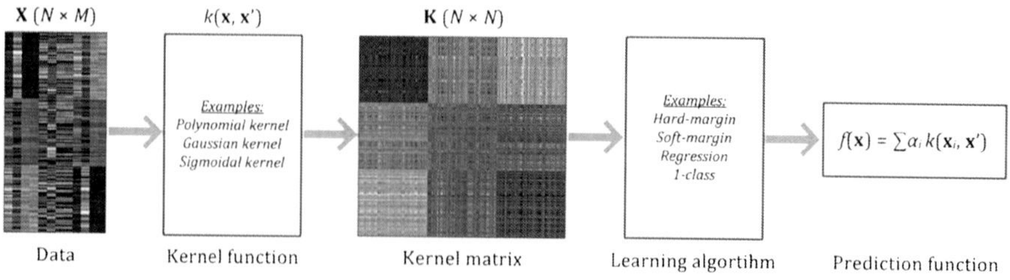

Fig. 4.11 Statistical learning with kernel methods

The kernel trick induces the modularity of kernel methods (Shawe-Taylor and Cristianini 2004): Any learning algorithm which operates on kernel matrices can be used in conjunction with a wide variety of kernel functions (see Fig. 4.11).

4.5 Support Vector Machines

Support vector machines (SVM) arose from the application of the kernel trick to linear margin classifiers (Boser et al. 1992; Cortes and Vapnik 1995). The dual form of the optimization problem, as well as the final decision function of a linear margin classifier, is expressed in terms of dot products of the training input data, $\mathbf{x}^{\mathrm{T}}\mathbf{x}'$. This linear kernel function can be replaced by any other kernel function k, to generalize the linear margin classifier in the input space to a linear margin classifier in some implicit nonlinear high-dimensional space.

The optimization problem for a hard margin support vector machine is then:

$$\max_{\alpha}\left(\sum_{i=1}^{N}\alpha_i - \frac{1}{2}\sum_{i,j=1}^{N}\alpha_i\alpha_j y_i y_j k(\mathbf{x}_i,\mathbf{x}_j)\right)$$
$$\text{subject to}\,\alpha_i \geq 0;\quad i = 1,\ldots,N \tag{4.59}$$
$$\text{and}\,\sum_{i=1}^{N}\alpha_i y_i = 0.$$

The final decision function is expressed in terms of the chosen kernel function:

$$y = f(\mathrm{x}) = \mathrm{sgn}\left(\sum_{j=1}^{N_{\mathrm{SV}}}\alpha_j y_j k(\mathbf{x},\mathbf{x}_j) + b\right). \tag{4.60}$$

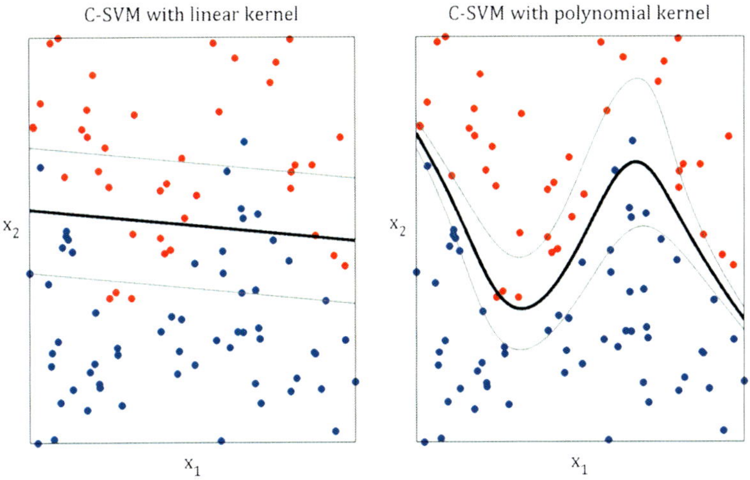

Fig. 4.12 Example of C-SVM with linear and polynomial kernels. *Blue* and *red points* indicate data of different groups. *Thick black lines* indicate the decision boundary, while *thin grey lines* indicate the margin

The bias term in the decision function above is also expressed in terms of the chosen kernel function:

$$b = \frac{1}{N_{SV}} \sum_{j=1}^{N_{SV}} \left(y_j - \sum_{i=1}^{N} \alpha_i y_i k(\mathrm{x}_i, \mathrm{x}_j) \right); \quad j = 1, \ldots, N_{SV}. \tag{4.61}$$

The kernel trick can also be used to create soft margin support vector machines, which leads to the C-SVM and ν-SVM algorithms. An example of the solutions to a soft margin SVM classification problem with linear and polynomial kernel functions is given in Fig. 4.12. The linear kernel approach cannot capture the nonlinearity of the separation between the two classes. The polynomial kernel approach shows a better approximation to the class boundary. Margin infringements (training data occurring in the margin) as well as training errors (training data on the wrong side of the decision boundary) can be seen, which are characteristic of C-SVM. Even though these infringements occur, the decision boundary delivers good separation between the two classes.

An example of highly nonlinear structure that can be captured with kernel functions is given in Fig. 4.13. The intertwined red and blue spirals can be successfully segregated with a Gaussian kernel and SVM classifier.

However, the nonlinear mapping strength of kernel functions comes at a price. Figure 4.13 also shows the spiral data classification result with a different Gaussian kernel parameter. Although no mislabelled points occur for the training data set, the highly localized class regions do not represent the general spiral pattern well.

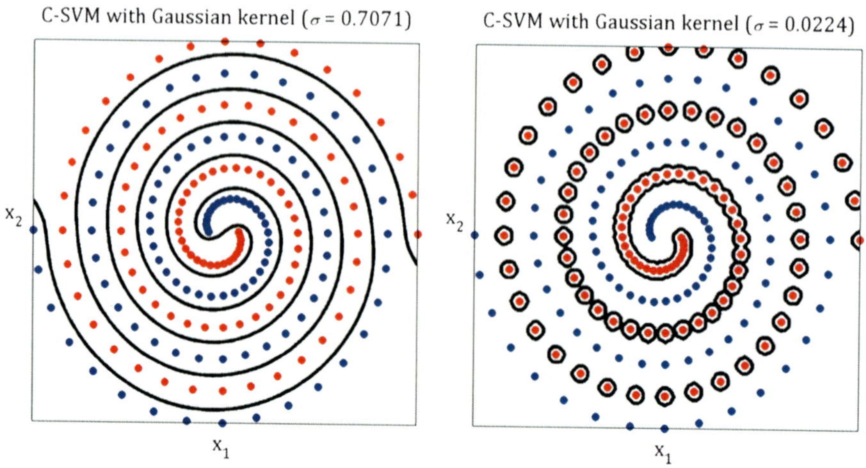

Fig. 4.13 Example of C-SVM with Gaussian kernels with different kernel width parameters. *Blue* and *red points* indicate data of different groups. *Thick black lines* indicate the decision boundary

4.5.1 Parameter Selection with Cross-Validation

As seen in Fig. 4.13, kernel parameter selection is an important consideration in SVM implementation. For soft margin applications (C-SVM and ν-SVM), the capacity trade-off parameter, C or ν, is also free. A popular and successful approach to determining free parameters for SVMs is through cross-validation:

Cross-Validation for Parameter Selection

- Select a set of Q parameters to vary, e.g. $C_1, \ldots, C_Q$
- Partition training data set $(\mathbf{X}, \mathbf{y})$ of N samples into K partitions:

$$\left\{\left(\mathbf{X}^{(1)}, \mathbf{y}^{(1)}\right), \left(\mathbf{X}^{(2)}, \mathbf{y}^{(2)}\right), \ldots, \left(\mathbf{X}^{(K)}, \mathbf{y}^{(K)}\right)\right\}.$$

- For $k = 1, \ldots, K$:
  - Train a learning machine for each q of the Q parameters on the training data set, excluding the k partition:

$$f^{(k,q)} \rightarrow (\mathbf{X}, \mathbf{y}) / \left(\mathbf{X}^{(k)}, \mathbf{y}^{(k)}\right).$$

  - Estimate the partition test error/loss function of the classifier $f^{(k,q)}$ from $(\mathbf{X}^{(k)}, \mathbf{y}^{(k)})$:

$$R_e\left(f^{(k,q)}\right) \rightarrow \left(\mathbf{X}^{(k)}, \mathbf{y}^{(k)}\right).$$

(continued)

(continued)

- Determine the overall estimate of the test error for each parameter C_q as the average of the k test partitions.
- Choose the parameter C_p^* associated with the lowest overall test error, and retrain the learning machine on the entire training data set $(\mathbf{X}, \mathbf{y})$.

4.5.2 VC Dimension of Support Vector Machines

As shown previously, the VC dimension of a linear separating hyperplane is related to the input dimensionality of the data set. When kernels are used to obtain a nonlinear mapping to some dot product space $\mathcal{H}$, the dimensionality of this space determines the VC dimension of a linear hyperplane of a kernel-based linear classifier.

A dth-order homogeneous polynomial kernel $(\mathbf{x}^{\mathrm{T}}\mathbf{x}')^d$ for a data set with original input space dimensionality of M results in a mapping to a space $\mathcal{H}$ with a dimensionality of (Burges 1998):

$$\dim(\mathcal{H}) = \binom{M+d-1}{d} \tag{4.62}$$

This dimensionality (and the associated VC dimension of a linear hyperplane classifier in $\mathcal{H}$) increases very quickly with increase in d or M.

The VC dimension of linear hyperplane classifiers in dot product spaces defined by Gaussian kernels can be very large, possible infinite (Burges 1998). If the Gaussian kernel width parameter is chosen to be small enough, each training data point will form the centre of its own small Gaussian, and an infinite number of points could be shattered by such a classifier. An example of small Gaussians creating small neighbourhoods is shown in Fig. 4.13.

The possibly very large VC dimension of polynomial and Gaussian kernels do not bode well for bounding risk and structural risk minimization. However, margin classifiers may have VC dimensions much smaller than the dimensionality of the dot product space in which the margin classifier is constructed (see (4.34)). The margin property of support vector machines ensures considerably better risk bounds than would be associated with simple linear classifiers in kernel-derived dot product spaces. Connection of the construction of the maximal margin with the estimation of the bound on the generalization performance according to (4.63) is therefore the important issue (Belousov et al. 2002):

$$R(f) \le R_e(f) + \sqrt{\frac{1}{Nm^2}(\ln 2Nm^2 + 1) + \frac{\ln\frac{4}{\delta}}{N}}, \tag{4.63}$$

where N is the number of training examples and m is the margin.

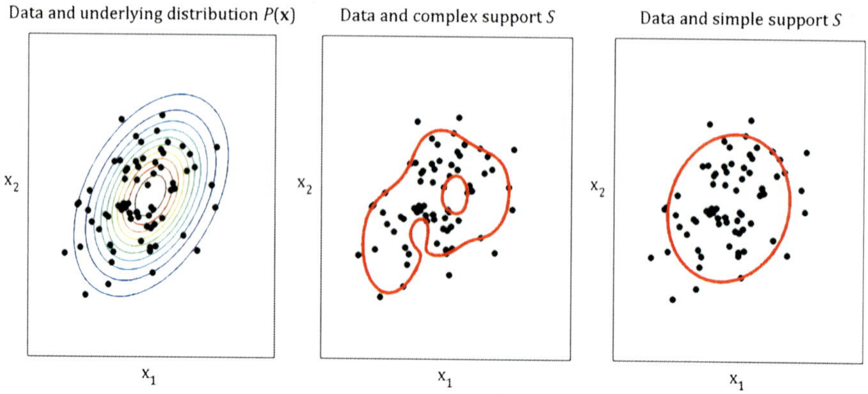

Fig. 4.14 Example of estimating the support for a data set with some underlying (generally unknown) distribution. Data are shown in *black*, the underlying multivariate distribution $P(x)$ is indicated on the *left plot*, while two possible support subsets S are shown with *red lines* in the *centre* and *right plots*

4.5.3 Unsupervised Support Vector Machines

The supervised learning problem of classification with support vector machines has been considered thus far. Another statistical learning application employing the maximal margin concept and kernel trick is unsupervised learning for novelty detection. In unsupervised learning, the training data set consists of only unlabelled input training examples, $\mathbf{x}_i \in \mathbb{R}^M, i = 1, \ldots, N$. The goal of the learning algorithm considered here is to estimate the support of the training data set $\mathbf{X}$ (Schölkopf et al. 2001), where the support is a simple subset S characterizing the underlying probability distribution $P(\mathbf{x})$. The support is defined according to a parameter that controls the probability of a test point (sampled from $P(\mathbf{x})$) falling within the simple subset S.

Figure 4.14 gives an example of a random set of samples drawn from a multivariate distribution $P(\mathbf{x})$. Two possible support subsets are shown, as regions enclosed by red lines. The one subset is complex, with a long and intricate edge and an inclusion, while the other subset is simpler; an ellipse. An upper bound on the number of points excluded from the support subset can be specified as an algorithmic parameter.

One-class support vector machines (1-SVM) aim to learn a function $f(\mathbf{x})$ that is positive inside a support subset S and negative outside S. As with the SVM algorithms for classification encountered before, the functional form of f is given in terms of a kernel expansion (which gives nonlinear capacity), and its complexity is controlled by maximizing a margin. The 1-SVM formulation can also be solved through primal or dual form convex optimization. The dual form is stated explicitly in terms of a kernel function, which allows the use of a variety of implicit nonlinear mappings through kernel choice.

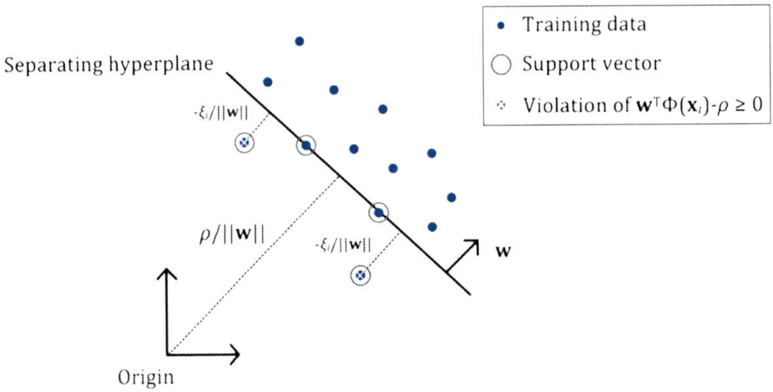

Fig. 4.15 Illustration of the separating hyperplane for a one-class support vector machine with slack variables ξ_i

The geometrical mechanics inspiring the 1-SVM algorithm involves mapping the training data to a (generally) nonlinear dot product space $\mathcal{H}$ and separating the data in this space from the origin by a maximal margin. Transformation to a nonlinear space allows a greater variety of support subsets to be learnt, while maximizing the margin reduces the capacity (and associated risks) of the function. As with the soft margin classifiers, a slack variable is introduced to allow margin infringements. These infringements are outliers to the support subsets, with the amount of infringements controlled by a model parameter. Figure 4.15 illustrates the 1-SVM separating hyperplane and margin infringements in some dot product space $\mathcal{H}$, where ρ is a bias defining the location of the hyperplane.

Given a set of unlabelled data $\mathbf{X} \in \mathcal{X}$, and an implicit mapping function $\Phi(\mathbf{x})$ to a dot product space $\mathcal{H}$ (encapsulated in some kernel function k), the 1-SVM decision function (returning $+1$ inside the support subset and -1 outside the support subset) is given by:

$$f(\mathbf{x}) = \operatorname{sgn}(\mathbf{w}^{\mathrm{T}}\Phi(\mathbf{x}) - \rho). \tag{4.64}$$

The primal optimization problem to solve for 1-SVM is given here:

$$\begin{aligned}
&\min_{\mathbf{w},\rho,\xi} \left(\frac{1}{2}\|\mathbf{w}\|^2 + \frac{1}{\nu N}\sum_{i=1}^{N}\xi_i - \rho \right) \\
&\text{subject to } \mathbf{w}^{\mathrm{T}}\Phi(\mathrm{x}_i) - \rho + \xi_i \geq 0; \quad i = 1, \ldots, N \\
&\text{and } \xi_i \geq 0; \quad i = 1, \ldots, N \\
&\text{and } \rho \geq 0.
\end{aligned} \tag{4.65}$$

Above, the ν-parameter (with a range from 0 to 1) controls the complexity of the support subset (by acting as a lower bound for the fraction of support vectors), and also controls the extent of the support subset (by acting as an upper bound for the fraction of outliers).

The dual formulation for 1-SVM employs the kernel formulation:

$$\min_{\alpha} \sum_{i,j=1}^{N} \alpha_i \alpha_j k(\mathbf{x}_i, \mathbf{x}_j)$$

$$\text{subject to } 0 \leq \alpha_i \leq \frac{1}{\nu N}; \quad i = 1, \ldots, N$$

$$\text{and } \sum_{i=1}^{N} \alpha_i = 1. \tag{4.66}$$

The bias is also expressed in terms of the kernel function:

$$\rho = \sum_{i} \alpha_i k\left(\mathrm{x}, \mathrm{x}_i\right). \tag{4.67}$$

Given a new test point, the value for the 1-SVM decision function $f(\mathbf{x})$ shows whether this test sample falls within ($f(\mathbf{x}) > 0$) or without ($f(\mathbf{x}) < 0$) the support subset. Where a test point falls outside the support, it is classified as a "novel" point with respect to the original underlying distribution $P(\mathbf{x})$, with some confidence related to ν. From this, 1-SVM is often referred to as a novelty detection learning algorithm.

Figure 4.16 shows 1-SVM results for Gaussian kernel functions (with two instances of the kernel width parameter σ), with three settings for ν. The larger kernel width gives smoother support subsets, while the smaller kernel width is more prone to overfitting. As the ν-parameter increases, the upper bound on the fraction of outliers increases, which translates to a larger number of data points excluded from the support subset.

4.5.4 *Support Vector Regression*

Another extension of support vector machines for classification is to the regression task. Where classification learns a function to determine class membership (e.g. $y \in \{-1, +1\}$), regression aims to learn output in the form of continuous variables (e.g. $y \in \mathbb{R}$). Beneficial aspects of SVM classification to be retained in support vector regression (SVR) include the nonlinear nature of certain kernel functions, sparseness of the solution (margins expressed in terms of a small number of support vectors), the deterministic nature of the convex optimization solution as well as capacity control through margin maximization.

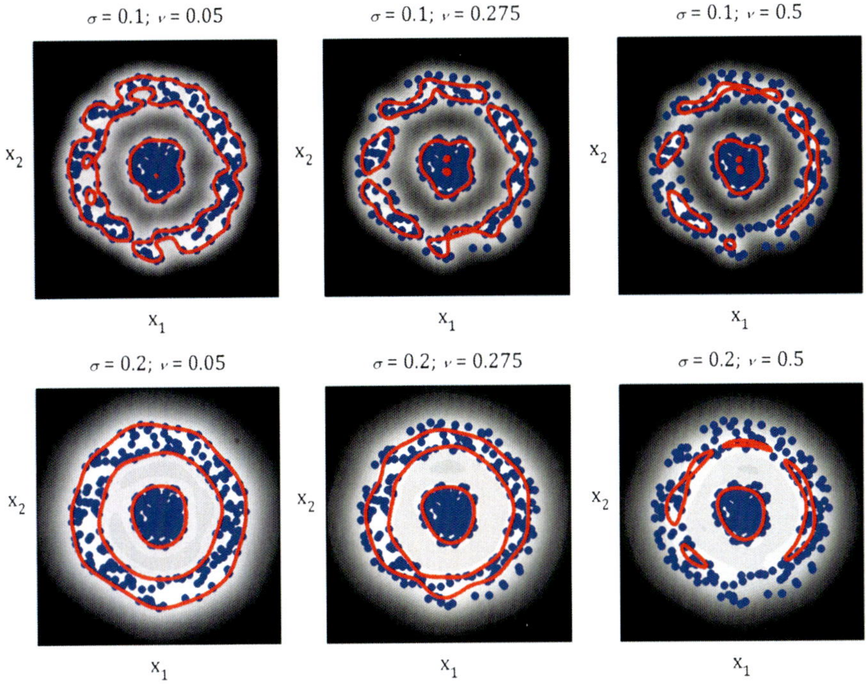

Fig. 4.16 Example of one-class SVM with Gaussian kernels with different kernel width parameters ($\sigma = 0.1$; 0.2) and trade-off parameters ($\nu = 0.05$; 0.275; 0.5). *Blue points* indicate data of one group. *Thick red lines* indicate the support boundaries. The *greyscale* background indicates the 1-SVM function value (*black* indicating negative values outside the support, *white* indicating positive values inside the support)

Sparseness in SVR can be enforced by employing a loss function that is insensitive to small residuals $(y - f(\mathbf{x}))$. An example of such a loss function is the ε-insensitive loss function, with $\varepsilon \geq 0$ a user-defined accuracy parameter (see Table 4.1, and below).

$$L(y, f(\mathbf{x})) = |y - f(\mathbf{x})|_\varepsilon - \max(0, |y - f(\mathbf{x})| \quad \varepsilon). \tag{4.68}$$

Support vector regression based on this loss function is known as ε-SVR (Smola and Schölkopf 2004). In effect, this insensitivity to deviances smaller than ε transpires to form a so-called ε-tube around the fitted function $f(\mathbf{x})$. Training data on and outside this ε-tube can be considered analogous to support vectors on and in the margin for SVM classification. This ε-tube for one-dimensional regression is illustrated in Fig. 4.17. As with SVM classification, slack variables can be introduced (and controlled by some parameter C) to allow for analogous margin infringements (in this case, ε-tube breakouts).

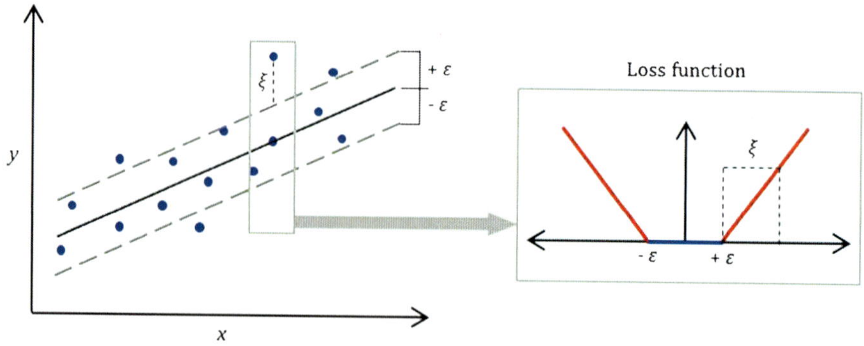

Fig. 4.17 Illustration of ε-SVR, with insert of ε-insensitive loss function and outlier with positive slack variable ξ_i

The ε-SVR regression function $f(\mathbf{x})$ is expressed in terms of a bias and a weight vector in some dot-product space $\mathcal{H}$, where the mapping to this dot-product space through $\Phi(\mathbf{x})$ is implicitly defined through some kernel function k:

$$y = f(\mathbf{x}) = \mathbf{w}^{\mathrm{T}}\mathbf{x} + b. \tag{4.69}$$

The primal formulation of the convex optimization problem for ε-SVR is given by:

$$\begin{aligned}
&\min_{\mathbf{w}} \left(\frac{1}{2}\|\mathbf{w}\|^2 + C\sum_{i=1}^{N}(\xi_i + \xi_i^*) \right) \\
&\text{subject to} - y_i + \mathbf{w}^{\mathrm{T}}\Phi(\mathbf{x}_i) + b \le \varepsilon + \xi_i; \quad i = 1, \ldots, N \\
&\text{and } y_i - \mathbf{w}^{\mathrm{T}}\Phi(\mathbf{x}_i) - b \le \varepsilon + \xi_i^*; \quad i = 1, \ldots, N \\
&\text{and } \xi_i, \xi_i^* \ge 0; \quad i = 1, \ldots, N.
\end{aligned} \tag{4.70}$$

The parameter C controls the trade-off between the "flatness" of the ε-tube (and thus the model complexity) obtained by minimizing $\|\mathbf{w}\|^2$ and the number of ε-tube breakouts, or outliers. The dual formulation of the above problem employs explicit kernel functions:

$$\begin{aligned}
&\min_{\alpha,\alpha^*} \begin{cases} -\frac{1}{2}\sum_{i,j=1}^{N}(\alpha_i^* - \alpha_i)\left(\alpha_j^* - \alpha_j\right)k\left(\mathbf{x}_i, \mathbf{x}_j\right) \\ -\varepsilon\sum_{i=1}^{N}(\alpha_i^* + \alpha_i) + \sum_{i=1}^{N} y_i(\alpha_i^* - \alpha_i) \end{cases} \\
&\text{subject to} \sum_{i=1}^{N}(\alpha_i - \alpha_i^*) = 0; \quad i = 1, \ldots, N \\
&\text{and } 0 \le \alpha_i, \alpha_i^* \le C; \quad i = 1, \ldots, N.
\end{aligned} \tag{4.71}$$

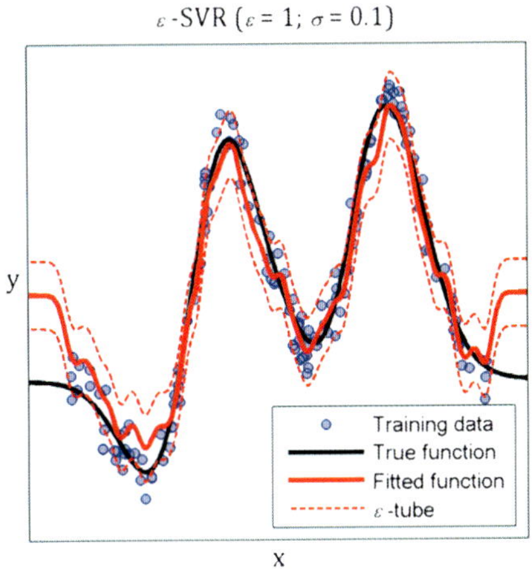

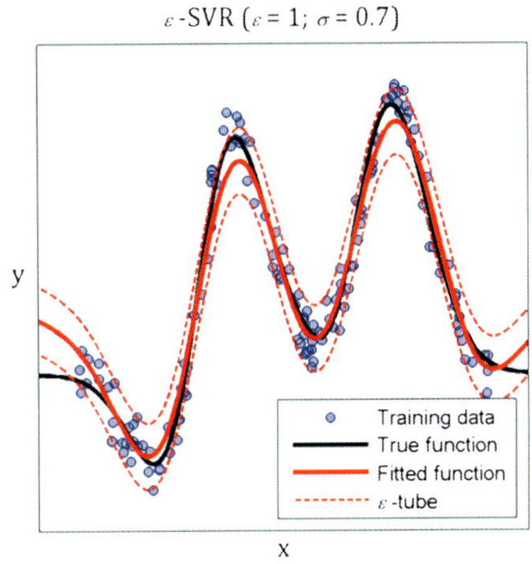

Fig. 4.18 Example of ε-SVR with Gaussian kernels with different kernel width parameters ($\sigma = 0.1;\ 0.7$) with a constant accuracy parameter ($\varepsilon = 1$). *Thick black lines* indicate the true underlying function, with *blue points* indicating noisy training data. *Thick red lines* indicate the fitted regression functions, while *thin red lines* indicate ε-tubes

Once the optimal dual variables are found through convex optimization, the final regression function is:

$$f(\mathbf{x}) = \sum_{i=1}^{N} \left(\alpha_i^* - \alpha_i\right) k\left(\mathbf{x}_i, \mathbf{x}\right) + b. \tag{4.72}$$

An example of the application of ε-SVR to a one-dimensional regression problem is shown in Fig. 4.18. For a constant ε of 1, two different Gaussian kernel widths were used with a Gaussian kernel function (σ of 0.1 and 0.7). As seen before, too small a Gaussian kernel width leads to overfitting and less smooth function approximation.

4.6 Transductive Support Vector Machines

Suppose one has available a set of independent and identically distributed labelled data,

$$Z_l = \{(x_1, y_1), (x_2, y_2), \ldots, (x_m, y_m)\}, \tag{4.73}$$

where $(x_i, y_i) \in \mathbb{R}^d \times \{\pm 1\}$, the *inductive learning* problem seeks to find a function F that describes statistical regularities in the data such that, given a test set of unlabelled data, i.e.

$$Z_u = \{x_{m+1}, x_{m+1}, \ldots, x_{m+k}\}, \tag{4.74}$$

the error in predicting test set labels using the learnt function F is minimized. In this inductive learning setup, the training data are used to adaptively adjust the parameters of a learning algorithm on the entire input domain represented by Z_l. The learning objective or cost function usually includes a smoothness or regularization constraint that guarantees the space of admissible functions the algorithm explores.

Transductive learning is a form of semi-supervised learning (SSL), where one gives up on finding the solution of a general learning rule as in inductive learning; instead, the goal is to learn the values of the *unknown* data generating function at specified points. Compared to induction, transduction is relatively simple since data points are mapped directly to the corresponding labels without solving an intermediate step of function estimation. Vapnik (2006) showed that transductive learning is more fundamental than induction, with better generalization bounds.

Large margin classifiers, such as support vector machines (SVMs), use the concept of the margin as a proxy for the model complexity measure. For a family of hyperplanes in a suitably defined vector space, the margin defines the largest separation between two classes. The margin is defined using the training data in inductive learning with SVMs, while both the training and test sets are used in the transductive SVM (TSVM). More specifically, TSVMs choose as an optimal hypothesis the class of functions that have minimal error on the labelled data and the largest margin on the joint data sets. The differences between SVM and TSVM margins are illustrated in Fig. 4.18.

Mathematically, TSVMs solve the following optimization problem

$$\min_{w,b,\{\xi_i\}_{i=1}^{m},\{\xi_i^*\}_{i=m+1}^{m+1}} \frac{1}{2}\|w\|_2^2 + C\sum_{i=1}^{m} q\left(y_i f_\theta(x_i)\right) + \tilde{C}\sum_{i=m+1}^{m+k} q\left(|f_\theta(x_i)|\right). \quad (4.75)$$

where (w, b) are hyperplane parameters, $q(z)$ is a hinge loss function with argument z, (ξ_i, ξ_i^*) are slack variables (introduced for nonseparable data) and $(C, \tilde{C})$ are specified misclassification costs that trade-off margin size against misclassifying labelled examples or excluding unlabelled ones. The standard SVM primal objective function is recovered by setting $\tilde{C} = 0$. TSVMs find the hyperplane that is maximally separate from the unlabelled points and, simultaneously, providing regularization on the labelled data. Interpreted differently, the large margin property is enforced on the labelled data, while ensuring the *cluster assumption* property holds on the unlabelled data. The cluster assumption captures the notion that a decision boundary traverses low-density regions that separate the different groups without cutting through the high-density regions. An efficient algorithm, the low-density separation (LDS), has been proposed to solve (3) and combines a graph-based representation of the data that exploits the cluster assumption and gradient descent optimization using the primal formulation of the transductive SVM

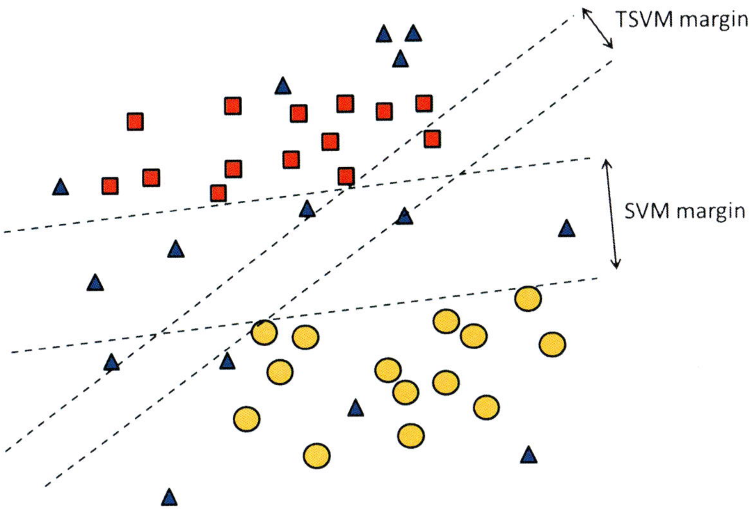

Fig. 4.19 Differences between transductive and inductive learning using SVMs. *Open squares* and *circles* denote "positive" and "negative" labelled examples, while the *solid triangles* are unlabelled examples (after Maulik and Chakraborty, 2013)

(Chapelle and Zien 2005). Solving in the primal directly enforces the decision boundary to traverse low-density regions. As in the standard SVM formulation, flexible decision functions are possible by using of kernel functions.

4.7 Example: Application of Transductive Support Vector Machines to Multivariate Image Analysis of Coal Particles on Conveyor Belts

The last few decades have seen growing interest in the industrial application of computer vision systems due to their non-intrusive nature, ease of use in potentially harsh environments as well as high sampling frequency not possible with manual methods. Furthermore, image-based sampling can be seamlessly integrated with existing plant-wide monitoring and control systems. In the mining industry, machine vision has been applied to many automation problems, including rock particle analysis (Jemwa and Aldrich 2012), froth monitoring on flotation circuits (Moolman et al. 1995; Kaartinen et al. 2006) and mineral composition analysis of ores (Tessier et al. 2007).

In the example considered here, the problem of characterizing the category of an observed image of coal particles, taken on a conveyor belt, with respect to the proportion of fines is investigated. The particle size analysis problem is considered as an image texture representation and classification problem. Image data of natural scenes contain geometrical structure or regularities that are clearly distinct from those of random patterns sampled from the general pixel space.

These localized nonlinear features can be represented statistically in terms of *textons* that compactly describe the textural properties. Although the use of machine vision for rock particle analysis has been considered in the past, most of the proposed approaches are based on image segmentation techniques, such as edge detection and morphological or watershed-like methods. Unfortunately, these techniques are prone to difficulties resulting from heterogeneous particle surface textures and variations in illumination induced by irregular reflection of light. This has the undesired effect of inconsistent or erroneous estimates, owing to spurious segmentation of image content. Moreover, segmentation methods are poorly suited to real-time PSD estimation due to inefficient processing when dealing with large volumes of image data.

Given a set of annotated image data (based on, e.g. PSD profile or fines content), the learning problem exploits statistical regularities in these data to derive a prediction function that can be used on new image data. Different forms of learning exist, depending on the available information. In general, annotated and representative data are difficult or expensive to obtain in many industrial setups. Therefore, the potential of *transductive inference* or *transduction* is investigated. Transduction is a particular form of the more general *semi-supervised learning* that seeks to learn the decision function values for unlabelled data without explicitly learning the decision function.

In the following, the basics of image representation using textons and transductive inference are described. A framework for classification of coal particles on a conveyor belt is then presented and evaluated using data from a pilot plant. The obtained results are discussed and compared with alternative supervised learning methods.

We propose a general procedure for texture-based classification of coal particles on a conveyor belt consisting of the following steps: (1) data acquisition (X_l, X_u) and annotation of labelled data (Y_l), (2) image representation in filter response space (X_l^*, X_u^*), and (3) prediction of the labels $\widehat{Y}_u$ for unlabelled data (X_u) via semi-supervised learning.

Coal particle classes with different proportion of fines (i.e. % passing 6 mm sieve size) were manually prepared by sieving a batch of industrial grade coal to generate three sets of groups: coarse (0–20 % fines), middlings (40–60 % fines) and fines (80–100 % fines). Samples from these groups are shown in Fig. 4.20. The images were converted to greyscale and rescaled using bicubic interpolation. In addition, intensity normalization was performed by scaling the image intensities to zero mean and unit variance. From each image, four non-overlapping patches (213 × 284) were sampled to form the data set used in the analysis.

The scatter plots in Fig. 4.21 show projections of the two leading principal components or scores computed on the raw image data, GLCM measurements and texton features as indicated, including the variance explained by each score. Superimposed on the plots is class information to aid in visualization of the distribution

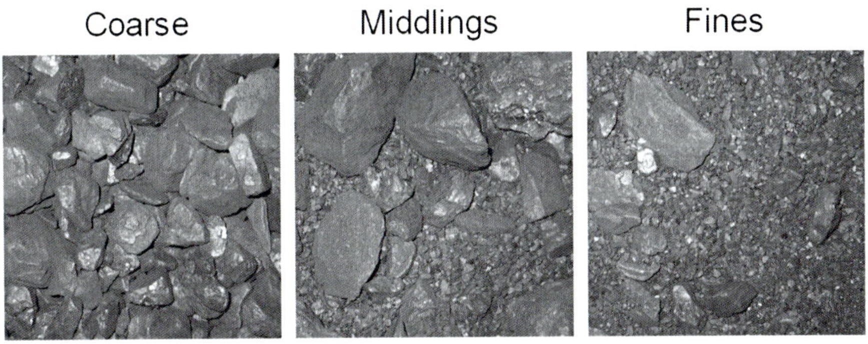

Fig. 4.20 Samples from each of the coarse, middlings and fines coal blends used in the experiments

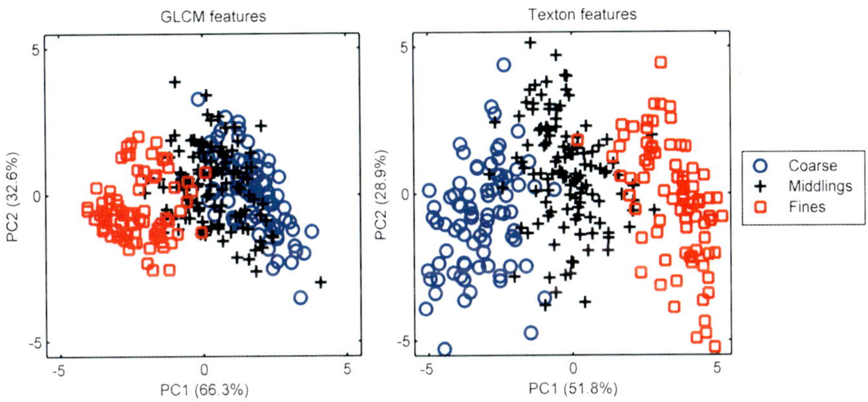

Fig. 4.21 Visualialization of the GLCM (*left*) and texton features (*right*) obtained from the raw image data after projecting into a 2D principal component space. Superimposed are the known class labels for each sample. The axes indicate the variance captured by the corresponding principal component

of the different groups in the respective (low-dimensional) feature space. Exploiting second-order correlations (GLCM) results in the separation between image data sampled from different textural groups, as indicated in Fig. 4.21 (left), and this is further enhanced when higher-order correlations (textons) are used, as is evident in Fig. 4.21 (right).

To better explain the foregoing observation, a more quantitative analysis using semi-supervised learning was performed. For each of group of features, the data were split into a training (labelled) and test (unlabelled) set. Different simulations were performed by varying the number of labelled samples but using the same size of labelled samples per group. These were selected randomly from the available data, and the remaining samples were used as the test or unlabelled

Table 4.2 Summary of classification performance for different types of image features with SSL

Data split		Classification error	
Training	Testing	GLCM	Textons
5/5/5	75/115/75	0.333 ± 0.003	0.147 ± 0.014
10/10/10	70/110/70	0.334 ± 0.004	0.138 ± 0.017
15/15/15	65/105/65	0.333 ± 0.005	0.132 ± 0.017
20/20/20	60/100/60	0.333 ± 0.005	0.135 ± 0.013
25/25/25	55/95/55	0.332 ± 0.006	0.131 ± 0.013
30/30/30	50/90/50	0.335 ± 0.008	0.136 ± 0.013
40/40/40	40/80/40	0.333 ± 0.004	0.148 ± 0.016

Table 4.3 Comparison of supervised and semi-supervised classification using coal image features

Features	Classification Scheme		
	k-NN (%)	SVM (%)	LDS (%)
GLCM	63.8	79.3	66.7
Textons	85.7	90.7	86.2

data. Specifically, denoting by $\{X_{\text{coarse}}/X_{\text{middling}}/X_{\text{fines}}\}$ the labelled data set, with X_i the number of samples from group i, the unlabelled data set becomes $\{X^*_{\text{coarse}}/X^*_{\text{middling}}/X^*_{\text{fines}}\}$, where X_i^* denotes the samples from group i not in the training data set. For each training sample size, random sampling was repeated ten times. Each labelled and unlabelled data set realization served as input into the LDS learner and the accuracy of the predictions computed using the 0/1-loss function. The results are summarized in Table 4.2.

The classification error of the data shown in Fig. 4.21 was approximately 33 % when using GLCM features and 13–14 % with texton features. Interestingly, the variation in the misclassification rate seems little affected by the size of the labelled data set, indicating the robustness of the *cluster assumption*-based LDS learning algorithm used. For comparative purposes, supervised models were also trained and the results are shown in Table 4.3.

In this case, half the samples were used as training data and the rest as test data. It can be observed that supervised learning yields slightly better performance. Although not shown here, observed performance significantly deteriorates with a decrease of the training data set size, unlike the case with semi-supervised learning.

The significant difference in classification performance observed across the set of features considered emphasizes the importance of feature extraction prior to using so-called *black box* algorithms, at least in this particular application. All non-parametric learning techniques require tuning of various parameters before a set of discriminative features can be identified. In most cases, this requires solving a combinatorial problem, with further dependency on the training set size. For the coal size analysis problem, these results indicate the huge benefits of using locally oriented feature representation in distinguishing between images with varying textural properties.

4.8 Kernel Principal Component Analysis

In a previous section, it was shown that kernel functions, together with margin maximization, can be used in the 1-SVM algorithm to estimate the support of an unlabelled data set. Another unsupervised learning approach that exploits kernel functions is kernel principal component analysis (Schölkopf et al. 1998).

Kernel principal component analysis (KPCA) applies principal component analysis (PCA) in a (often nonlinear) dot product space $\mathcal{H}$, allowing the extraction of nonlinear features. Figure 4.22 illustrates kernel principal component analysis in terms of mapping from an input space $\mathcal{X}$ to a dot-product space $\mathcal{H}$, linear projection in $\mathcal{H}$ to the KPCA feature space $\mathcal{T}$ and approximate reconstruction from $\mathcal{T}$ back to $\mathcal{X}$ (via $\mathcal{H}$). Finding an appropriate reconstruction is also known as pre-image problem.

4.8.1 *Principal Component Analysis*

Before venturing forth into the dot product space $\mathcal{H}$, standard principal component analysis in the input space ($\mathcal{X}$) is considered. Principal component analysis (PCA) is a linear feature extraction technique, which calculates orthogonal vectors in the measurement space $\mathcal{X}$ onto which data are projected (Hotelling 1933). The criterion for determining these principal components is to maximize variance of data projected onto said components. Each principal component is a linear combination of measured variables, uncorrelated to all other components. The first component accounts for the most variance, the second component for the next most variance, etc. The determination of these components is an optimality guaranteed problem in the form of spectral decomposition of the covariance matrix of the data.

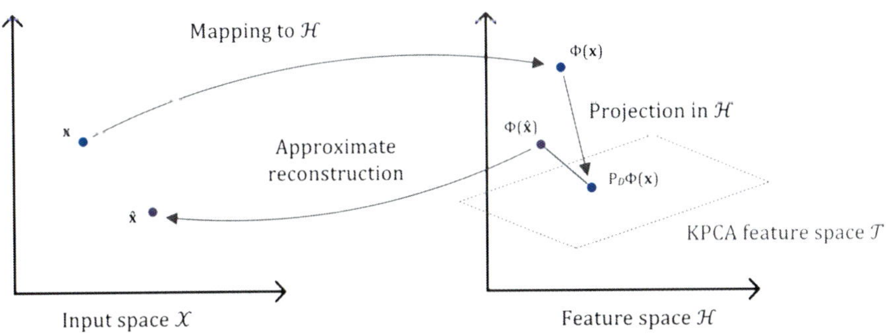

Fig. 4.22 Illustration of kernel principal component analysis the pre-image problem: data are mapped from an input space $\mathcal{X}$ to a feature space $\mathcal{H}$ through a nonlinear mapping function $\Phi(x)$. Linear projection through principal component analysis is done to the KPCA feature space $\mathcal{T}$ by means of D projection vectors $\mathbf{p}$. Approximate solutions exist to reconstruct an input space image $\hat{\mathbf{x}}$ from a feature space projected sample $P_D\Phi(x)$, such that the feature space distance between $\Phi(\hat{\mathbf{x}})$ and $P_D\Phi(x)$ is minimized

The PCA feature extraction algorithm consists of training and application stages. During the training stage, a training data set $\mathbf{X} \in \mathcal{X}$ (N samples in M dimensions) is employed to determine D projection vectors (principal components) $\mathbf{P}^*$. The projection of the training data onto the retained principal components gives the features **T**in the d-dimensional feature space $\mathcal{T}$. During the application stage, new (test) data $\mathbf{X}^{\text{test}} \in \mathcal{X}$ (N^{test} samples in M dimensions) are projected onto the D retained principal components to give test features $\mathbf{T}^{\text{test}} \in \mathcal{T}$. The test data can also be reconstructed from the retained features, to give the reconstructed test data set $\hat{\mathbf{x}}^{\text{test}}$. The PCA feature extraction algorithm is given here:

Principal Component Analysis Feature Extraction Algorithm

Training stage:

- Center the samples of the training data set **X:**

$$\hat{\mathbf{x}}_i = \frac{\mathbf{x}_i}{\boldsymbol{\mu}}, i = 1, \ldots, N.$$

Where $\boldsymbol{\mu}$ is the mean vector of the training data set:

$$\boldsymbol{\mu} = \frac{1}{N}\sum_{i=1}^{N} \mathbf{x}_i.$$

- The covariance matrix **C** based on the centred training data $\tilde{X}$ is calculated:

$$\mathbf{C} = \frac{1}{N}\sum_{i=1}^{N} \tilde{x}_i \tilde{x}_i^j.$$

- Eigendecomposition of **C** returns the ordered eigenvectors **P** and eigenvalues **Λ** such that:

$$\mathbf{CP} = \mathbf{\Lambda P}.$$

- The first D principal components are retained as the projection vectors $\mathbf{P}^*$, where the selection of D is based on some selection criteria (e.g. certain fraction of variance accounted for).

Application stage:

- Centre the samples of the test data set $\mathbf{X}^{\text{test}}$, using the mean vector of the training data set:

$$\hat{\mathbf{x}}_i^{\text{test}} = \frac{\mathbf{x}_i^{\text{test}}}{\boldsymbol{\mu}}, \quad i = 1, \ldots, N^{\text{test}}.$$

(continued)

(continued)

- d-dimensional test features $\mathbf{T}^{\text{test}}$ are calculated by projecting the centred test data $\tilde{\mathbf{X}}^{\text{test}}$ onto the first D principal components $\mathbf{P}^*$:

$$\mathbf{T}^{\text{test}} = \hat{\mathbf{X}}^{\text{test}}\mathbf{P}^*.$$

- Reconstruct the centred test data in the input space from the d-dimensional test features $\mathbf{T}^{\text{test}}$:

$$\widehat{\mathbf{X}}^{\text{test}} = \mathbf{T}^{\text{test}}(\mathbf{P}^*)^{\mathrm{T}}.$$

4.8.2 *Principal Component Analysis in Kernel Feature Space*

Where standard PCA determines orthogonal vectors (principal components) in the input space $\mathcal{X}$ which (in decreasing order) account for the most variance in a training data set, kernel PCA finds orthogonal vectors in an implicit dot-product space $\mathcal{H}$ (Schölkopf et al. 1998). This dot-product space, and the mapping Φ from $\mathcal{X}$ to $\mathcal{H}$, is defined implicitly through the use of kernel functions.

Let $\Phi(\mathbf{x})$ be a mapping from the input space $\mathcal{X}$ to a larger (possibly infinite) dimensional feature space $\mathcal{H}$, implicitly encapsulated by a kernel function k. Assume that the features of a N-sample training set are centered:

$$\sum_{i=1}^{N} \Phi(\mathbf{x}_i) = 0. \tag{4.76}$$

Diagonalization of Covariance Matrix

The covariance matrix in the dot product space $\mathcal{H}$ for the N training samples is given by:

$$\mathbf{C} = \frac{1}{N}\sum_{i=1}^{N} \Phi(\mathbf{x}_i)\Phi(\mathbf{x}_i)^{\mathrm{T}}. \tag{4.77}$$

Linear transformation, such as PCA projection, in a nonlinear implicit feature space $\mathcal{H}$ translates to nonlinear transformations in the original input space $\mathcal{X}$. The linear transformation in the case of KPCA involves the diagonalization of the covariance matrix $\mathbf{C}$ in $\mathcal{H}$ (through eigendecomposition):

$$\mathbf{C}\mathbf{p}^m = \lambda^m \mathbf{p}^m. \tag{4.78}$$

In the equation above, $\mathbf{C}$ is the covariance matrix in $\mathcal{H}$ and λ^m is the mth eigenvalue associated with the mth eigenvector $\mathbf{p}^m$. The solution for each of the r eigenvectors $\mathbf{p}^m(m = 1,\ldots,r)$ associated with r non-zero eigenvalues ($\lambda^m \neq 0$, $m = 1, \ldots, r$) is a weighted linear combination of the training data mapped points $\Phi\left(\mathbf{x}_j\right)$ $(j = 1,\ldots,N)$:

$$\mathbf{p}^m = \sum_{j=1}^{N} \alpha_j^m \Phi\left(\mathbf{x}_j\right). \tag{4.79}$$

Where $\boldsymbol{\alpha}^m$ consists of α_j^m $(j = 1, \ldots, N)$, the coefficients for the linear combination that produces $\mathbf{p}^m$.

The eigenvalue problem in (4.75) can be reconsidered in the following form:

$$\Phi(x_n) \cdot (\mathbf{C}\mathbf{p}^m) = \lambda^m \Phi(\mathbf{x}_n) \cdot \mathbf{p}^m. \tag{4.80}$$

The term $\mathbf{C}\mathbf{p}^m$ above can be written as:

$$\begin{aligned}
\mathbf{C}\mathbf{p}^m &= \frac{1}{N}\sum_{i=1}^{N} \Phi(\mathbf{x}_i)\Phi(\mathbf{x}_i)^{\mathrm{T}}\mathbf{p}^m \\
\therefore \mathbf{C}\mathbf{p}^m &= \frac{1}{N}\sum_{i=1}^{N} \Phi(\mathbf{x}_i)(\Phi(\mathbf{x}_i)\cdot \mathbf{p}^m) \\
\therefore \mathbf{C}\mathbf{p}^m &= \frac{1}{N}\sum_{i=1}^{N} \Phi(\mathbf{x}_i)\left(\Phi(\mathbf{x}_i)\cdot \sum_{j=1}^{N}\alpha_j^m \Phi(\mathbf{x}_j)\right) \\
\therefore \mathbf{C}\mathbf{p}^m &= \frac{1}{N}\sum_{i=1}^{N} \Phi(\mathbf{x}_i)\left(\sum_{j=1}^{N}\alpha_j^m \Phi(\mathbf{x}_i)\cdot \Phi(\mathbf{x}_j)\right) \\
\therefore \mathbf{C}\mathbf{p}^m &= \frac{1}{N}\sum_{i=1}^{N}\alpha_j^m \sum_{j=1}^{N} \Phi(\mathbf{x}_i)\Phi(\mathbf{x}_i)\cdot \Phi(\mathbf{x}_j).
\end{aligned} \tag{4.81}$$

Substituting (4.78) in (4.77) gives:

$$\Phi(\mathbf{x}_n)\cdot\left(\frac{1}{N}\sum_{j=1}^{N}\alpha_j^m\sum_{i=1}^{N}\Phi(\mathbf{x}_i)\Phi(\mathbf{x}_i)\cdot\Phi(\mathbf{x}_j)\right) = \lambda^m\Phi(\mathbf{x}_n)\cdot\left(\sum_{j=1}^{N}\alpha_j^m\Phi(\mathbf{x}_j)\right)$$

$$\therefore \frac{1}{N}\sum_{j=1}^{N}\alpha_j^m\Phi(\mathbf{x}_n)\cdot\left(\sum_{i=1}^{N}\Phi(\mathbf{x}_i)\Phi(\mathbf{x}_i)\cdot\Phi(\mathbf{x}_j)\right) = \lambda^m\sum_{j=1}^{N}\alpha_j^m\Phi(\mathbf{x}_n)\cdot\Phi(\mathbf{x}_j). \tag{4.82}$$

Above, $n = 1, \ldots, N$. The kernel matrix **K** entries for the training data set are defined as:

$$K_{ij} = \Phi(\mathbf{x}_i) \cdot \Phi(\mathbf{x}_j). \tag{4.83}$$

Substituting (4.80) in (4.79) yields:

$$\frac{1}{N} \sum_{j=1}^{N} \alpha_j^m \Phi(\mathbf{x}_n) \cdot \left(\sum_{i=1}^{N} \Phi(\mathbf{x}_i) K_{ij} \right) = \lambda^m \sum_{j=1}^{N} \alpha_j^m K_{nj}$$

$$\therefore \frac{1}{N} \sum_{j=1}^{N} \alpha_j^m \left(\sum_{i=1}^{N} \Phi(\mathbf{x}_n) \cdot \Phi(\mathbf{x}_i) K_{ij} \right) = \lambda^m \sum_{j=1}^{N} \alpha_j^m K_{nj}$$

$$\therefore \frac{1}{N} \sum_{j=1}^{N} \alpha_j^m \sum_{i=1}^{N} K_{ni} K_{ij} = \lambda^m \sum_{j=1}^{N} \alpha_j^m K_{nj}. \tag{4.84}$$

In matrix form, the above result can be rewritten as:

$$\mathbf{K}^2 \boldsymbol{\alpha}^m = N \lambda^m \boldsymbol{\alpha}^m \mathbf{K}. \tag{4.85}$$

Above, $\boldsymbol{\alpha}^m = (\alpha_1^m, \ldots, \alpha_N^m)$ is the column vector of weights for eigenvector $\mathbf{p}^m$. It can be shown (Schölkopf et al. 1998) that, for non-zero eigenvalues λ^m, the diagonalization solution of the above equation is equal to the diagonalization solution of the equation below:

$$\mathrm{K} \boldsymbol{\alpha}^m = N \lambda^m \boldsymbol{\alpha}^m = \tilde{\lambda}^m \boldsymbol{\alpha}^m. \tag{4.86}$$

Above, $\tilde{\lambda}^m$ represents the scaled eigenvalues ($\tilde{\lambda}^m = \lambda^m / N$).

Normalization of Weight Vectors

It is desirable to have eigenvectors of unit length, as this simplifies calculations. To ensure that the eigenvectors $\mathbf{p}^m$ are of unit length, the weight vectors $\boldsymbol{\alpha}^m$ must be normalized. Enforcing unit length eigenvectors requires the following:

$$\mathbf{p}^m \cdot \mathbf{p}^m = 1. \tag{4.87}$$

Rewriting the above in terms of the weight vectors $\boldsymbol{\alpha}^m$ yields (substituting (4.74) and (4.83) in (4.84)):

$$\mathbf{p}^m \cdot \mathbf{p}^m = \left(\sum_{i=1}^{N} \alpha_i^m \Phi(\mathbf{x}_i)\right) \cdot \left(\sum_{j=1}^{N} \alpha_j^m \Phi(\mathbf{x}_j)\right) = 1$$

$$\therefore \sum_{i=1}^{N}\sum_{j=1}^{N} \alpha_i^m \alpha_j^m \Phi(\mathbf{x}_i) \cdot \Phi(\mathbf{x}_j) = \sum_{i=1}^{N}\sum_{j=1}^{N} \alpha_i^m \alpha_j^m K_{ij} = 1$$

$$\therefore \boldsymbol{\alpha}^m \cdot (\mathrm{K}\boldsymbol{\alpha}^m) = \tilde{\lambda}^m \boldsymbol{\alpha}^m \cdot \boldsymbol{\alpha}^m = 1. \tag{4.88}$$

From above, the normalized weight vector $\tilde{\boldsymbol{\alpha}}^m$ is given by:

$$\tilde{\boldsymbol{\alpha}}^m = \frac{\boldsymbol{\alpha}^m}{\sqrt{\tilde{\lambda}^m}}. \tag{4.89}$$

After normalizing the weight vectors, the unit-length eigenvectors are now:

$$\mathbf{p}^m = \sum_{j=1}^{N} \tilde{\alpha}_j^m \Phi(\mathbf{x}_\mathrm{j}) = \frac{1}{\sqrt{\tilde{\lambda}^m}} \sum_{j=1}^{N} \alpha_j^m \Phi(\mathbf{x}_\mathrm{j}). \tag{4.90}$$

Calculation of Feature Scores

The solution of the eigenproblem (for a training data set) stated in (4.83) yields the weight vectors $\boldsymbol{\alpha}^m$ ($m = 1, \ldots, r$) associated with r non-zero eigenvalues. To calculate the mth feature score t^m for an input space sample $\mathbf{x}$, the mapping $\Phi(\mathbf{x})$ is projected onto the eigenvector $\mathbf{p}^m$, characterized by the weight vector $\boldsymbol{\alpha}^m$:

$$t^m = \mathbf{p}^m \cdot \Phi(\mathbf{x}) = \frac{1}{\sqrt{\tilde{\lambda}^m}} \sum_{j=1}^{N} \alpha_j^m \Phi\left(\mathrm{x}_j\right) \cdot \Phi(\mathrm{x}) = \frac{1}{\sqrt{\tilde{\lambda}^m}} \sum_{j=1}^{N} \alpha_j^m k\left(\mathbf{x}_j, \mathbf{x}\right). \tag{4.91}$$

As with standard (linear) PCA, a subset of $D < r$ components can be retained for the calculation of the feature scores. This d-dimensional feature space $\mathcal{T}$ is then the KPCA feature space. The selection of the number of components to retain can be determined by some form of selection criteria; e.g. the fraction of variance retained by the features of the training set.

4.8.3 Centering in Kernel Feature Space

The above derivation assumes that the mapped training samples $\Phi(\mathbf{x}_i)$ for $i = 1, \ldots, N$ are centered in $\mathcal{H}$. Since kernel matrices represent the implicit mapping of samples to $\mathcal{H}$, a centering operation must be applied to these matrices directly. The centered kernel matrix $\tilde{\mathbf{K}}$ for the training data is calculated as follows: $\hat{\mathbf{x}}$

$$\tilde{\mathbf{K}} = \mathbf{K} - \mathbf{K}\frac{\mathbf{1}_{N\times N}}{N} - \frac{\mathbf{1}_{N\times N}}{N}\mathbf{K} + \frac{\mathbf{1}_{N\times N}}{N}\mathbf{K}\frac{\mathbf{1}_{N\times N}}{N}. \tag{4.92}$$

In the above equation, $\mathbf{1}_{a\times b}$ is an a by b matrix of ones.

The weight vectors obtained from training data with KPCA can be used to calculate KPCA features of new (test) data, similar to 4.88. The kernel matrix for $\mathbf{K}^{\text{test}}$ test data (relative to the training data) is required:

$$K_{ij}^{\text{test}} = k(\mathbf{x}_i, \mathbf{x}_j^{\text{test}}). \tag{4.93}$$

Above, $\mathbf{x}_i (i = 1, \ldots, N)$ are the samples of the training set and $\mathbf{x}_j^{\text{test}} (j = 1, \ldots, N^{\text{test}})$ are samples of the test set.

The centering operation on a kernel matrix $\mathbf{K}^{test}$ derived from test data is then given by:

$$\tilde{\mathbf{K}}^{\text{test}} = \mathbf{K}^{\text{test}} - \frac{\mathbf{1}_{N^{\text{test}}\times N}}{N}\mathbf{K} - \mathbf{K}^{\text{test}}\frac{\mathbf{1}_{N\times N}}{N} + \frac{\mathbf{1}_{N^{\text{test}}\times N}}{N}\mathbf{K}\frac{\mathbf{1}_{N\times N}}{N}. \tag{4.94}$$

The derivation of the KPCA weight vectors and feature scores given in the previous Sect 4.8.1 is valid for the centered matrices calculated with (4.90) and (4.91). From (4.88), the mth feature score t_j^m for a test sample $\mathbf{x}_j^{\text{test}}$ can be calculated from the centered test kernel matrix $\mathbf{K}^{\text{test}}$ entries:

$$t_j^m = \frac{1}{\sqrt{\tilde{\lambda}^m}} \sum_{i=1}^{N} \alpha_i^m \tilde{K}_{ij}^{\text{test}}. \tag{4.95}$$

4.8.4 *Effect of Kernel Type and Kernel Parameters*

An example of the nonlinear mapping directions that can be achieved with kernel PCA is shown in Fig. 4.23. Three types of kernels were used; a linear kernel, a homogeneous polynomial kernel with $d = 3$ and a Gaussian kernel with $\sigma = 1$. The plots show contour lines for principal component values (projected features).

4.8.5 *Reconstruction from Kernel Principal Component Analysis Features*

With standard (linear) PCA, it is straightforward to determine the reconstruction $\hat{\mathbf{x}}$ in the M-dimensional input space $\mathcal{X}$ from a sample $\mathbf{t}$ in the d-dimensional PCA feature space $\mathcal{T}$. Reconstruction with standard PCA is a one-to-one function: Given a feature vector $\mathbf{t}$ and the D retained principal components $\mathbf{P}^*$, the reconstructed vector $\hat{\mathbf{x}}$ in the input space is given by:

$$\hat{\mathbf{x}} = \mathbf{t}(\mathbf{P}^*)^{\mathrm{T}}. \tag{4.96}$$

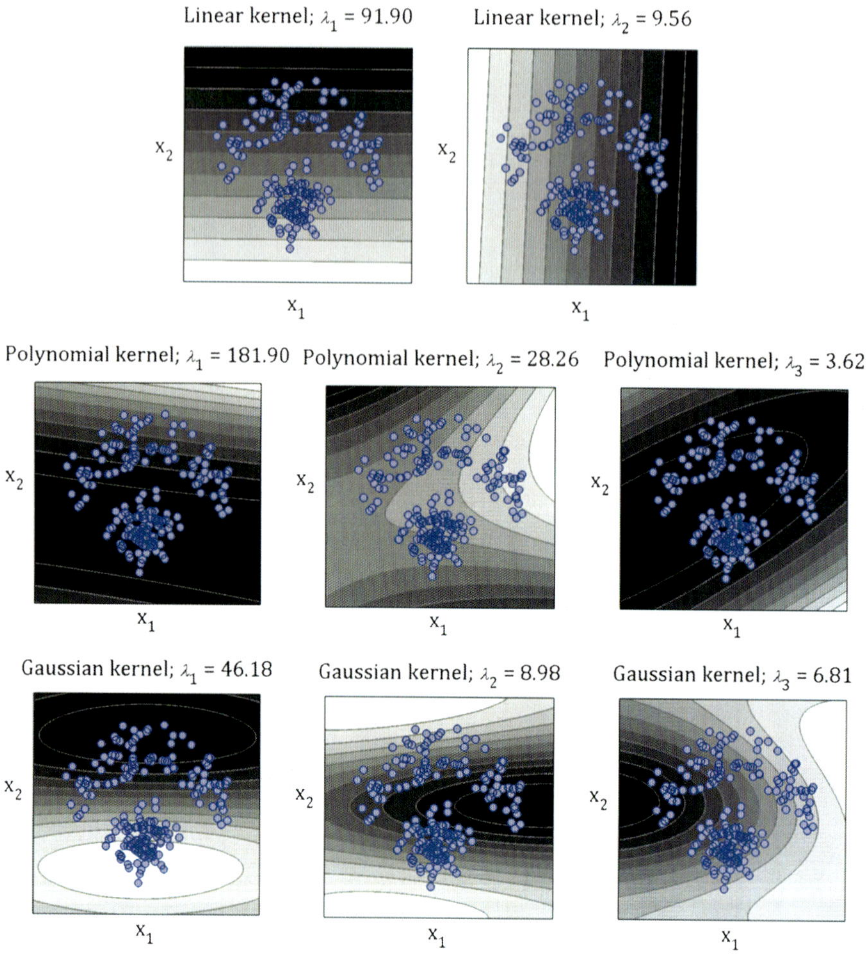

Fig. 4.23 Illustration of kernel principal component contours for three kernel types: linear, polynomial and Gaussian

Reconstruction in an input space from a feature space is known as the pre-image problem. Where the feature samples are originally projections from an input space with the ultimate goal of noise removal in this same input space, this reconstruction operation is known as denoising. Denoising is achieved by retaining $D < M$ principal components for projection, with the assumption that the discarded components represent noise only. An indication of the accuracy of reconstruction is the squared Euclidian distance (squared error) $d_{\mathcal{X}}^2$ in the input space:

$$d_{\mathcal{X}}^2(\mathbf{x}, \hat{\mathbf{x}}) = \|\mathbf{x} - \hat{\mathbf{x}}\|^2 \tag{4.97}$$

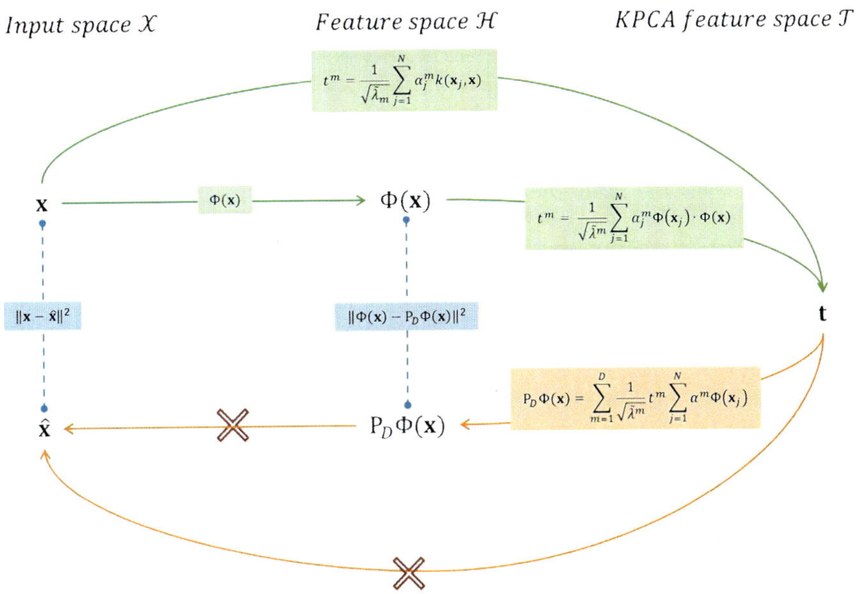

Fig. 4.24 Illustration of mapping and demapping in KPCA showing the pre-image problem

For standard PCA, the overall reconstruction error for the training data set is guaranteed to be minimal, given the number of retained components D.

Reconstruction, denoising, or the pre-image problem is not as straightforward in KPCA as in PCA. In KPCA, the linear PCA reconstruction shown above is valid for the reconstruction of a mapped sample in $\mathcal{H}$, $\Phi(\mathbf{x})$, from the projection of said sample in the reduced PCA feature space represented by $\mathcal{T}$. This PCA reconstruction approach cannot be used to reconstruct a sample in the original input space $\mathcal{X}$. In fact, due to the properties of $\mathcal{H}$ defined by certain kernel functions, an exact pre-image might not necessarily exist, and if it exists, it might not be unique (Mika et al. 1999).

Figure 4.24 illustrates the KPCA pre-image problem. Given a mapping function $\Phi(\mathbf{x})$, explicit functions exist to map an input space sample $\mathbf{x}$ to the feature space $\mathcal{H}$, to map a feature space sample $\Phi(\mathbf{x})$ to the KPCA feature space $\mathcal{T}$, and to map $\mathbf{x}$ directly to $\mathcal{T}$ (with kernel functions corresponding to $\Phi(\mathbf{x})$) to give the KPCA feature space sample $\mathbf{t}$.

To reconstruct a sample from the KPCA feature space $\mathcal{T}$ to the feature space $\mathcal{H}$, a projection operator $P_D\Phi(\mathbf{x})$ is defined:

$$P_D\Phi(\mathbf{x}) = \sum_{m=1}^{D} \frac{1}{\sqrt{\tilde{\lambda}^m}} t^m \mathbf{p}^m. \tag{4.98}$$

From (4.76), the projection operator equation can be stated in terms of training data feature space samples $\Phi(\mathbf{x}_j)$, with $j = 1, \ldots, N$:

$$P_D\Phi(\mathbf{x}) = \sum_{m=1}^{D} \frac{1}{\sqrt{\tilde{\lambda}^m}} t^m \sum_{j=1}^{N} \alpha_j^m \Phi(\mathbf{x}_j). \tag{4.99}$$

The error of reconstruction $d_{\mathcal{H}}^2$ for $\mathbf{x}$ in the feature space can be expressed in terms of the squared Euclidian distance (squared error) between a feature space sample $\Phi(\mathbf{x})$ and the projection operator result:

$$d_{\mathcal{H}}^2(\Phi(\mathbf{x}), \mathrm{P}_D\Phi(\mathbf{x})) = \|\Phi(\mathbf{x}) - \mathrm{P}_D\Phi(\mathbf{x})\|^2 \tag{4.100}$$

Minimizing the above result does not necessarily imply that the input space reconstruction error is minimal. Furthermore, due to the lack of uniqueness and/or possible non-existence of pre-images mentioned before, no explicit functions exist to determine $\hat{\mathbf{x}}$ from either $P_D\Phi(\mathbf{x})$ or $\mathbf{t}$.

Another complication is that the mapping function $\Phi(\mathbf{x})$ may only be defined implicitly, through the expression of a kernel function k. In the KPCA framework, the reconstruction solution must be expressed in terms of kernel functions or kernel matrices. Conveniently, the error of reconstruction stated in (4.97) results in dot-products in $\mathcal{H}$, which can be expressed in terms of kernel functions:

$$d_{\mathcal{H}}^2(\Phi(\mathbf{x}), \mathrm{P}_D\Phi(\mathbf{x})) = \|\Phi(\mathbf{x})\|^2 - 2\Phi(\mathbf{x}) \cdot \mathrm{P}_D\Phi(\mathbf{x}) + \|\mathrm{P}_D\Phi(\mathbf{x})\|^2$$

$$\begin{aligned}
\therefore\ & d_{\mathcal{H}}^2(\Phi(\mathbf{x}), \mathrm{P}_D\Phi(\mathbf{x})) \\
&= \Phi(\mathbf{x}) \cdot \Phi(\mathbf{x}) - 2\Phi(\mathbf{x}) \cdot \left(\sum_{m=1}^{D} \frac{1}{\sqrt{\tilde{\lambda}^m}} t^m \sum_{j=1}^{N} \alpha_j^m \Phi(\mathbf{x}_j) \right) \\
&\quad + \left(\mathrm{P}_D\Phi(\mathbf{x})\right) \cdot \left(\mathrm{P}_D\Phi(\mathbf{x})\right)
\end{aligned}$$

$$\begin{aligned}
\therefore\ & d_{\mathcal{H}}^2\left(\Phi(\mathbf{x}), \mathrm{P}_D\Phi(\mathbf{x})\right) \\
&= k(\mathbf{x}, \mathbf{x}) - 2 \sum_{m=1}^{D} \sum_{j=1}^{N} \frac{1}{\sqrt{\tilde{\lambda}^m}} t^m \alpha_j^m \Phi(\mathbf{x}) \cdot \Phi(\mathbf{x}_j) \\
&\quad + \left(\sum_{m=1}^{D} \sum_{j=1}^{N} \frac{1}{\sqrt{\tilde{\lambda}^m}} t^m \alpha_j^m \Phi(\mathbf{x}_j) \right) \\
&\quad \cdot \left(\sum_{m=1}^{D} \sum_{n=1}^{N} \frac{1}{\sqrt{\tilde{\lambda}^m}} t^m \alpha_n^m \Phi(\mathbf{x}_n) \right)
\end{aligned} \tag{4.98}$$

$$\therefore d^2_{\mathcal{H}}(\Phi(\mathbf{x}), \mathrm{P}_D\Phi(\mathbf{x}))$$
$$= k(\mathbf{x},\mathbf{x}) - 2\sum_{m=1}^{D}\sum_{j=1}^{N}\frac{1}{\sqrt{\tilde{\lambda}^m}}t^m\alpha_j^m k(\mathbf{x},\mathbf{x}_j)$$
$$+ \sum_{m=1}^{D}\sum_{j=1}^{N}\sum_{n=1}^{N}\frac{1}{\tilde{\lambda}^m}(t^m)^2\alpha_j^m\alpha_n^m\Phi(\mathbf{x}_j)\cdot\Phi(\mathbf{x}_n)$$

$$\therefore d^2_{\mathcal{H}}(\Phi(\mathbf{x}), \mathrm{P}_D\Phi(\mathbf{x}))$$
$$= k(\mathbf{x},\mathbf{x}) - 2\sum_{m=1}^{D}\sum_{j=1}^{N}\frac{1}{\sqrt{\tilde{\lambda}^m}}t^m\alpha_j^m k(\mathbf{x},\mathbf{x}_j)$$
$$+ \sum_{m=1}^{D}\sum_{j=1}^{N}\sum_{n=1}^{N}\frac{1}{\tilde{\lambda}^m}(t^m)^2\alpha_j^m\alpha_n^m k(\mathbf{x}_j,\mathbf{x}_n)$$

Reconstruction by Fixed-Point Iteration

The above result allows the reconstruction error in feature space to be expressed in terms of kernel functions. This allows the formulation of approximate solutions to the KPCA pre-image problem, assuming that a result that ensures minimal reconstruction error in the feature space translates to sufficiently low reconstruction error in the input space.

An illustration of an approximate solution strategy is shown in Fig. 4.25. Let $\hat{\mathbf{x}}_e$ represent an estimate of the KPCA reconstruction of a sample $\mathbf{x}$. This estimated pre-image can be mapped to the feature space $\mathcal{H}$, as $\Phi(\hat{\mathbf{x}}_e)$ and compared with the KPCA projection of the original sample, $P_D\Phi(\mathbf{x})$. The error of deconstruction now compares the feature space mapping of the estimated pre-image, $\Phi(\hat{\mathbf{x}}_e)$, to the projection of the original sample $\mathbf{x}$ in the KPCA feature space, $P_D\Phi(\mathbf{x})$:

$$d^2_{\mathcal{H}}(\Phi(\hat{\mathbf{x}}_e), \mathrm{P}_D\Phi(\mathbf{x})) = \|\Phi(\hat{\mathbf{x}}_c) - \mathrm{P}_D\Phi(\mathbf{x})\|^2 \tag{4.101}$$

Minimizing the reconstruction error $d^2_{\mathcal{H}}(\mathbf{x}, \hat{\mathbf{x}}_e)$ with respect to $\hat{\mathbf{x}}_e$ represents an approximate solution to the pre-image problem.

Schölkopf et al. (1999) solved this optimization problem (for Gaussian kernels) by setting the gradient of (4.98) (with respect to $\mathbf{x}$) to zero. With fixed-point iteration, the approximate reconstruction in input space ($\hat{\mathbf{x}}_e$) can be calculated using the following iteration formula (Schölkopf et al. 1999):

$$\hat{\mathbf{x}}_{e,n+1} = \frac{\sum_{j=1}^{N}\gamma_j \exp\left(-\frac{\|\hat{\mathbf{x}}_{e,n}-\mathbf{x}_j\|^2}{2\sigma^2}\right)\mathbf{x}_j}{\sum_{j=1}^{N}\gamma_j \exp\left(-\frac{\|\hat{\mathbf{x}}_{e,n}-\mathbf{x}_j\|^2}{2\sigma^2}\right)};\ \gamma_j = \sum_{m=1}^{D}\sum_{j=1}^{N}\frac{1}{\sqrt{\tilde{\lambda}^m}}t^m\alpha_j^m. \tag{4.102}$$

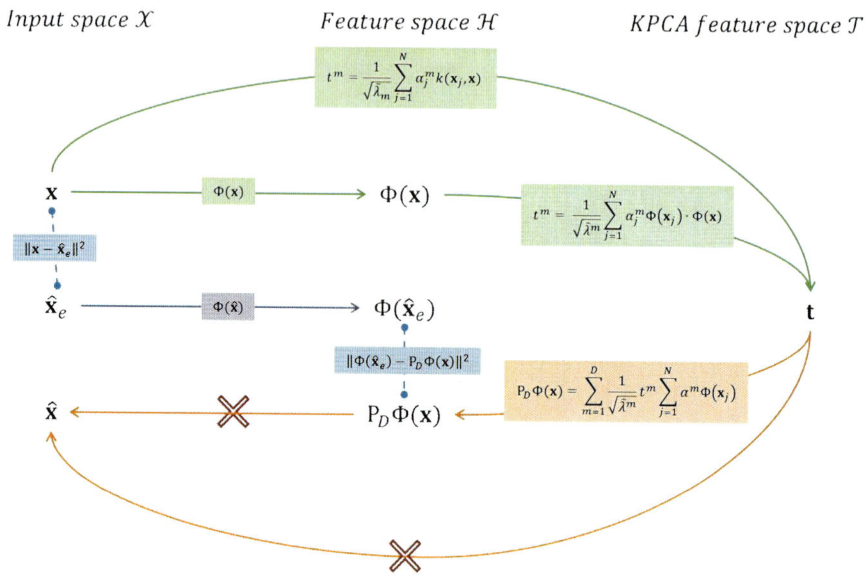

Fig. 4.25 Illustration of an approximate solution to the KPCA pre-image problem

Above, n is the iteration counter. The iteration of the above formula continues until the change in estimation between iteration steps $|\hat{\mathbf{x}}_{e,n+1} - \mathbf{x}_{e,n}|$ falls below a certain threshold, or a set number of iteration steps are completed.

Reconstruction by Distance Constraints

Another approach to approximate KPCA reconstruction makes use of the fact that knowing the distances between a new sample and a base set of other samples geometrically constrains the possible coordinates of the new sample. The rigidity of this constraint depends on the dimensionality of the space in which the coordinates are to be determined, the number of samples in the base set, and the consistency of the distances. Where distances are consistent, $m + 1$ distances (relative to $m + 1$ points in a base set) is required to determine the coordinates of a new point in a m-dimensional coordinate system.

As an example, Fig. 4.26 shows two consistent distant constraint situations, where the coordinates of a new point should be determined based on its distances from three reference points. In a two-dimensional coordinate space, a unique solution exists; while in a three-dimensional space, two possible solutions exist. If the distances of a new point to points in a base set are not consistent, an exact solution may not exist, but an approximate solution minimizing some strain function may be found.

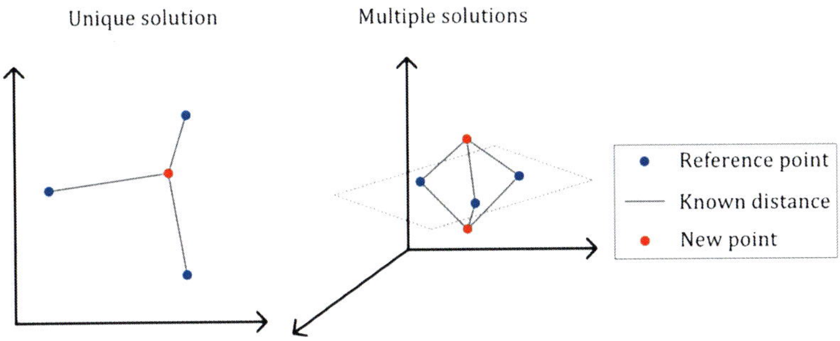

Fig. 4.26 Illustration of unique or multiple solutions to a distance constraint problem (with consistent distances)

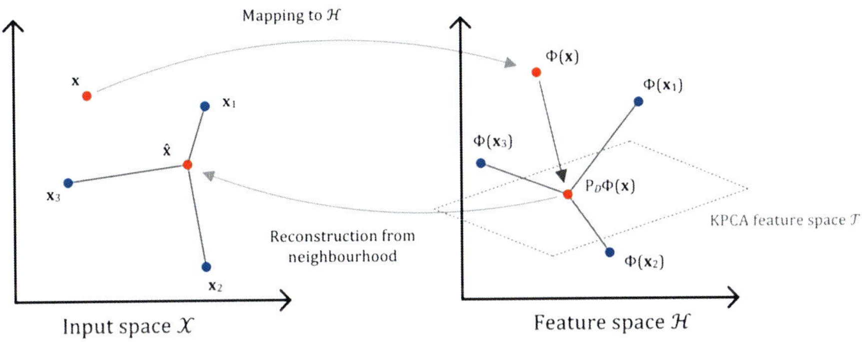

Fig. 4.27 Illustration of distance constraint approach to KPCA reconstruction

An approach using distance constraints to find an approximate solution to the KPCA pre-image problem was suggested by Kwok and Tsang (2004). Their approach is illustrated in Fig. 4.27. For a given sample $\mathbf{x}$ to be reconstructed, its projection $P_D\Phi(\mathbf{x})$ in KPCA feature space is determined. After calculating the feature space distances between this projection $P_D\Phi(\mathbf{x})$ and all other training sample features, n closest neighbours are retained for further calculations. (In Fig. 4.27, three neighbours in feature space are considered). The corresponding input space distances between these n neighbours and the reconstructed sample $\hat{\mathbf{x}}$ can be calculated analytically for certain kernel functions. Once the neighbour input space distances are known, the calculation of the location of $\hat{\mathbf{x}}$ becomes a simple coordinates-by-distance constraints problem.

The distance constraint reconstruction algorithm is expanded here in more detail. Given a point $\mathbf{x}$ for which a KPCA reconstruction must be calculated, its projection $P_D\Phi(\mathbf{x})$ in the KPCA is calculated. This KPCA projection will be at a squared feature space distance $d^2_{\mathcal{H}}(\Phi(\mathbf{x}_i), P_D\Phi(\mathbf{x}))$ from each sample $i = 1, \ldots, N$ in the training data set:

$$d^2_{\mathcal{H}}(\Phi(\mathbf{x}_i), \mathrm{P}_D\Phi(\mathbf{x})) = \|\Phi(\mathbf{x}_i) - \mathrm{P}_D\Phi(\mathbf{x})\|^2 \qquad (4.103)$$

These squared feature space distances are collected in a vector, $\mathbf{d}^2_{\mathcal{H}}$. For certain kernel functions (e.g. Gaussian kernels and polynomial kernels with odd degree), there is a direct functional relation between kernel feature space distances and input space distances. For example, for a Gaussian kernel function k with kernel width σ, the relation between a squared kernel feature space distance ($d^2_{\mathcal{H}}$) and its corresponding squared input space distance ($\mathbf{d}^2_{\mathcal{X}}$) is:

$$d^2_{\mathcal{X}}(\mathbf{x}_i,\mathbf{x}_j) = -2\sigma^2 \ln\left(\frac{\left(k(\mathbf{x}_i,\mathbf{x}_i) + k(\mathbf{x}_j,\mathbf{x}_j) - d^2_{\mathcal{H}}\left(\Phi(\boldsymbol{x}_i),\Phi(\boldsymbol{x}_j)\right)\right)}{2}\right) \tag{4.104}$$

The input space distance vector $d^2_{\mathcal{X}}$ associated with $\mathbf{d}^2_{\mathcal{H}}$ can be calculated from relations such as the above. To determine the coordinates of $\hat{\mathbf{x}}$, only a subset of the input samples that correspond with n nearest neighbours in the feature space $\mathcal{H}$ are considered. These n samples are collected in the nearest neighbour input matrix $\mathbf{X}^{(n)}$, which, together with the corresponding nearest neighbour input space distances $\mathbf{d}^{2(n)}_{\mathcal{X}}$, is used to estimate the coordinates of $\hat{\mathbf{x}}$.

A least-squares approximation to the distance constraint coordinate problem stated above is given here. The nearest neighbour inputs $\mathbf{X}^{(n)}$ are first centred to $\mathbf{X}^{(n)}_c$:

$$x^{(n)}_{c,i} = \mathbf{x}^{(n)}_i - \bar{\mathbf{x}}^{(n)}; \quad \text{with} \quad \bar{\mathbf{x}}^{(n)} = \frac{1}{n}\sum_{i=1}^{n}\mathbf{x}^{(n)}_i. \tag{4.105}$$

A new coordinate system defined through singular value decomposition is obtained, with $\mathbf{W}^{(n)}$ the coordinates of the n nearest neighbours projected to the first q coordinate eigenvectors ($\mathbf{U}$) with non-zero eigenvalues.

$$\mathbf{X}^{(n)}_c = \mathbf{U}\boldsymbol{\Lambda}\mathbf{V}' = \mathbf{U}\mathbf{W}^{(n)}. \tag{4.106}$$

The n nearest neighbour squared distances in this new coordinate space is represented by $\mathbf{d}^{2(n)}_{\mathcal{W}}$. The least-squares solution of the estimated reconstruction in the new coordinate system $\widehat{\mathbf{w}}$, which minimizes the difference between $\mathbf{d}^{2(n)}_{\mathcal{X}}$ and $\mathbf{d}^{2(n)}_{\mathcal{W}}$ is given by:

$$\widehat{\mathbf{w}} = -\frac{1}{2}\boldsymbol{\Lambda}^{-1}\mathbf{V}'\left(\mathbf{d}^{2(n)}_{\mathcal{X}} - \mathbf{d}^{2(n)}_{\mathcal{W}}\right) \tag{4.107}$$

The estimated reconstruction $\widehat{\mathbf{w}}$ is defined in terms of the new coordinate system given by $\mathbf{U}$. Transforming back the input space $\mathcal{X}$, with decentring, gives the coordinates of the reconstructed input sample $\hat{\mathbf{x}}$:

$$\hat{\mathbf{x}} = \mathbf{U}\widehat{\mathbf{w}} + \bar{\mathbf{x}}^{(n)}. \tag{4.108}$$

Reconstruction by Learning

While the above approaches are commendable in that kernel function information is exploited, these methods require an iterative (fixed-point iteration method) or least-squares (distance constraints method) solution for each reconstruction prediction. Another approach to the KPCA reconstruction problem is to learn a single mapping function, $\Gamma(\mathbf{t})$, which maps vectors in KPCA feature space $\mathcal{T}$ to the input space $\mathcal{X}$. The input data $\mathbf{X}$ used to train KPCA, with their associated KPCA features $\mathbf{T}$, serves as a labelled training data set to learn the function $\Gamma(\mathbf{t})$.

Bakır and co-workers have suggested using kernel ridge regression as the learning approach for KPCA reconstruction (Bakır et al. 2004, 2005). Since the map $\Gamma(\mathbf{t})$ outputs M-dimensional input space vectors $\mathbf{x}$, multivariate kernel ridge regression is appropriate to learn the map from $\mathcal{T}$ to $\mathcal{X}$, given training data.

Ridge regression is similar to normal linear regression, except that the sizes of the model coefficients are penalized. Where the loss function for normal linear regression (with coefficients $\mathbf{w}$) is given by the quadratic loss between the dependent variable $\mathbf{y}$ and the regression predictions $\mathbf{w}^{\mathrm{T}}\mathbf{X}$, ridge regression further imposes a penalty on the model coefficients.[3] The contribution of this penalty to the overall loss function is regulated by the ridge parameter, λ:

$$L(\mathbf{y}, \mathbf{w}^T\mathbf{X}) = \sum_{i=1}^{N} \left(y_i - \mathbf{w}^T\mathbf{x}_i\right)^2 + \lambda\|\mathbf{w}\|^2. \tag{4.109}$$

The introduction of the ridge parameter reduces high variance of coefficients where independent variables in $\mathbf{X}$ are correlated (Hastie et al. 2009). Cross-validation can be used to determine an optimal ridge parameter for a specific application.

The ridge regression predicted label $\hat{y}$ of a new sample $\mathbf{x}$ is given by:

$$\hat{y} = \mathbf{w}^T\mathbf{x} = \mathbf{y}^{\mathrm{T}}(\mathbf{K} + \lambda \mathrm{I})^{-1}\boldsymbol{\kappa}. \tag{4.110}$$

Above, $\hat{y}$ is the predicted dependent variable value, $\mathbf{y}$ is the vector of dependent variable training samples, $\mathbf{K}$ is a matrix containing the dot products of the training set input data vectors ($K_{ij} = \mathbf{x}_i^{\mathrm{T}}\mathbf{x}_j$; $i = 1, \ldots, N$; $j = 1, \ldots, N$) and $\boldsymbol{\kappa}$ is a vector containing the dot products of the new input sample with the training set input data vectors ($\kappa_i = \mathbf{x}^{\mathrm{T}}\mathbf{x}_i$; $i = 1, \ldots, N$).

Multiple ridge regression, where more than one dependent variable (collected as $\mathbf{Y}$) are to be predicted from dependent variables $\mathbf{X}$, can be achieved by extending the coefficients and loss function of ridge regression:

[3] For the progression of the ridge regression explanation, the general statistical nomenclature of $\mathbf{x}$ for independent variables and $\mathbf{y}$ for dependent variables will be used. KPCA reconstruction by learning has the input space $\mathcal{X}$ as output, and the KPCA feature space $\mathcal{T}$ as input. The KPCA nomenclature will be returned to once the ridge regression explanation is completed.

$$L(\mathbf{Y}, \mathbf{W}^{\mathrm{T}}, \mathbf{X}) = \sum_{i=1}^{N} \|y_i - \mathbf{W}^{\mathrm{T}}\mathbf{x}_i\|^2 + \lambda\|\mathbf{W}\|^2. \tag{4.111}$$

Multiple kernel ridge regression follows from above by replacing input samples ($\mathbf{x}$) by nonlinear mappings of these input samples, $\varphi(\mathbf{x})$. If these nonlinear mapping functions are associated with kernel functions, the kernel trick can be exploited to avoid explicit calculation of $\varphi(\mathbf{x})$. Predicted dependent variable values $\hat{\mathbf{y}}$ can then be calculated from:

$$\hat{y} = \mathrm{Y}(\mathrm{K} + \lambda \mathrm{I})^{-1}\kappa. \tag{4.112}$$

Above, $\hat{\mathbf{y}}$ is the vector of predicted dependent variable values, $\mathbf{Y}$ is the matrix of dependent variable training samples, $\mathbf{K}$ is the kernel matrix of the training set input data vectors ($K_{ij} = k(\mathbf{x}_i, \mathbf{x}_j)$; $i = 1, \ldots, N$; $j = 1, \ldots, N$) and $\boldsymbol{\kappa}$ is a vector containing the kernel function results of the new input sample with the training set input data vectors ($\kappa_i = k(\mathbf{x}_i, \mathbf{x})$;$i = 1, \ldots, N$).

Multiple kernel ridge regression applied to KPCA reconstruction takes the training features $\mathbf{T}$ as independent variables, and the training input samples $\mathbf{X}$ as dependent variables. For a given choice of kernel function, kernel parameters and ridge parameter, a reconstructed input space vector can be predicted from its projection in KPCA feature space:

$$\hat{\mathbf{x}} = \mathrm{T}(\mathrm{K} + \lambda \mathrm{I})^{-1}\kappa. \tag{4.113}$$

Above, the kernel matrix $\mathbf{K}$ is calculated for the KPCA features of the training data set ($K_{ij} = k(\mathbf{t}_i, \mathbf{t}_j)$; $i = 1, \ldots, N$; $j = 1, \ldots, N$), and the kernel vector $\boldsymbol{\kappa}$ contains the kernel function evaluations of a new KPCA feature sample $\mathbf{t}$ (to be reconstructed in the input space $\mathcal{X}$) against the training data KPCA features ($\kappa_i = k(\mathbf{t}_i, \mathbf{t})$;$i = 1, \ldots, N$).

The design process for multiple kernel ridge regression learning involves selecting a kernel function, kernel parameters for said kernel function, and an optimal ridge parameter. For the case where multiple kernel ridge regression is to be used for KPCA reconstruction, the kernel function and kernel parameters need not be the same as those used in KPCA. Since learning a map from the KPCA feature space to the input space is a supervised approach (with training data available), cross-validation approaches can be used to select kernel parameters and the ridge parameter. An efficient approach for selecting the ridge parameter for multiple kernel ridge regression is given by An et al. (2007).

Advantages of the learning approach to KPCA reconstruction include its avoidance of difficult and potentially unstable numerical optimization, faster evaluations and better suitability to high-dimensional input spaces (Bakır et al. 2004).

4.8.6 *Kernel Principal Component Analysis Feature Extraction Algorithm*

Considering the previous subsections, a general framework for feature extraction with kernel principal component analysis can now be summarized. The KPCA feature extraction algorithm consists of training and application stages. During the training stage, a training data set $\mathbf{X} \in \mathcal{X}$ (N samples in M dimensions) is transformed to a kernel feature space, where D projection vectors $\mathbf{P}^*$ are determined by PCA. The kernel transformation and subsequent projection of the training data onto the retained principal components in the kernel feature space gives the features $\mathbf{T}$ in the d-dimensional KPCA feature space $\mathcal{T}$. During the application stage, new (test) data $\mathbf{X}^{\text{test}} \in \mathcal{X}$ (N^{test} samples in M dimensions) are implicitly transformed to the kernel feature space, relative to the training data kernel matrix. These implicit test data are projected onto the D retained kernel principal components to give test features $\mathbf{T}^{\text{test}} \in \mathcal{T}$. The test data can also be reconstructed from the retained features (using some approximate reconstruction method), to give the reconstructed test data set $\hat{\mathbf{x}}^{\text{test}}$. The KPCA feature extraction algorithm is given here:

Kernel Principal Component Analysis Feature Extraction Algorithm

Training stage:

- Centre the samples of the M-dimensional training data set $\mathbf{X}$ to give $\tilde{\mathbf{X}}$.
- Given a specific kernel function k, calculate the kernel $\mathbf{K}$ and centred kernel $\tilde{K}$ matrices of the centred training data. (Kernel function parameters can be determined by cross-validation, where the cross-validation criterion may be mean squared reconstruction error.)
- Eigendecomposition of $\tilde{K}$ returns r ordered eigenvectors $\mathbf{P}$ and with non-zero eigenvalues $\mathbf{\Lambda}$.
- The first D kernel principal components are retained as the projection vectors $\mathbf{P}^*$, where the selection of D is based on some selection criteria (e.g. certain fraction of variance accounted for).

Application stage:

- Centre the samples of the test data set $\mathbf{X}^{\text{test}}$, using the mean vector of the training data set, to give the centred test data set $\tilde{X}^{\text{test}}$.
- d-dimensional test features $\mathbf{T}^{\text{test}}$ are calculated by projecting the centred test data $\tilde{X}^{\text{test}}$ onto the first D kernel principal components $\mathbf{P}^*$.
- Reconstruct the centred test data in the input space from the d-dimensional test features, using some approximate reconstruction method (e.g. fixed-point iteration or distance constraint approach).

4.9 Example: Fault Diagnosis in a Simulated Nonlinear System with Kernel Principal Component Analysis

The use of kernel-based methods to detect faults in nonlinear systems can be illustrated by the following example based on a simulated system, previously considered by Dong and McAvoy (1992) and Jemwa and Aldrich (2006). The behaviour of the systems is determined by a single variable (t) that is inaccessible. The only information available is that derived from three measurements satisfying

$$\begin{aligned} x_1 &= t + e_1 \\ x_2 &= t^2 - 3t + e_2 \\ x_3 &= -t^3 + 3t^2 + e_3. \end{aligned} \tag{4.114}$$

t is a sample from the uniform distribution on [0.01 1], and e_i ($i = 1, 2$ and 3) are Gaussian random noise variables, each with a mean of zero and variance of 0.02. A fault condition is introduced by inducing small changes to the measured variable x_3 such that

$$\begin{aligned} x_1 &= t + e_1 \\ x_2 &= t^2 - 3t + e_2 \\ x_3 &= -1.1t^3 + 3.2t^2 + e_3. \end{aligned} \tag{4.115}$$

Hundred samples were collected from each of the two systems described by equations (4.107) and (4.108). The two sets of data are shown in Fig. 4.28. It is difficult to detect the difference between the two systems by visual inspection

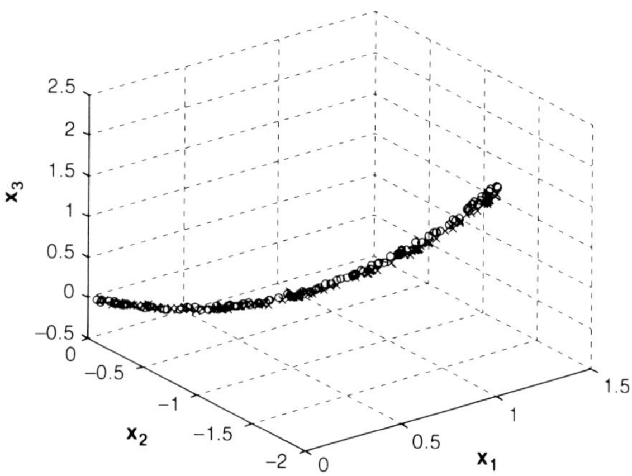

Fig. 4.28 3D scatter plot of the two data sets represented by equations (4.107) ('x') and (4.108) ('o'), simulating normal and faulty operating conditions respectively

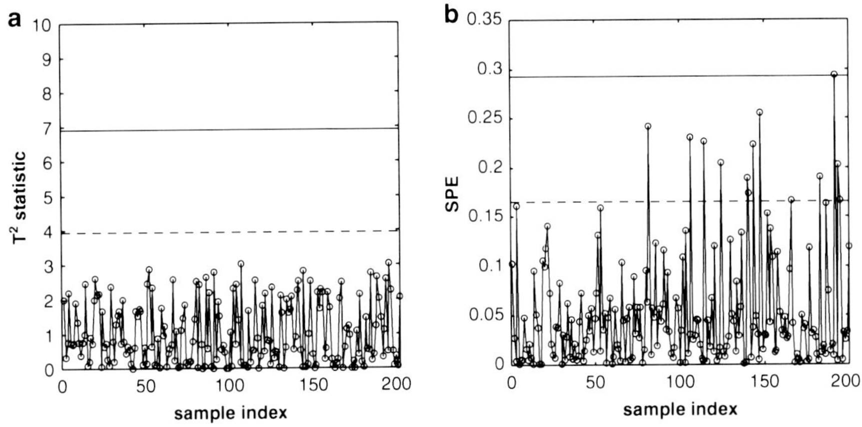

Fig. 4.29 (**a**) Hotelling's T^2 and (**b**) Squared Prediction Error (Q) statistics for both the normal data (first 100 samples) and the fault condition data (last 100 samples) calculated for the simulated system (4.114 and 4.115). The superimposed horizontal lines indicate the 95 % (*dashed line*) and 99 % (*solid line*) confidence limits for each statistic. In both cases, the fault condition remains undetected at the 99 % confidence level

of the data. Under these circumstances, monitoring methods based on classical principal component analysis are also inadequate, as indicated in Fig. 4.29, since the correlations between the variables are nonlinear.

Dong and McAvoy (1992) have considered a nonlinear diagnostic approach based on the use of principal curves and autoassociative multilayer perceptrons. Principal curves as proposed by Hastie and Stuetzle (1989) were used to find the nonlinear scores, while a multilayer perceptron was used to find both the forward and reverse mappings between the scores and original data, since a nonlinear principal loading is not explicitly defined for the principal curves. The approach is effective, as indicated in Fig. 4.30, which is a plot of the squared prediction error or Q statistic derived for their method.

Similar results are obtained with kernel PCA on the same data, as indicated in Fig. 4.31, where kernel PCA was used to extract nonlinear features from the data. Both the T^2 and SPE or Q-statistics indicate a process shift in the data sampled in the presence of the fault condition. The two methods can be compared further in Fig. 4.32. Figure 4.32 shows the leading principal direction (dashed line) and the one-dimensional principal curve (solid line). For comparative purposes, Fig. 4.32 shows the kernel-based principal component (solid line), which is virtually identical to the principal curve in Fig. 4.32.

The advantage of the proposed method compared to the principal curves approach is that nonlinear optimization is avoided in kernel PCA, and hence potentially suboptimal solutions. Also, principal curves require prior specification

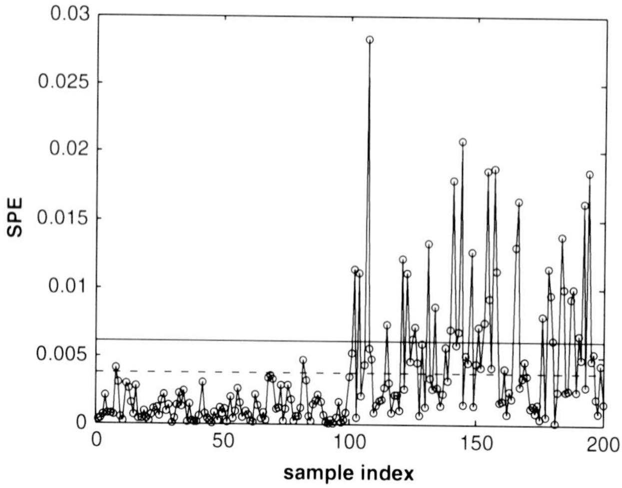

Fig. 4.30 SPE monitoring chart using principal curves-multilayer perceptron method. The onset of the fault condition after the 100th sample is clearly indicated

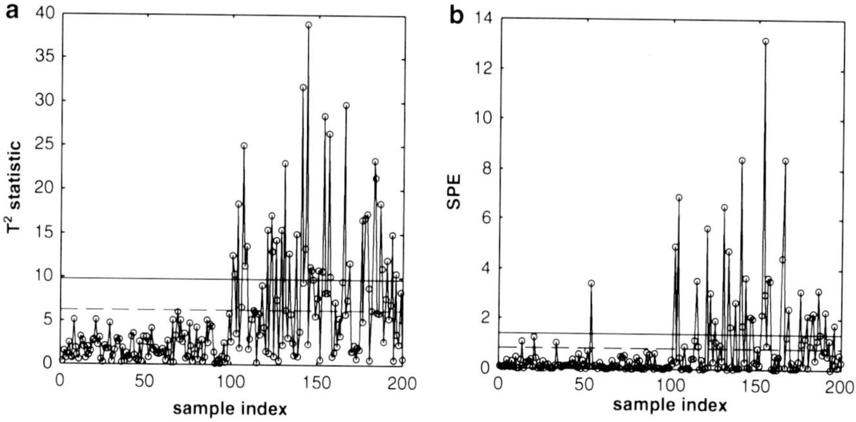

Fig. 4.31 (**a**) Hotelling's T^2 and (**b**) SPE statistics for the simulated system after extracting dominant nonlinear features using kernel PCA, and performing linear PCA on the residuals. Similar to Fig. 4, the onset of the fault condition can be distinguished using either of the statistics

of the number of features to be extracted. The main disadvantage of the kernel PCA method is the lack of clear geometric interpretation of the features extracted in the high-dimensional space. In our experiments the approximate pre-images were sensitive to the reconstruction method used. Further work will be directed towards refining the approach to reduce sensitivity to the way in which reconstruction is done, e.g. by developing novel statistical criteria for the detection of process deviations in the feature space.

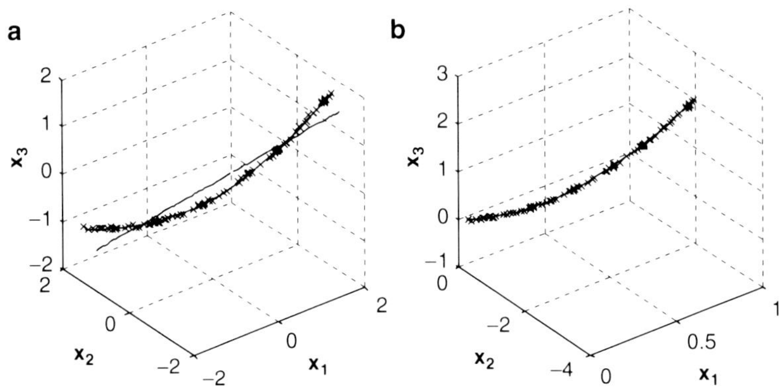

Fig. 4.32 (**a**) Nonlinear PCA using principal curves. The *solid line* is the 1st principal curve and the *dashed line* is the first linear principal component analysis. (**b**) Nonlinear principal component obtained on performing kernel PCA with a Gaussian kernel of unit width. The four leading features were retained in the feature space, and input space reconstruction was done using a multidimensional scaling approach (Kwok and Tsang 2004)

4.10 Concluding Remarks

Support vector machines and other kernel-based algorithms are popular statistical learning methods, for a number of reasons. One of these is their solid theoretical basis (in terms of structural risk minimization). SVMs are also fairly intuitive on a geometrical basis. Much work has been done on understanding the generalization ability of SVMs (i.e. their ability to do well on unseen data by not overfitting training data). SVM training algorithms are guaranteed to achieve a global optimum (as opposed to locally friable neural networks and decision trees). The kernel trick allows any algorithm that can be expressed in terms of dot products to be extended via nonlinear kernel functions, unleashing a cornucopia of possibilities. A caveat of this cornucopia is the selection of kernel parameters; this task can be guided through cross-validation. The practicalities of computational optimization are not discussed here, but provide fertile ground for research on and improvement in kernel algorithms.

As was discussed in Chap. 2, kernel methods have been used extensively in fault diagnosis in recent years. This includes kernel-based versions of multivariate methods, such as principal and independent components, as well as the use of one-class support vector machines to derive confidence limits of arbitrary complexity. Figure 4.33 gives a conceptual summary of the generalized framework for constructing fault diagnostic systems with kernel methods. In addition to forward and reverse mapping models as indicated, one-class support vector machines are also used for the estimation of confidence limits where the data (features, $\boldsymbol{F}$ and residuals $\boldsymbol{E}$) are not normally distributed.

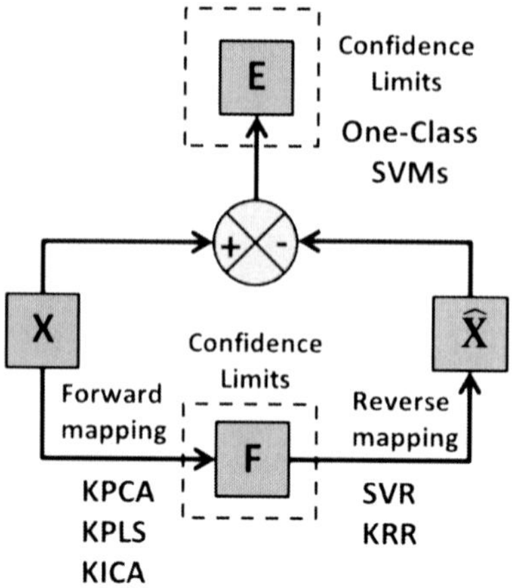

Fig. 4.33 Fault diagnosis with kernel-based models, including kernel principal component analysis (*KPCA*), kernel partial least squares (*KPLS*), and kernel independent component analysis (KICA) for feature extraction, support vector regression (*SVR*) or kernel ridge regression (*KRR*) for reverse mapping and one-class support vector machines for the estimation of confidence limits in non-Gaussian data in the feature and residual spaces

References

Bakır, G. H. (2005). *Extension to kernel dependency estimation with applications to robotics*. PhD dissertation. Berlin: Technical University of Berlin.

Bakır, G. H., Weston, J., & Schölkopf, B. (2004). Learning to find pre-images. In *Advances in neural information processing systems* (pp. 449–456). Cambridge, MA: MIT Press.

Belousov, A. I., Verzakov, S. A., & von Frese, J. (2002). Applicational aspects of support vector machines. *Journal of Chemometrics, 16*(8–10), 482–489.

Berk, R. A. (2008). *Statistical learning from a regression perspective* (1st ed.). New York: Springer.

Boser, B. E., Guyon, I. M., & Vapnik, V. N. (1992). A training algorithm for optimal margin classifiers. In *Proceedings of the fifth annual workshop on Computational Learning Theory – COLT'92. The 5th annual workshop* (pp. 144–152), Pittsburgh, PA, USA. Available at: http://portal.acm.org/citation.cfm?doid=130385.130401. Accessed 27 May 2011.

Boyd, S., & Vandenberghe, L. (2004). *Convex optimization*. Cambridge/New York: Cambridge University Press.

Burges, C. J. C. (1998). A tutorial on support vector machines for pattern recognition. *Data Mining and Knowledge Discovery, 2*(2), 121–167.

Chapelle, O., & Zien, A. (2005). Semi-supervised classification by low-density separation. In *Proceedings of the 10th international workshop on Artificial Intelligence and Statistics* (pp. 57–64). The Savannah Hotel: Barbados.

Cortes, C., & Vapnik, V. (1995). Support vector networks. *Machine Learning, 20*(3), 273–297.

Cover, T. M. (1965). Geometrical and statistical properties of systems of linear inequalities with applications in pattern recognition. *IEEE Transactions on Electronic Computers, EC-14*(3), 326–334.

Dong, D., & McAvoy, T. J. (1992). Nonlinear principal component analysis – Based on principal curves and neural networks. *Computers and Chemical Engineering, 16*, 313–328.

Fisher, R. A. (1936). The use of multiple measurements in taxonomic problems. *Annals of Eugenics, 7*, 179–188.

Franc, V., Schlesinger, M. I., & Hlavac, V. (2008). *Statistical pattern recognition toolbox for Matlab*. Available at: http://cmp.felk.cvut.cz/cmp/software/stprtool/. Accessed 12 Dec 2011.

Hastie, T., & Stuetzle, W. (1989). Principal curves. *Journal of the American Statistical Association, 84*, 502–516.

Hastie, T., Tibshirani, R., & Friedman, J. (2009). *The elements of statistical learning – Data mining, inference and prediction* (2nd ed.). New York: Springer.

Hotelling, H. (1933). Analysis of a complex of statistical variables into principal components. *Journal of Educational Psychology, 24*(6), 417–441.

Hsieh, W. (2009). *Machine learning methods in the environmental sciences: Neural networks and kernels*. Cambridge/New York: Cambridge University Press.

Jemwa, G. T., & Aldrich, C. (2006). Kernel-based fault diagnosis on mineral processing plants. *Minerals Engineering, 19*(11), 1149–1162.

Jemwa, G. T., & Aldrich, C. (2012). Estimating size fraction categories of coal particles on conveyor belts using image texture modelling methods. *Expert Systems with Applications, 39*(9), 7947–7960.

Kaartinen, J., Hätönen, J., Hyötyniemi, H., & Miettunen, J. (2006). Machine-visionbasedcontrol of zinc flotation – A case study. *Control Engineering Practice, 14*, 1455–1466.

Kwok, J. T.-Y., & Tsang, I. W.-H. (2004). The pre-image problem in kernel methods. *IEEE Transactions on Neural Networks, 15*(6), 1517–1525.

Maulik, U., & Chakraborty, D. (2013). Learning with transductive SVM for semisupervised pixel classification of remote sensing imagery. *ISPRS Journal of Photogrammetry and Remote Sensing, 77*, 66–78.

Mika, S., Schölkopf, B., Smola, A., Müller, K.-R., Scholz, M., & Rätsch, G. (1999). Kernel PCA and de-noising in feature spaces. In *Advances in neural information processing systems 11* (pp. 536–542). Cambridge: MIT Press.

Moolman, D. W., Aldrich, C., van Deventer, J. S. J., & Stange, W. W. (1995). The classification offroth structures in a copper flotation plant by means of a neural net. *International Journal of Mineral Processing, 43*, 23–30.

Müller, K.-R., Mika, S., Ratsch, G., Tsuda, K., & Scholkopf, B. (2001). An introduction to kernel-based learning algorithms. *IEEE Transactions on Neural Networks, 12*(2), 181–201.

Schölkopf, B., & Smola, A. J. (2001). *Learning with kernels: Support vector machines, regularization, optimization, and beyond* (1st ed.). Cambridge: MIT Press.

Schölkopf, B., Smola, A., & Müller, K.-R. (1998). Nonlinear component analysis as a kernel eigenvalue problem. *Neural Computation, 10*(5), 1299–1319.

Schölkopf, B., Mika, S., Burges, C. J. C., Knirsch, P., Muller, K.-R., Ratsch, G., & Smola, A. J. (1999). Input space versus feature space in kernel-based methods. *IEEE Transactions on Neural Networks, 10*(5), 1000–1017.

Schölkopf, B., Platt, J. C., Shawe-Taylor, J., Smola, A. J., & Williamson, R. C. (2001). Estimating the support of a high-dimensional distribution. *Neural Computation, 13*(7), 1443–1471.

Shawe-Taylor, J., & Cristianini, N. (2004). *Kernel methods for pattern analysis*. Cambridge: Cambridge University Press.

Smola, A. J., & Schölkopf, B. (2000). Sparse greedy matrix approximation for machine learning. In *Proceedings of the seventeenth International Conference on Machine Learning. ICML'00* (pp. 911–918). San Francisco: Morgan Kaufmann Publishers Inc.. Available at: http://dl.acm.org/citation.cfm?id=645529.657980.

Smola, A. J., & Schölkopf, B. (2004). A tutorial on support vector regression. *Statistics and Computing, 14*(3), 199–222.

Smola, A. J., Mangasarian, O. L., & Schölkopf, B. (1999). *Sparse kernel feature analysis*. Madison: Data Mining Institute.

Tessier, J., Duchesne, C., & Bartolacci, G. (2007). A machine vision approach to on-line estimation of run-of-mine ore composition on conveyor belts. *Minerals Engineering, 20*(12), 1129–1144.

Tipping, M. (2001). Sparse kernel principal component analysis. In T. K. Leen, T. G. Dietterich, & V. Tresp (Eds.), *Advances in neural information processing systems* (Neural Information Processing Systems 13 (NIPS 2000), pp. 633–639). Cambridge, MA: MIT Press.

Vapnik, V. (2006). Transductive inference and semi-supervisedlearning. In O. Chapelle, B. Schölkopf, & A. Zien (Eds.), *Semi-supervised learning* (pp. 453–472). Cambridge, MA: MIT Press.

Nomenclature

Symbol	Description
$\tilde{x}$	Point satisfying all inequality constraints of an optimization problem
K_{ij}	Element of Gram matrix in *i*th row and *j*th column
N_{SV}	Number of support vectors in a training data set
$R_e(f)$	Empirical risk of overfitting function *f*
$f_0(\mathbf{x})$	Objective function in an optimization problem
$f_0(\tilde{\mathbf{x}})$	Objective function value at point where all inequality constraints of optimization problem are satisfied
f_0^*	Optimal value of objective function
$m_G(\mathbf{x}, y)$	Geometrical margin, i.e. the distance of a point **x** with associated label *y* from a separating hyperplane defined by parameters **w** and *b*
ξ_i	*i*th slack variable of an optimization problem
$C_p{}^*$	Optimal parameter in a kernel function
C_q	Parameter in a kernel function
D	Diameter of sphere enclosing a set of (training) data
D_s	Diameter of smallest sphere enclosing a set of (training) data
$\mathcal{F}$	Function space, class of functions
f^*	Function associated with lowest risk bound
h	Capacity parameter of a model, such as VC dimension
$\mathcal{H}$	Feature space
$\mathbf{K}$	Gram matrix
$L(y, f(\mathbf{x}))$	Loss function
M	Dimensionality of input space
m	Margin or shortest distance between two separating hyperplanes
$P(\mathbf{x}, y)$	Joint probability distribution between **x** and *y*
Q	Arbitrary number of parameters
δ	Confidence
$\boldsymbol{\xi}$	Vector of slack variables
ρ	Bias defining location of a hyperplane
σ	Width of Gaussian kernel
$\Phi(\mathbf{x})$	Mapping function from space $\mathcal{X}$ to space $\mathcal{H}$
C	Covariance matrix of mean-centred data matrix **X**
v	Arbitrary vector
$L(\tilde{\mathbf{x}}, \propto)$	Lagrangian function value at point where all inequality constraints of optimization problem are satisfied
$L(\mathbf{x}, \propto)$	Lagrangian function

(continued)

(continued)

Symbol	Description
N	Number of samples in a training data set
$R(f)$	Risk of overfitting function f
$C(h, N, \delta)$	Capacity of a model
$g(\propto)$	Lagrangian dual function
$k(\mathbf{x}, \mathbf{x}')$	Kernel function
$m(\mathbf{w}, b)$	Margin of a separating hyperplane
$\mathcal{T}$	Principal component score space
$\mathcal{X}$	Vector space of $\mathbf{x}$
$\propto$	Vector of Lagrangian multipliers
θ	Angle
κ	Sigmoidal kernel parameter
ν	Parameter in the optimization of a soft margin classifier
ρ	Parameter in the optimization of a soft margin classifier
ϑ	Sigmoidal kernel parameter

Chapter 5
Tree-Based Methods

5.1 Generalized Framework for Data-Driven Fault Diagnosis by the Use of Tree-Based Methods

Unlike neural networks or kernel methods, tree-based methods have figured prominently in the development of data-driven fault diagnostic methods, except perhaps for fault classification, which is not considered in this book. Although tree-based models, and regression and decision trees in particular, have been around since the 1970s, these methods do not generally lend themselves to unsupervised learning and hence feature extraction from data. It is only with the advent of random forests in the 1990s that unsupervised learning with trees has become possible, but even so, this aspect has not received much attention to date, much less so in the context of fault diagnosis. Nonetheless, the ability of random forests to facilitate both unsupervised and supervised learning means that they, together with other tree-based approaches, constitute a potentially important class of analytical methods as far as the construction of fault diagnostic models is concerned. These approaches are explored in this chapter.

5.2 Decision Trees

A decision tree is a statistical model learnt from a given training data set to perform a classification or regression task. The training data set consists of a number of input vectors $\mathbf{X}$ (N samples and M rows) and a corresponding response vector $\mathbf{y}$ (N samples). Where $\mathbf{y}$ can have a discrete class membership, the tree model is known as a classification tree, while continuous response values are obtained from a regression tree.

Decision trees refer to a class of learning algorithms that recursively partition an input data space ($\boldsymbol{x}$) in order to obtain subspaces with increasingly purer output

C. Aldrich and L. Auret, *Unsupervised Process Monitoring and Fault Diagnosis with Machine Learning Methods*, Advances in Computer Vision and Pattern Recognition,
DOI 10.1007/978-1-4471-5185-2_5, © Springer-Verlag London 2013

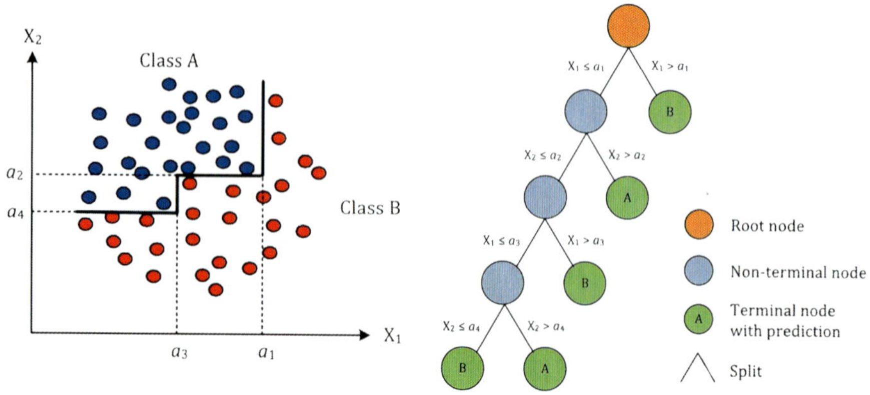

Fig. 5.1 Classification problem with decision boundaries and corresponding classification tree

distributions. Here, purity is defined by some impurity measure (e.g. information entropy for classification or least-squares deviation for regression). Once these data-adaptive subspaces have been found, simple local models can be fitted. A decision tree can be seen as a collection of non-overlapping local models, where the regions spanned by each local model are determined from the distribution of the training data.

Popular tree-growing algorithms, such as CART (Breiman et al. 1984) and C4.5 (Quinlan 1993), create subspaces by recursively searching for a partition on a single input variable that results in the largest reduction in impurity of the output variable. The algorithm continues partitioning until no improvement in output impurity is possible or some stopping criterion is satisfied. The result is a collection of non-overlapping hyperrectangles in the input data space, with partitions parallel to the input variable axes.

A decision tree model can be visualized as a collection of if–then–else rules, specifying the split variables and split positions. This visual presentation of data partitioning is an attractive feature of decision trees, aiding interpretation of the model (see Fig. 5.1).

The power of decision tree algorithms relates to their ability to fit nearly any data distribution. This can also be seen as their downfall, as decision trees may erroneously overfit noisy distributions. Overfitting can be counteracted by restricting the complexity of the tree structures or by combining decision tree models in ensembles, such as with random forest and boosted tree ensembles.

Whereas neural network algorithms (such as multilayer perceptrons) define model structures and then optimize the model parameters to achieve an explicit global fitness function, decision tree algorithms make use of greedy optimization of local fitness functions, resulting in emergent model structures.

5.2.1 Development of Decision Trees

One of the earliest references to recursive partitioning of input variables to aid prediction of an output variable is Benson's algorithm (Belson 1959). Benson's algorithm caters for classification (i.e. categorical output variable) based on categorical input variables, with binary splits. Belson devised partitions on categorical input variables by comparing the prior $P(\mathbf{y})$ and conditional probabilities $P(\mathbf{y}|\mathbf{x})$ of a categorical output variable. Where the prior and conditional probabilities differ by the largest margin, the input variable is considered relevant to the output variable and used to partition the input data space (Freund and Schapire 1997; Strobl et al. 2009).

An automated decision tree-growing algorithm for the regression task (i.e. continuous output variable), automatic interaction detection (AID), was developed by Morgan and Sonquist (1963). The prediction for a particular subspace is the average of the output samples in the said subspace, while suitable partitions are selected by optimizing the least-squares deviations of predictions on either side of the partition. Both categorical and continuous input variables can be handled, and binary splits are made.

In subsequent research, the AID algorithm was modified to allow multiple output variables (MAID-M: modified AID for multiple output variables) (Gillo and Shelly 1974), to function as a classification model (THAID: theta AID) (Messenger and Mandell 1972) and to introduce significance testing to restrict overfitting (CHAID: chi-square statistic AID) (Kass 1980).

One of the most widely used decision tree algorithms, the classification and regression tree (CART) algorithm, was developed by Breiman et al. (1984). CART provides classification and regression modelling for both categorical and continuous variables, as well as a suggested solution to overfitting: cost-complexity pruning. Cost-complexity pruning attempts to balance model complexity (size of tree) and model generalization (robustness to overfitting) by generating a succession of subtrees and assigning trade-off parameter-dependent cost-complexities to these subtrees. The trade-off parameter can be determined by cross-validation, and the optimal subtree defined.

Another prominent family of decision tree algorithms was developed by Quinlan: ID3 (Quinlan 1986), C4.5 (Quinlan 1993) and the commercial implementation of C4.5, C5.0/See5 (RuleQuest Research 2011), first available in 1998. The early algorithms were restricted to categorical input variables and provided no automatic pruning. A further difference from the CART algorithm is that ID3 and C4.5 generate multiple partitions per split, and not binary partitions. Later versions are more similar to CART, including continuous input variables and the so-called pessimistic pruning. The C4.5 algorithm provides a scheme for deriving a simplified set of rules from the decision tree partitions, to simplify interpretation and allow extension to expert systems.

5.2.2 Construction

Theoretically, a decision tree model could be found by generating all possible decision tree structures, and through exhaustive search selecting the model with the best predictive performance. However, for large data sets, this is computationally infeasible (Quinlan 1986). Instead, most tree construction algorithms make use of greedy, stepwise induction. This top-down induction consists of three tasks: selecting the split position (partition) at each new node, determining whether a node is terminal or not and assigning a predict response to terminal nodes.

The essence of the top-down tree induction algorithms is recursive partitioning of the input space.[1] Starting with the entire input space $\boldsymbol{x}$ (represented by a training input matrix $\mathbf{X}$), the tree-growing algorithm attempts to find a binary partition to increase the response ($\mathbf{y}$) purity in the subspaces formed by the partition. The partition is defined as a hyperplane perpendicular to one of the coordinate axes of $\boldsymbol{x}$. The purity of the resulting subspaces depends on the purity of the response classes (for classification) or the squared error from the response subspace average (for regression). Binary partitioning is repeated in each new subspace until subspace response purity, or some other stopping criterion, is achieved. Where input variables are categorical (nominal), partitions can be defined by creating two sets partitioning the unique values in the categorical variable.

The successive binary partitions can be expressed as a tree diagram. When training commences, the entire input space $\boldsymbol{x}$ is represented as a root node. The first binary split is shown as two branches leading to left and right children nodes. The split is defined as a specific value of a certain variable: If an input for a specific sample has a value smaller than the split, it reports to the left child node; if it is larger, it reports to the right child node. These children nodes now contain only the samples relevant to their defined subspaces. Repeated partitioning produces a tree with one root node and a number of non-terminal and terminal nodes. Figure 5.1 gives an example of a simple classification problem and a classification tree structure associated with this problem.

Split Selection

The partition at each non-terminal node is optimized to result in the largest increase in child node purity or, conversely, the largest decrease in impurity. For the classification task, the impurity of the categorical output variable is based on an impurity measure $i(\eta)$ calculated from the class proportions for class k, $p(k|\eta)$, present at node η. The Gini index is used as a default impurity measure in the CART algorithm:

$$i(\eta) = \sum_{k=1}^{C} p(k|\eta)(1 - p(k|\eta)) = 1 - \sum_{k=1}^{C} p(k|\eta)^2 . \quad (5.1)$$

[1] Binary splitting is considered here; extension to multiple splits is trivial.

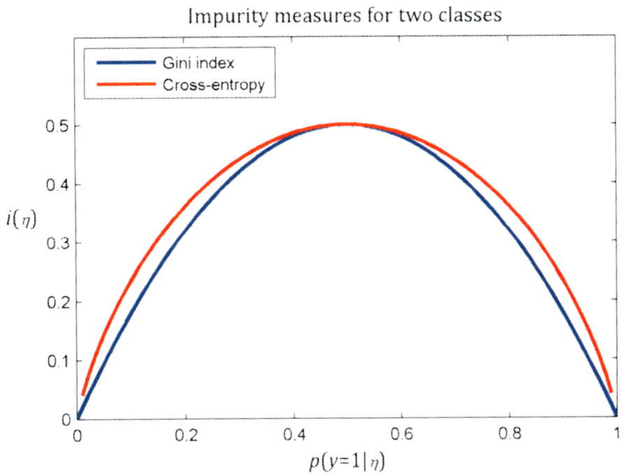

Fig. 5.2 Example of the Gini and cross-entropy impurity measures for two-class problem: the more mixed the classes (i.e. proportion close to 0.5), the higher the impurity measure

Cross-entropy is used as a default impurity measure in the ID3, C4.5 and C5.0 algorithms:

$$i\ (\eta) = -\sum\nolimits_{k=1}^{C} p\ (k\,|\,\eta) \log p(k\,|\,\eta) \tag{5.2}$$

where C is the number of classes in the response vector.

The impurity measure must be at a maximum where all classes are present in equal proportion and a minimum where only one class is present. The distribution of the Gini index and cross-entropy for a two-class problem is shown in Fig. 5.2.

The decrease in impurity $\Delta i(\varsigma, \eta)$ for a candidate split position ς depends on the impurity measures of the current node, the candidate child nodes and the proportion of samples reporting to the left and right child nodes (p_L and p_R):

$$\Delta i\ (\varsigma, \eta) = i\ (\eta) - p_R i\ (\eta_R) - p_L i\ (\eta_L). \tag{5.3}$$

At each node, the best split position ς^* is selected from the best split positions from each input variable that results in the largest decrease in impurity.[2] As the training sample is of a finite size, there is a finite number of possible split positions for each variable, but only a maximum of $N-1$ options per variable, where N is the sample size of the training data set. The actual number of viable split positions is often less than this for classification, as only split positions that correspond with a change in the response (class membership) need to be considered (Fayyad and Irani 1992).

[2]The C4.5 algorithm (Quinlan 1993) scales the decrease in impurity for categorical input variables, as a bias favouring multilevel variables exists in the cross-entropy impurity function. This corrected impurity decrease is known as the gain ratio.

Regression trees do not utilize the Gini index or cross-entropy for split selection, but rather minimize the mean squared error in the children nodes, where the error is the departure from the child node predictions (c_L for the left node η_L, and c_R for the right node η_R):

$$\Delta i\,(\varsigma, \eta) = \sum\nolimits_{\mathbf{X}_i \in \eta_L} (y_i - c_L)^2 + \sum\nolimits_{\mathbf{X}_i \in \eta_R} (y_i - c_R)^2. \tag{5.4}$$

Termination of Splitting

There are three motives for declaring a node a terminal node. If all samples in a node have the same response value, there is no rationale to create further partitions. A similar situation arises when only one sample is present in a node. One could also specify a minimum number of samples to be present in all terminal nodes. The rationale for a minimum leaf node size is to prevent overfitting. Heuristically, the minimum node size is often specified as five samples for regression. Another termination criterion requires splitting to stop if a specified minimum decrease in impurity is no longer possible.

Terminal Node Prediction

For classification models, each terminal node prediction will be that of the class majority present. For regression models, the average value of learning sample responses in terminal nodes is assigned as predictions.

To predict the response of a new input, the sample is simply "fed" down the tree from the root node, reporting to successive child nodes according to the learnt splits. When the new input reaches a terminal node, the terminal node prediction is assigned.

5.2.3 *Decision Tree Characteristics*

Decision trees can be seen as models that partition the input space into rectangular subspaces and assign a simple prediction (subspace class majority or response average) to each subspace. In this sense, a regression tree can be thought of as a histogram estimate of the response regression surface (Breiman et al. 1984). This discontinuous nature can be beneficial in terms of capturing of nonlinear and local effects, but detrimental when smooth function approximation is required (Hastie et al. 2009). Figure 5.3 illustrates the discontinuous nature of decision tree predictions.

Decision tree models are conceptually simple, as only an impurity measure and stopping criteria have to be specified. No distributional assumptions are made regarding either the input or output variables. The downside of this is the lack of

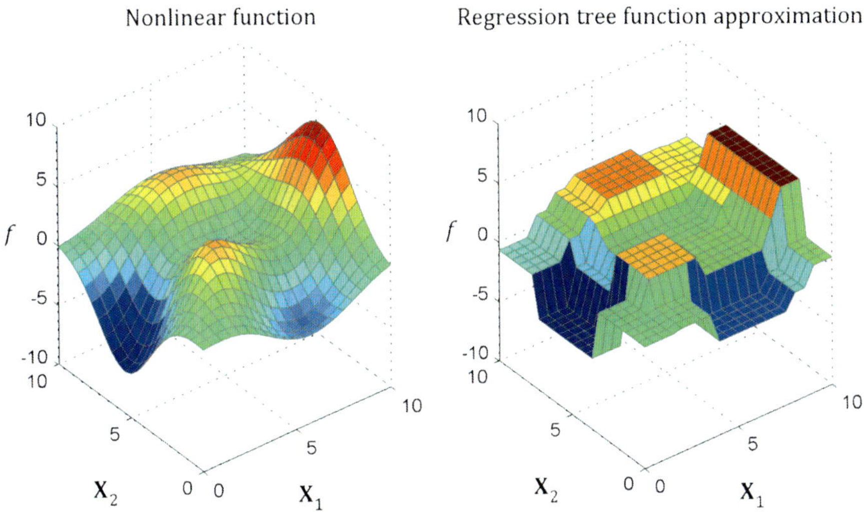

Fig. 5.3 Example of regression tree prediction for a nonlinear function: $f(\mathbf{X}) = 3(1-X_1)^2 \exp[-(\mathbf{X}_1^2) - (X_2+1)^2] - 10(X_1/5 - \mathbf{X}_1^3 - \mathbf{X}_2^5)\exp(-\mathbf{X}_1^2 - \mathbf{X}_2^2) - \exp[-(X_1+1)^2 - \mathbf{X}_2^2]/3$

confidence intervals for decision tree predictions. The confidence in new decision tree predictions is then solely based on previous results of the specific model on unseen testing data.

Training is computationally inexpensive, and assigning predictions for new inputs is even faster, as the tree structure can be summarized as a set of simple if–then–else rules. Due to their low computational expense, decision trees are suitable for inductive learning of large data sets. Decision tree rules and tree diagrams also make the models easily interpretable.

As splitting is only rank dependent, decision tree training is invariant to monotone transformations of variables. Both continuous ordered variables and categorical discrete variables can be accommodated as input. Procedures are also incorporated in some decision tree algorithms to allow for training and testing records with missing data.

The greedy learning approach has drawbacks: As terminal nodes can be defined up to a single training point, a large tree might overfit the data. Another drawback is the instability of decision trees. A small change in the learning sample can result in a completely different tree. The reason for this is its hierarchical approach to learning. An error early in the tree construction can change the entire tree structure (Hastie et al. 2009). As will be seen below, this instability is desirable when creating voting committees of decision trees.

As partitions are defined parallel to the coordinate axes, capturing additive structure requires a complicated tree with many levels, with repetition of splits across different subtrees (Hastie et al. 2009). Other "diagonal" partitions cannot be specified in single splits and may require more complex tree structures. This decision tree characteristic is illustrated in Fig. 5.4.

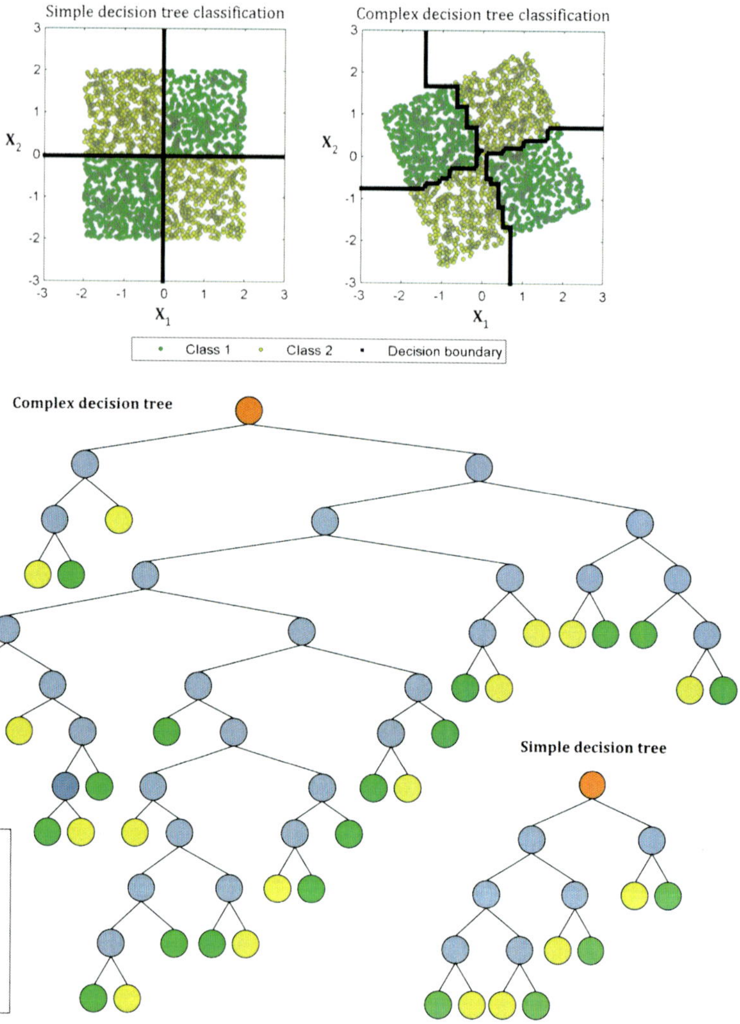

Fig. 5.4 Illustration of parallel decision boundary characteristics of decision trees, showing the complex nature of a decision tree trained on data with diagonal decision boundaries (tree structure and terminal node classification are given, with split decisions omitted)

The tree structure of the simple decision tree in Fig. 5.4 further shows the limitations of greedy splitting. As impurity decrease is optimized locally over each new split, the most parsimonious tree structure will not necessarily be found with the greedy tree-growing algorithm. Selecting a single partition separating the four class groups into two groups of two classes provides only marginal impurity increase, which is solely a function of the random distribution of data created for this toy problem. This difficulty for greedy tree algorithms in splitting so-called checkerboard patterns has been mentioned in early work on decision trees (Morgan and Sonquist 1963).

5.3 Ensemble Theory and Application to Decision Trees

Decision trees, similar to neural networks and subset selection linear regression, are unstable statistical models, in the sense that a small change in the learning set or model initialization can dramatically affect the final model (Breiman 1996). This instability can be exploited using ensemble theory.

5.3.1 *Combining Statistical Models*

An ensemble algorithm constructs a set of statistical models and typically uses majority voting (for classification) or averaging (for regression) to make predictions (Dietterich 2000b). For an ensemble to be more accurate than its members, two necessary and sufficient conditions must be met: The ensemble members must have individual accuracies better than random guessing and the ensemble members must be diverse (Hansen and Salamon 1990). These conditions are also known as the strength and diversity of an ensemble (Breiman 2001). Two models are considered diverse if the errors they make on unseen data are uncorrelated. Generally, ensemble algorithms do not explicitly optimize for diversity, but attempt to introduce and increase diversity heuristically through model parameter randomization (Polikar 2006).

Unstable statistical models are ideally suited to being collected in an ensemble, as a great variety of different models can be constructed given a finite learning data set. This can be expressed in terms of statistical, computational and representational advantages in hypothesis space (Dietterich 2000b). Figure 5.5 illustrates these concepts.

A learning algorithm attempts to find the best approximation of the true function, given a specific training set, in a space of many possible hypotheses. Statistically, without sufficient data, the learning algorithm may find many different hypotheses

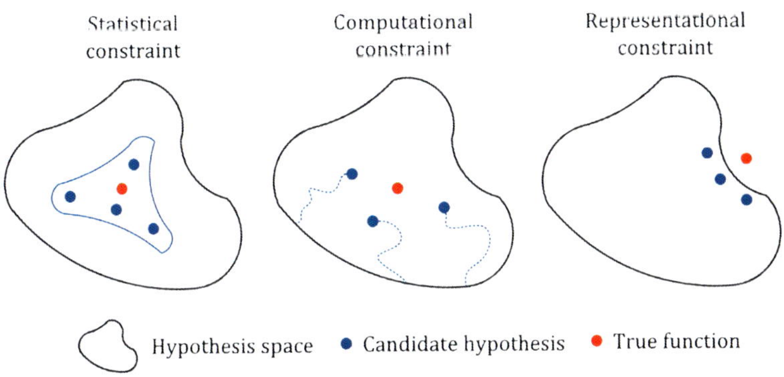

Fig. 5.5 Ensemble motivation based on hypothesis space concepts (Dietterich 2000b)

that have identical accuracies on the training set. By using all these hypotheses in an ensemble, the risk of choosing the wrong hypothesis is reduced. Computationally, iterative learning may get stuck in local optima in the hypothesis space. By repeating learning from different initial positions, local optima may be circumvented. Finally, the hypothesis space of a single model might fundamentally be unable to represent the true function. This representational problem may be skirted by combining models in an ensemble (Dietterich 2000b).

Decision trees are now considered in light of these hypothesis space constraints. Certain learning problems may require a large decision tree to give an accurate prediction, due to the complexity of the decision boundaries. The larger the decision tree, the larger the training data set required to ensure a good fit. If the sample size of the learning data set is insufficient, a number of trees may show the same training performance, although their generalization performance may differ. The rationale of combining models is then to reduce the risk of selecting a poor model (Polikar 2006). This represents the statistical constraint.

In terms of computational constraints, decision trees are unstable and may arrive at different hypothesis due to perturbations of the training data set. Multiple trees may then reflect multiple optima, with a combination of these models preventing the narrow view of a single locally optimal model.

The last constraint is the representational capacity of decision trees. The hyper-rectangular partitions of decision trees may hinder the successful representation of complex (especially diagonal) decision surfaces. However, combining many trees allows approximation of such structures (Dietterich 2000b); see Fig. 5.6.

5.3.2 *Ensembles of Decision Trees*

An early example of combining decision trees (albeit modified diagonal partition trees) employs different randomly selected subspaces of the original learning set to construct an ensemble of diverse trees (Ho 1995). Each tree is given only a subset of all input variables available in the learning set to use in split selection, and splits are defined perpendicular to vectors connecting class centres. Another approach (Breiman 1996) manipulates the learning set for each tree through bagging. Bagging is the term for bootstrap aggregating: taking a different bootstrapped sample with replacement from the learning set as modified learning set for each new tree and aggregating their predictions through voting or averaging.

Amit and Geman (1997) described a tree ensemble algorithm for shape recognition, where tree diversity is created by restricting splits to random subsets, as well as increasing the complexity of the input space with increasing tree depth. An alternative method for introducing diversity to trees is through split variable randomization (Dietterich 2000a): The top splits for each node are calculated, with the split variable randomly selected from a subset of the top-ranked variables. Breiman (2001) combined bagging and aspects of split randomization to create the random forest algorithm.

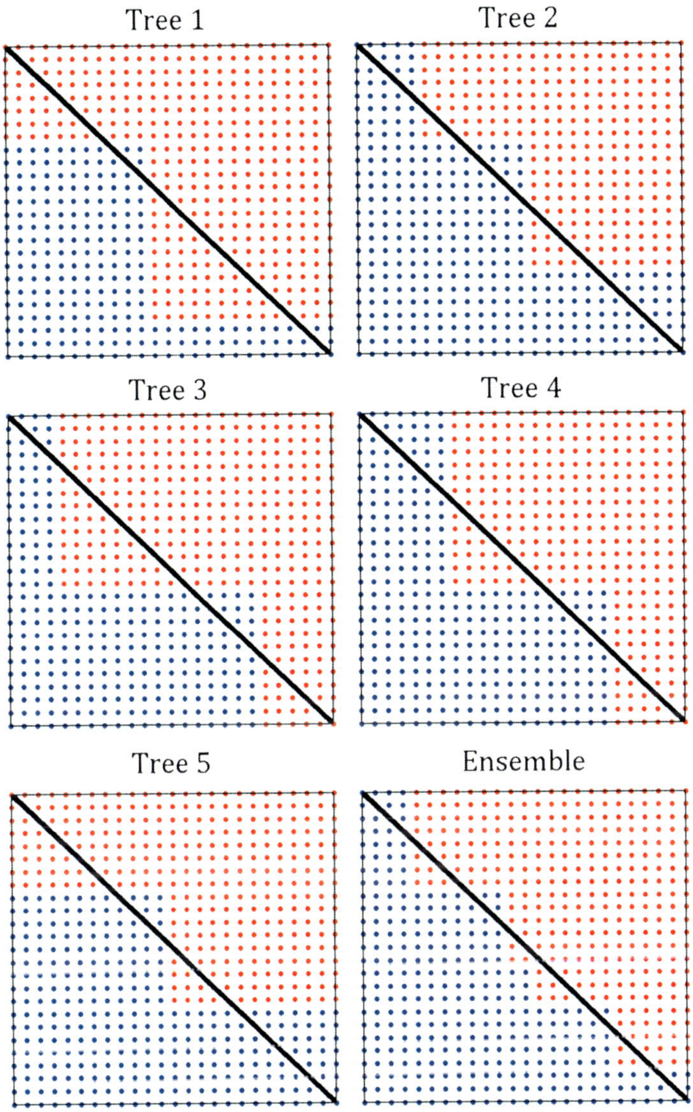

Fig. 5.6 Combination of classification decision boundaries in a majority-voting ensemble (*red* and *blue regions* represent model prediction, and *diagonal line* represents true decision boundary)

All these procedures can be expressed as combinations of decision trees such that the structure of each tree t_k depends on a random vector $\boldsymbol{\theta}_k$ that is independent of past $\boldsymbol{\theta}$ but belongs to the same distribution. The kth tree predictor is then $t_k = t(\boldsymbol{\theta}_k, \mathbf{X}, \mathbf{y})$ (Breiman 2001). For bagging trees, $\boldsymbol{\theta}$ represents the bootstrap sample identifier; for random subspace trees, $\boldsymbol{\theta}$ represents the selection of input vectors available for training.

It has been shown that the generalization error for a decision tree ensemble converges asymptotically to a limit as the number of trees in the ensemble increases, subject to accuracy and diversity constraints as mentioned before. This ensemble limit can be improved by minimizing correlation among members while maintaining their accuracy (Breiman 2001).

5.4 Random Forests

Breiman (2001) suggested an improvement on bagging decision tree ensembles by decreasing correlation through further randomization: employing random split selection. This ensemble, known as a random forest, is constructed by not only using different bootstrapped training sets for each tree but also by restricting the available split variables at each node to m randomly drawn input variables (Breiman 2001).

5.4.1 *Construction*

The construction of a random forest follows the following algorithm (Breiman 2001; Hastie et al. 2009):

Random Forest Classification and Regression Algorithm

- For $k = 1$ to K (size of ensemble)
  - Construct a bootstrap sample with replacement $\mathbf{X}^k$ from the learning set $\mathbf{X}$, of the same size as the learning set.
  - Grow a random forest tree t_k on $\mathbf{X}^k$ by employing the CART tree-growing algorithm, with the following modification at each node:
    - Select m random input variables from $\mathbf{X}^k$ to use as possible split variables.
    - Calculate best split position ς^* from these m variables.
    - Split the node into two child nodes.
  - Repeat the above tree-growing algorithm until the following stopping criteria are achieved:
    - A node size of one (for classification) or five (for regression)
    - A node with homogeneous class membership or response values
- Output the ensemble of trees $T = \{t_k\}_1{}^K$.
- A new prediction is made as follows:
  - The prediction of each tree $t_k(\mathbf{x})$ is calculated.
  - For classification, the majority vote over all K trees is assigned.
  - For regression, the average response value over all K trees is assigned.

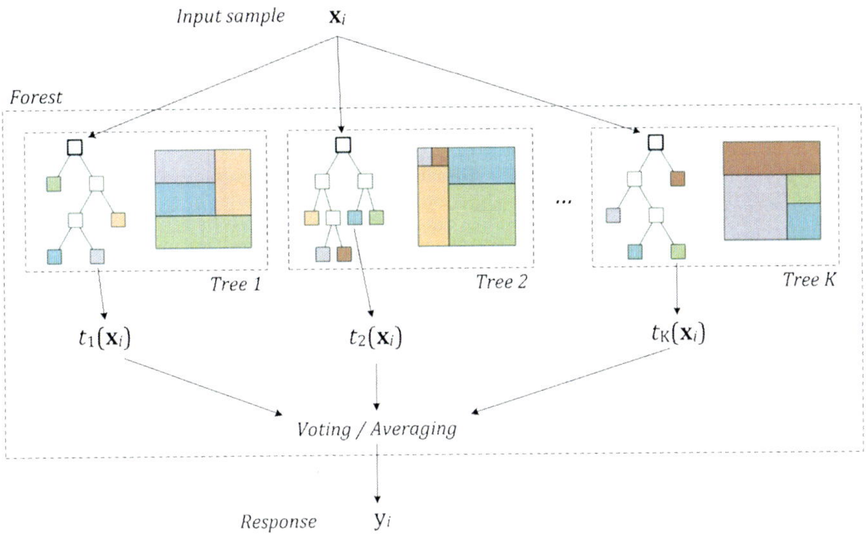

Fig. 5.7 Schematic of decision trees collected in an ensemble model

The only model parameters to be specified is the number of trees to grow (K) and the number of random split variables to consider at each split (m).

Figure 5.7 shows a schematic of a random forest model. Tree structures indicate if–then rules at each branching, with the associated subspace partitioning of a hypothetical two-dimensional space shown. Individual predictions from all trees are collected and combined as a single ensemble prediction by voting (for classification) or averaging (for regression).

5.4.2 *Model Accuracy and Parameter Selection*

The accuracy of a random forest can be candidly estimated using so-called out-of-bag data. As each tree is grown on a bootstrapped sample $\mathbf{X}^k$ of the original training set $\mathbf{X}$, a certain portion (about one-third) of samples will not have been used in training tree t_k. Aggregating votes over tree for only this out-of-bag (OOB) data can be used to calculate an honest estimate of generalization error (Breiman 2001).

Breiman (2001) derives measures of the strength and correlation of trees in a forest based on the margin function of a tree and forest (as applied to classification problems). The strength of a forest relates to the average tree accuracy, while correlation is a measure of tree diversity. The larger the correlation among trees, the lower their diversity.

Generally, two model parameters are required when constructing random forests: the number of trees K and the number of random split variables m. Other adjustable parameters include limitations on tree sizes, by specifying minimum sample sizes in leaf nodes, and tree depth.

As the generalization error generally approaches a limit asymptotically in terms of increasing number of trees (Breiman 2001), this parameter can be selected to be sufficiently large so that OOB error no longer improves significantly with additional trees. Breiman and Cutler (2003) suggest using large numbers of trees (1,000–5,000), especially if additional model properties are to be assessed.

In terms of the number of split variables, it is intuitive that choosing smaller values for m will increase the diversity (at least structural, if not in predictions) of trees (Hastie et al. 2009). However, the accuracy of a tree may be decreased for small m. Liaw and Wiener (2002) suggests using m equal to the floored square root of the number of inputs M for classification and $m/3$ for regression.

5.4.3 Model Interpretation

Although decision trees are easily interpreted due in part to their rule-based construction, this characteristic does not extend to random forests. As a forest consists of many different decision trees, a simple model structure is not evident (Breiman 2001).

However, random forests do contain a wealth of information that can be inspected. This includes variable importance, partial dependence and proximities.

Variable Importance

An approach to quantifying the importance of variables in function approximation is to permute the values of the predictor variables, one at a time, and determine the decrease in model accuracy for each variable. The permutation of a variable destroys its association with the response. If the prediction accuracy of the new model is significantly lower than for the original model, it implies that the association between predictor and response is significant. For random forests, the OOB samples can be permuted, without having to train new forests (Breiman and Cutler 2003; Archer and Kimes 2008). The variable importance measure based on permutation $\omega_j(t_k)$ is calculated according to Eq. 5.5 (where mse is the mean squared error of the model and $\mathbf{X}^k{}_{\mathrm{OOB}(j)}$ is the OOB input learning data for the kth tree with variable j permuted):

$$\omega_j(t_k) = mse\left(\mathbf{y}^k_{\mathrm{OOB}}, t_k\left(\mathbf{X}^k_{\mathrm{OOB}}\right)\right) - mse\left(\mathbf{y}^k_{\mathrm{OOB}}, t_k\left(\mathbf{X}^k_{\mathrm{OOB}(j)}\right)\right). \tag{5.5}$$

These variable importance measures can be expanded to ensembles of trees by averaging individual tree importance measures:

$$\omega_j = \frac{1}{k}\sum\nolimits_{k=1}^{K} \omega_j(t_k). \tag{5.6}$$

The advantage of permuted variable importance measures such as ω_j is that it considers multivariate interactions with other input variables, as permutation destroys association with other input variables as well. However, this also implies that variable importance measures are sensitive to the correlation structure of the predictor variables (Archer and Kimes 2008; Strobl et al. 2008; Nicodemus and Malley 2009).

Figure 5.8 shows a schematic of variable importance calculation based on random forest regression models. The calculation of variable importance for input variable $\mathbf{X}_j$ is demonstrated. The greyscale matrices represent the other input variables (excluding $\mathbf{X}_j$), while the greyscale column represents the response vector $\mathbf{y}$. The coloured vector represents the range of values of $\mathbf{X}_j$. A random forest model has been constructed on the original training data set, $\{\mathbf{X}, \mathbf{y}\}$. For each tree k, another learning set $\{\mathbf{X}^k{}_{\mathrm{OOB}(j)}, \mathbf{y}\}$ is constructed by randomly permuting the variable values of $\mathbf{X}^k{}_j$. Tree predictions of the response are then made for the original and permuted training sets, for all K trees of the ensemble and for all N samples. Per-tree variable importance for $\mathbf{X}_\mathrm{j}$ is calculated as the difference in mean squared error (*mse*) of the N samples of the original and permuted data. The average of the per-tree variable importance yields the overall forest variable importance for variable $\mathbf{X}_j$.

Partial Dependence

Once important variables have been identified, the nature of its influence on the response is considered. Friedman has suggested an approach to investigating the influence of a variable on a response (given any predictive model): the partial dependence plot (Friedman 2001). It is useful to inspect a plot of the predicted response for the range of values of a specific input variable, as averaged over all training values of the other input variables (known as the marginal average for a specific variable). Identification of important variables, combined with partial dependence plots for these important variables, provides a power visualization and interpretation tool for random forest models.

Nonlinear models do not lend themselves to simple direction of influence analysis as with linear models. The reason for this is inherent in the definition of a nonlinear model: Interactions and transformations of variables are taken into consideration. One cannot then simply say (as for linear models) that as a certain variable increases, the response would necessarily increase at a constant rate, for all possible values of all other variables. Correlation and interactions of variables add intricacies to the interpretation of influence.

Friedman has suggested an approach to investigating the influence of a variable on a response (given any predictive model): the partial dependence plot (Friedman 2001). It is useful to inspect a plot of the predicted response for the range of values of a specific input variable, as averaged over all training values of the other input variables.

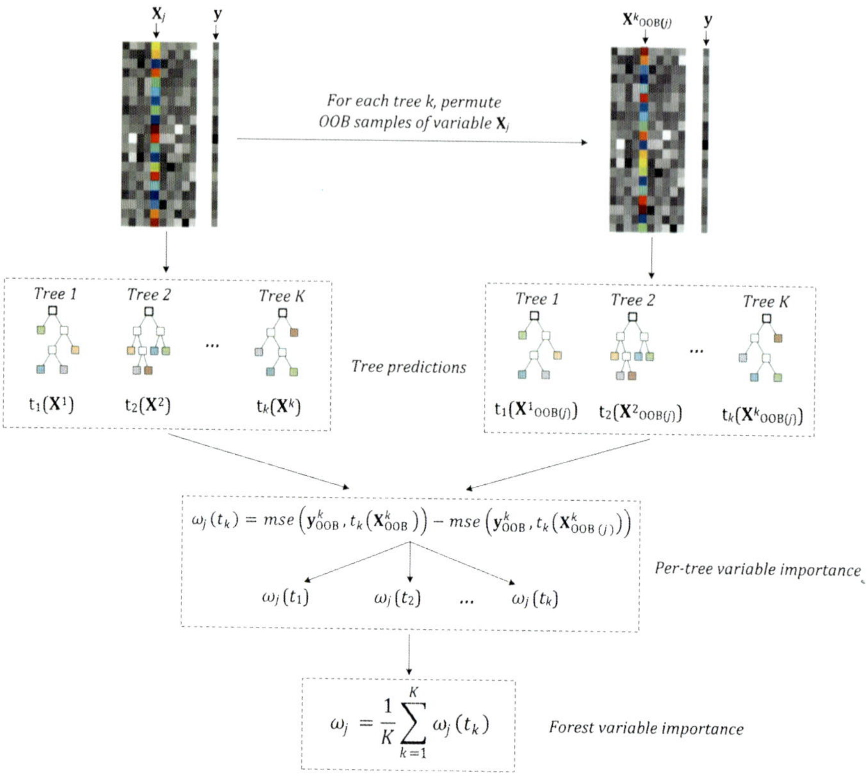

Fig. 5.8 Schematic of variable importance calculation

Suppose $\mathbf{X}$ is the input data and $f(\mathbf{X})$ is the predictor model for the response. Let $\mathbf{X}_S$ be a subset of the input variables of interest, and $\mathbf{X}_C$ all other input variables not included in $\mathbf{X}_S$. The model depends on all variables, $f(\mathbf{X}) = f(\mathbf{X}_S, \mathbf{X}_C)$, but the partial dependence of the predicted response to a subset of interest can be defined as the marginal average of the approximation function over $\mathbf{X}_C$ (Friedman 2001):

$$\bar{f}(\mathbf{X}_S) = \frac{1}{N}\sum_{i=1}^{N} f(\mathbf{X}_S, X_{i,C}) \tag{5.7}$$

with $X_{i,C}$ the values of samples in $\mathbf{X}_C$ in the training sample.

Figure 5.9 shows a schematic of partial dependence calculation for an input variable of interest. The partial dependence function is calculated for 30 equispaced values (b_1, b_2, … b_{30}) of a one-dimensional subset of interest $\mathbf{X}_S$. The coloured $\mathbf{X}_S$ vector represents the range of training data values for $\mathbf{X}_S$, while the greyscale $\mathbf{X}_C$ matrix represents the subset of input variables not of interest for this particular partial dependence calculation. To calculate the partial dependence function value at, for example, $\mathbf{X}_S = b_1$, forest predictions are made for every combination of

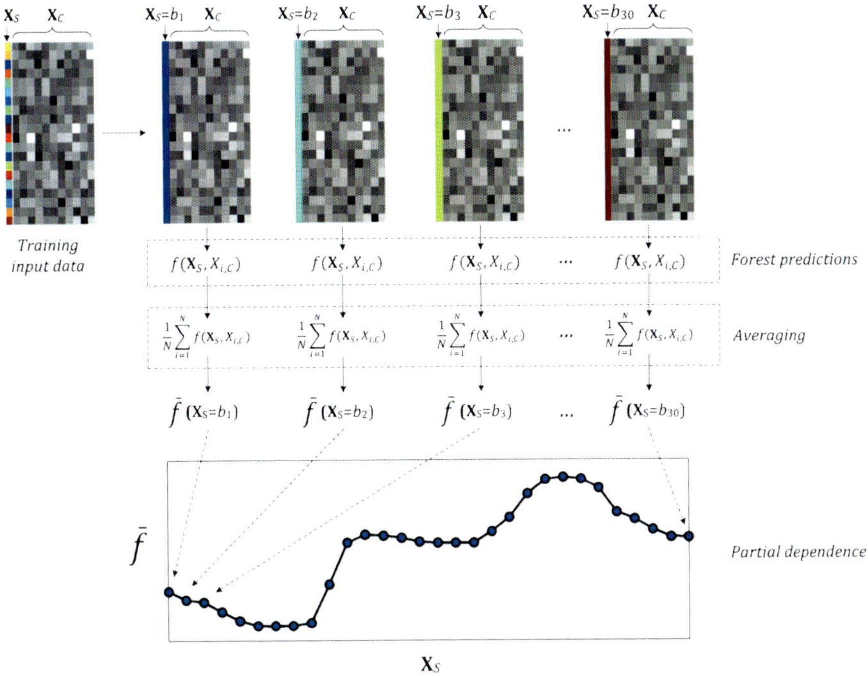

Fig. 5.9 Schematic showing partial dependence calculations for one variable of interest

$\mathbf{X}_S = b_1$ with the N original training samples in $\mathbf{X}_C$. The average of these forest predictions then represents the partial dependence function of $\mathbf{X}_S$ at b_1. These calculations are repeated for $\mathbf{X}_S = b_2$ to b_{30}, with the final result being the partial dependence function for $\mathbf{X}_S$ over its range of training data input values.

Partial dependence functions based on random forests are essentially marginal averages, of tree-averaged predictions, of local subspace averages of training samples. The ability of partial dependence functions to reflect the nonlinear influence of input variables on the response is then closely associated with the ability of the model to capture the structure of the input data.

Proximities

Another aid to data interpretation arising from random forests is found by inspecting the terminal node characteristics of all trees. By investigating whether samples report to the same leaf nodes, one can construct a proximity measure for all points. A leaf node essentially spans a hyperrectangle in the input space; if two samples report to the same hyperrectangle, they must be proximal. The algorithm for constructing a proximity matrix is summarized below (Breiman and Cutler 2003):

Random Forest Proximity Algorithm

- Construct a random forest model on learning set **X** with response **y**.
- Create an empty similarity (proximity) matrix **S**.
- For each tree $k = 1$ to K
  - For each sample combination (i, j), determine the terminal nodes to which they report.
  - If a sample combination (i, j) reports to the same terminal node, increase $S_{i,j}$ by one.
  - Repeat for all possible sample combinations.
- Scale the similarity matrix **S** by dividing by the number of trees K; the similarity matrix is symmetric and positive definite, with entries ranging from 0 to 1 and diagonal elements equal to 1.

This similarity matrix can be converted to a dissimilarity matrix:

$$\mathbf{D} = 1 - \mathbf{S} \quad \text{(Breiman and Cutler 2003)} \tag{5.8}$$

or

$$\mathbf{D} = \sqrt{(1 - \mathbf{S})} \quad \text{(Shi and Horvath 2006)} \tag{5.9}$$

The dissimilarity matrix may be considered as Euclidean distances in high-dimensional space and multidimensional scaling employed to obtain a lower-dimensional representation of the distances between data points (Breiman and Cutler 2003).

Given a dissimilarity matrix, multidimensional scaling (Cox and Cox 2001) attempts to find coordinates for data points that preserve the dissimilarity as Euclidian distances in the found coordinate space. The preservation of dissimilarities is measured by a stress function, the sum of squared differences between point dissimilarities and the new coordinate distances (Hastie et al. 2009). Classical multidimensional scaling finds an explicit solution to the stress function by utilizing eigenvalue decomposition. The eigenvectors are the MDS coordinates, while the eigenvalues give an indication of every coordinate's contribution to the squared distance between points. Smaller eigenvalues thus indicate less significant coordinates and can be inspected in order to select only significant coordinates (Cox and Cox 2001). Classical MDS is not considered a manifold technique, as it explicitly attempts to minimize all pairwise distances (Hastie et al. 2009). These MDS coordinates can be considered as random forest features to represent the data in low dimensions.

Hastie et al. (2009) have noted that random forest proximity plots look very similar for different classification data sets. Each class seem to be represented by one arm of a star, with points in pure class regions mapping to the extremities of the star, while points closer to the decision surface map near the centre of the star. They deduce that this is due to data points in pure class regions reporting to the same leaf nodes, while the uncertainty close to a decision surface reduces the likelihood of proximity between points. Even so, proximity plots are useful in classification and regression problem data visualization, especially in identifying cluster structure and separation (Cutler 2009).

Proximities generated using random forests do not only measure similarity but incorporate variable importance as well. As an example, samples with different values on certain input variables may still report to the same terminal nodes, if the said differences occur only in unimportant variables (Cutler 2009).

Decision trees in random forests are not pruned. This implies that proximities based on all data may overfit, as samples from different classes will have proximities of zero for a tree where they served in the training data set. A manner to circumvent this problem may be to only use OOB data to calculate proximities (Cutler 2009).

5.4.4 *Unsupervised Random Forests for Feature Extraction*

A feature extraction technique must be able to compute significant features for data sets where no response variables are present. This forms part of unsupervised learning, whereas supervised learning attempts to fit input variables to a given response. Random forest classification and regression are supervised learning methods and require some adaptation to be employed to the unsupervised case.

Unsupervised Learning as Supervised Learning

The application of supervised methods to unsupervised problems is discussed by Hastie et al. (2009). In their discussion, they use density estimation as an example of an unsupervised learning. The concepts discussed can be extended to the characterization of multivariate structure as an unsupervised problem.

If g is the unknown data density to approximate, let g_0 be any known reference distribution. A synthetic data set can be constructed by sampling g_0. A supervised task can now be assigned to build a model that can distinguish between data generated by g and g_0. The model predictions can be inverted to provide an estimate for g.[3] In essence, the supervised model provides estimates on the departure of g from g_0. The selection of g_0 is dictated by what type of departure is considered interesting. Departures from uniformity or multivariate normality can be investigated by choosing the corresponding distributions for g_0, while departures from independence between variables can be ascertained by choosing the product of marginal distributions for g_0. (The product of marginal distributions is simply the resampling, for each variable, from its own values, independent from other input variables.)

[3] See "The Elements of Statistical Learning" (Hastie et al. 2009) for details.

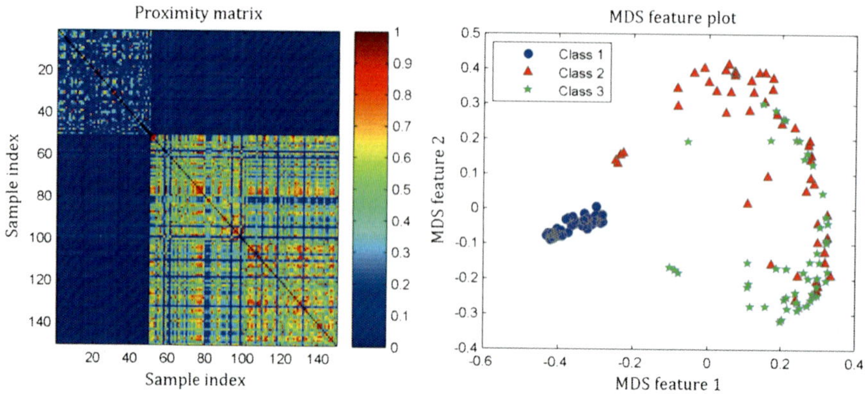

Fig. 5.10 Example of random forest proximity matrix and corresponding multidimensional scaling features for unsupervised three-class data set (class information added post priori to features)

Unsupervised Random Forests

The concept of unsupervised learning is further extended to random forests (Breiman and Cutler 2003), with a further study (Shi and Horvath 2006) investigating its properties extensively.[4] The unlabeled original data is given a class label, say 1. A synthetic contrast data set is created and given a different class label, say 2. A classification random forest can now be constructed which attempts to separate the two classes. From this forest, proximities can be calculated for the original unlabeled data, using the aforementioned proximity procedure. Obtaining dissimilarities from the similarity matrix, multidimensional scaling is applied to produce a set of features (coordinates) and their corresponding eigenvalues. These features now represent a projection of the original unlabeled data. An example of a proximity matrix and random forest features plot can be found in Fig. 5.10.

The random forest feature extraction algorithm is summarized below (Shi and Horvath 2006):

Random Forest Feature Extraction Algorithm

- For an unlabeled learning set $\mathbf{X}$, create a synthetic data set $\mathbf{X}^0$ by random sampling from the product of marginal distributions of $\mathbf{X}$
- Label the response of $\mathbf{X}$ inputs as class 1 and the response of $\mathbf{X}^0$ inputs as class 2

(continued)

[4]Shi and Horvath (2006) focused on the clustering utility of random forest proximities, a subtle difference to general feature extraction applications. Here, clustering refers to the ability of a feature extraction method to generate projections where known clusters are separate, without using cluster information in training.

(continued)

- Concatenate input matrices $\mathbf{X}$ and $\mathbf{X}^0$ as $\mathbf{Z}$, with concatenated response $\mathbf{y}$
- Construct a random forest classification model to predict $\mathbf{y}$ given $\mathbf{Z}$
- Create an empty similarity (proximity) matrix $\mathbf{S}$
- For each tree $k = 1$ to K
  - For each sample combination of $\mathbf{x}_{i.}$ and $\mathbf{x}_{j.}$, determine the terminal nodes to which they report
  - If a sample combination of $\mathbf{x}_{i.}$ and $\mathbf{x}_{j.}$, report to the same terminal node, increase $S_{i,j}$ by one
  - Repeat for all possible sample combinations
- Scale the similarity matrix $\mathbf{S}$ by dividing by the number of trees K; the similarity matrix is symmetric and positive definite, with entries ranging from 0 to 1, and diagonal elements equal to 1
- Determine the dissimilarity matrix $\mathbf{D} = 1 - \mathbf{S}$
- Use the dissimilarity matrix $\mathbf{D}$ as input to classical multidimensional scaling, retaining d scaling coordinates $\mathbf{T}$ as random forest features

Specifying Random Forest Feature Extraction Models

A number of decisions must be made when constructing an unsupervised random forest in order to extract features: what type of contrast distribution to use, the size of the forest (K) and the number of random split variables (M).

As with unsupervised learning applied to density estimation, the choice of the reference distribution for the synthetic data set determines what is considered as interesting structure. Shi and Horvath (2006) considered the uniform and product of marginal densities. If a uniformly distributed synthetic class is used, variables that show great departure from uniformity are often selected during tree construction. If, however, the product of marginal densities is used, trees will select splitting variables that are dependent on other variables.

It was found that generating a synthetic class using the product of marginal densities provided the most insightful scaling coordinates for a variety of data sets. However, redundant collinear variables may be detrimental to interpreting scaling plots (Shi and Horvath 2006).

Using the product of marginal densities to create a synthetic class destroys the multivariate structure of the original data. If the original data had little multivariate structure to start with, the synthetic data will be similar to the original data, and a random forest model will be similar to random guessing, with an accuracy close to 50 %. Higher accuracies may be indicative that the original data contained dependent multivariate structure (Cutler and Stevens 2006). This leads to the observation that, in general, there is an inverse relationship between the OOB error rate of an unsupervised random forest and the cluster signal in its proximity plot (Shi and Horvath 2006).

The elements of the similarity matrix are averaged over all trees in the forest, suggesting intuitively that the more trees, the more statistically sound the similarity. As previously discussed, random forest performance does not deteriorate as the number of trees increase. The only constraint on forest size is determined by computational resources. Shi and Horvath (2006) typically used 2,000 trees per forest for random forest feature extraction. However, they recommended that more than one forest should be grown and the resulting similarity matrices averaged. This ensures that the similarities are not dependent on a single manifestation of the synthetic contrast.

The clustering performance of random forest feature extraction was shown to be robust to the number of random split variables. Very low and very high values for M were not conducive to good clustering, while selecting M which corresponds to a low OOB error rate did not improve clustering (Shi and Horvath 2006).

Interpreting Random Forest Features

The presence of clusters in a random forest feature plot may not be indicative of the presence of clusters in the original data. It is also difficult to provide a geometric interpretation of random forest features (Shi and Horvath 2006). This may be due, in part, to the disjoint nature of decision tree partitions.

As with feature extractive methods such as Sammon mapping (Sammon 1969), random forest feature extraction does not provide a direct mapping function from input variables to features. Separate mapping and demapping regression functions must be learnt to project new data points.

5.4.5 *Random Forest Characteristics*

Random forest classification models are a surprisingly powerful technique (Hastie et al. 2009). This is the paraphrased response of the writers (Hastie et al. 2009) of The Elements of Statistical Learning to the performance of random forests on the Neural Information Processing Systems (NIPS) classification competition. The challenge required five data sets with 500–100,000 variables and 100–6,000 samples to be classified. Random forests obtained an average rank of 2.7 for classification accuracy and 1.9 for computational time, competing with the likes of Bayesian neural networks, boosted trees, boosted neural networks and bagged neural networks (Hastie et al. 2009). The creator of random forests (Breiman 2001) is the first to acknowledge that random forest regression is less successful than its classification counterpart.

The strengths of random forests are not limited to its classification accuracy but also extend to the very little tuning required (Hastie et al. 2009), its automatic calculation of an generalization error, an ability to handle missing data, variable importance measures and the application possibilities of proximities (Breiman and Cutler 2003).

However, variable importance measures do not substitute entirely for model interpretability. It has also been shown that random forest performance may be reduced when a large number of noisy variable are present and a small number of random split variables are used (Hastie et al. 2009).

As the training of individual ensemble members is self-contained, the random forest algorithm is so called embarrassingly parallel. Trees can be trained and employed for prediction concurrently, with ensemble outputs generated by collecting the contributions of individual members. This parallelizability suggests considerable computational optimization as parallel and distributed computing continues to develop.

5.5 Boosted Trees

Random forests exploit the diversity of independently trained decision trees to create a stronger model through committee voting and averaging. In contrast to this, boosting algorithms (Schapire 1990; Freund and Schapire 1996; Friedman et al. 2000; Friedman 2001, 2002) construct models sequentially, with each additional model building on the existing ensemble, in order to reduce the training error of the entire ensemble. Weak models are "boosted" to perform better, by, for example, focusing the attention of each iteratively added model on specific training samples that show large residuals or misclassification.

The original idea for improving weak learners originates from probably approximately correct (PAC) learning (Valiant 1984) and strengthening weak learners (Schapire 1990). Valiant (1984) views modelling in the context of a teacher–learner interaction, where a learner (an inductive modelling algorithm) tries to learn a certain concept based on randomly chosen examples of the said concept. A weak learner performs slightly better than random guessing, while a strong learner achieves a low error with high confidence. Schapire (1990) describes a method for converting a weak learning algorithm into a stronger learning algorithm, by modifying the distribution of the concept examples to emphasize the harder-to-learn examples.

Compared to related ensemble techniques such as bagging, random input spaces and random forests, boosting models form a unique ensemble approach of forward, stagewise learning.

5.5.1 *AdaBoost: A Reweighting Boosting Algorithm*

As an example, boosting is now described within the framework of the popular AdaBoost algorithm for classification (Freund and Schapire 1997; Izenman 2008). The training data consists of N samples of M input variables in $\mathbf{X}$ and the associated classification response vector $\mathbf{y}$, with y_i $(i = 1, \ldots, N) \in \{-1, 1\}$ representing two classes. The goal of the boosting algorithm is to create a weighted ensemble of

K weak classifiers (weighted by the K weights vector $\boldsymbol{\beta}$), where each new weak classifier is trained on a differently weighted training set (samples weighted by N weights vector $\mathbf{w}$). Sample weights are adjusted so that emphasis is placed on misclassified samples in each new weak classifier.

AdaBoost Boosting Algorithm

- Initialize the sample weights:

$$w_i = \frac{1}{N}, \quad i = 1, 2, \ldots, N.$$

- For $k = 1$ to K (size of ensemble):
  - Fit a classifier $f_k(\mathbf{x})$ to the training data $(\mathbf{X}, \mathbf{y})$ using the weight vector $\mathbf{w}$.
  - Calculate the current ensemble error:

$$\epsilon_k = \frac{\sum_{i=1}^{N} w_i I\left(y_i \neq f_k\left(\mathbf{x}_i\right)\right)}{\sum_{i=1}^{N} w_i}$$

    where I is the indicator function (1 if statement is true, 0 if statement is false).
  - Calculate the ensemble member weight:

$$\beta_k = \frac{1}{2} \ln\left(\frac{1 - \epsilon_k}{\epsilon_k}\right).$$

  - Update the sample weights:

$$w_i \leftarrow \frac{w_i}{W_k} e^{2\beta_k I\left(y_i \neq f_k(\mathbf{x}_i)\right)}, \quad i = 1, 2, \ldots, N$$

    where W_k is a normalizing constant to ensure $\sum_{i=1}^{N} w_i = 1$.
- Output the ensemble:

$$F(\mathbf{x}) = sign\left[\sum_{k=1}^{k} \beta_k f_k(\mathbf{x})\right]$$

where *sign* returns $+1$ if its argument is positive or zero and -1 if its argument is negative.

5.5.2 Gradient Boosting

In their statistical view of boosting, Friedman et al. (2000) showed that the AdaBoost algorithm above is equivalent to a forward stagewise additive model with an exponential loss function, where the minimization of the loss function is achieved by a coordinate-descent algorithm. First, the loss function is minimized with respect to the weak classifier function f_k by fixing β_k, and then, given f_k, the loss function is minimized with respect to β_k.

Friedman et al. (2000) proposed that a variety of boosting algorithms can be created by varying, firstly, how lack of fit is quantified (through loss functions) and, secondly, how the settings for the next iteration are defined (minimization procedure). A loss function quantifies the difference between the true value of a response vector and its approximation by some predictive function. For example, the squared-error loss function is well known from its prevalence in least-squares regression:

$$L(y_i, f(\boldsymbol{x}_i)) = \frac{1}{2}(y_i - f(\boldsymbol{x}_i))^2. \tag{5.10}$$

Friedman (2001) presented algorithms employing squared-error, absolute-error and Humber loss functions for regression, with steepest-descent minimization, and further extended their application to decision trees. For classification, the exponential (as in the original AdaBoost), binomial and squared-error loss functions can be used. These loss functions are continuous and convex and thus suitable for gradient boosting.

Given a loss function $L(y, f(\mathbf{x}))$ for the regressor/classifier $f(\mathbf{x})$, the direction of the optimization step at iteration k is determined from the gradient $g_k(\mathbf{x})$ at the current iteration, while the step size ρ_k is determined by a line search along the gradient to minimize the loss function (Izenman 2008):

$$g_k(\mathbf{x}_i) = \left.\frac{\partial L(y_i, f(\mathbf{x}_i))}{\partial f(\mathbf{x}_i)}\right|_{f(\mathbf{x}_i)=f_{(k-1)}(\mathbf{x}_i)}, i = 1, 2, \ldots, N \tag{5.11}$$

$$\rho_k = \arg\min_\rho \sum_{i=1}^{N} L\left(y_i, f_{(k-1)}(\mathbf{x}_i) - \rho g_k(\mathbf{x}_i)\right). \tag{5.12}$$

The regression/classification function $f(\mathbf{x})$ is then updated:

$$f_k(\mathbf{x}) = f_{k-1}(\mathbf{x}) - \rho_k g_k(\mathbf{x}). \tag{5.13}$$

An algorithm for gradient boosting is shown here (Izenman 2008):

Gradient Boosting Algorithm

- For a training set $\{\mathbf{x}_i, y_i\}$ with $i = 1, 2, \ldots, N$ and loss function $L(y, \mathbf{x})$:
  - Initialize the ensemble model as a constant:

$$f_0(\mathbf{x}) = \arg min_\rho \sum_{i=1}^{N} L(y_i, \rho).$$

- For $k = 1$ to K (size of ensemble):
  - Determine the working response (the negative gradient of the loss function)

$$z_i = -g_k(\mathbf{x}_i) = -\left.\frac{\partial L(y_i, f(\mathbf{x}_i))}{\partial f(\mathbf{x}_i)}\right|_{f(\mathbf{x}_i)=f_{k-1}(\mathbf{x}_i)}, i = 1, 2, \ldots, N.$$

  - Use least-squares minimization to determine the best parameters $\boldsymbol{\theta}_k$ for the current member model $h(\mathbf{x}, \boldsymbol{\theta}_k)$ and its weight in the ensemble, β_k

$$(\boldsymbol{\theta}_k, \beta_k) = \arg min_{\boldsymbol{\theta},\beta} \sum_{i=1}^{N} (z_i - \beta h(\mathbf{x}_i, \boldsymbol{\theta}))^2.$$

  - Determine the optimal step size along the gradient ρ_k

$$\rho_k = \arg min_\rho \sum_{i=1}^{N} L(y_i, f_{k-1}(\mathbf{x}_i) + \rho h(\mathbf{x}_i, \boldsymbol{\theta}_k))$$

  - Update the ensemble model

$$f_k(\mathbf{x}) = f_{k-1}(\mathbf{x}) + \beta_k h(\mathbf{x}, \boldsymbol{\theta}_k).$$

- Output the ensemble:

$$F(\mathbf{x}) = f_k(\mathbf{x}) = \sum_{k=1}^{K} \beta_k h(\mathbf{x}, \boldsymbol{\theta}_k).$$

5.5.3 *Model Accuracy*

Boosting algorithms have been shown to be very accurate in learning problems with low noise, while prediction accuracy is lower when higher levels of noise are present

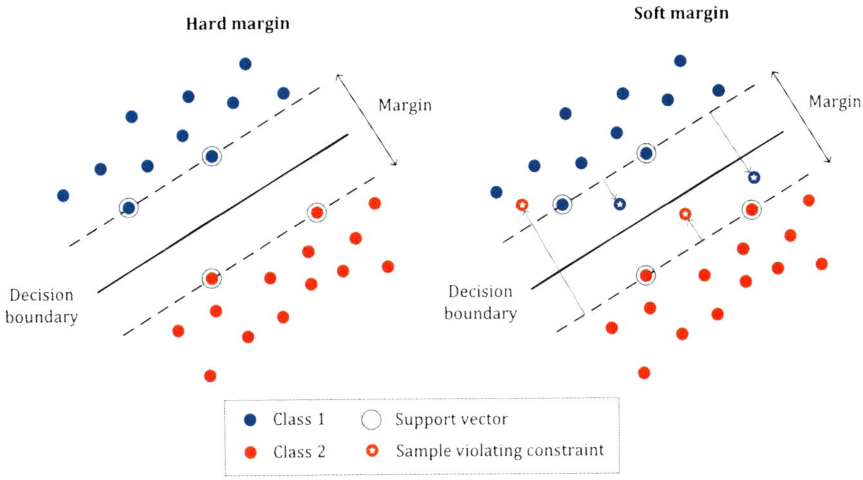

Fig. 5.11 Hard and soft margins for a classification problem

(Ratsch et al. 2001). An explanation for the good performance of classification boosting algorithms for low noise situations has been given in terms of margin theory (Schapire et al. 1998).

Boosting and Maximal Margins

Even when the boosting ensemble training error for a specific learning set has reached and maintained a minimum with the addition of ensemble members, certain data sets have shown continued decrease in the test (generalization) error. This characteristic relates to the classification margin of the trained boosting ensemble. A classification margin can be interpreted as the distance of an instance from the decision boundary between different classes. The greater the distance of a sample from the decision boundary, the more confidence a classification function can assign in its prediction.

It has been shown (Schapire et al. 1998) that the AdaBoost algorithm attempts to find a decision boundary that maximizes the margin for the training samples. The AdaBoost margin is defined in terms of the difference in number of votes for the correct class and the largest incorrect class but is analogous to the classification margin described above. In effect, the AdaBoost algorithm concentrates on hard-to-learn patterns close to the decision boundary, similar to support vector machines (see Chap. 5). However, this "hard margin" approach is not suitable for noisy data, where samples may be found on the "wrong" side of the margin due to overlapping distributions or mislabelling (see Fig. 5.11).

In Fig. 5.11, support vectors define the linear hyperplanes separating the classes. In the case of a hard margin, no training samples are allowed in the margin area.

However, mislabelling and overlapping distributions result in classes that are not linearly separable. By allowing a number of samples to violate the margin constraint, a soft margin can be defined.

Soft margin approaches, which introduce weighted margin maximization depending on the influence of training samples (Ratsch et al. 2001), have been introduced for AdaBoost to allow for better representation of noisy data sets.

Regularization

Regularization is another weapon against overfitting, which involves restricting the complexity of the base models, the size of the ensemble, or shrinking ensemble member weights (Izenman 2008; Hastie et al. 2009). Applying a bagging strategy for ensemble member training data sets can also improve robustness against overfitting (Friedman 2002).

5.5.4 *Model Interpretation*

As with random forests, variable importance and partial dependence methods can be employed to aid interpretation of boosted tree ensembles. It has been suggested (Friedman et al. 2000; Hastie et al. 2009) that the depth of trees employed in a boosting ensemble controls the level of variable interaction that is modelled. For example, tree stumps (only one level of splitting) can model only main effects, trees with a depth of two splits can model first-order interactions, etc. However, complex data sets may contain high-order interactions, and limiting tree depth may restrict the attainable accuracy of the entire ensemble (Izenman 2008). Larger trees may then allow more accurate models, at the cost of reduced interpretation through direct relation of tree depth to level of interaction.

5.6 Concluding Remarks

Decision trees are versatile, computationally inexpensive and data adaptive. Where single trees may overfit training data, this can be mitigated by combining trees in ensembles through bagging, random split selection, boosting and combinations thereof. Although the benefits of single tree interpretation are lost when they are collected in ensembles, tools such as variable importance and partial dependence calculations allow an appreciation of the relation between response and input variables. The wealth of information contained in an ensemble of tree structures can be further exploited through proximity measures and more, to extend these supervised learning approaches to unsupervised learning challenges.

5.7 Code for Tree-Based Classification

To demonstrate tree-based concepts introduced in this chapter, MATLAB code is given here for simple classification modelling. Functions are presented to calculate impurity changes when splits are made on a variable, for training of and prediction with tree stumps and the implementation of this tree stump for bagging, random forest and boosting ensemble training and prediction.

The impurity indices in the following code take a sample weight vector **w** into account, allowing the emphasis of classification performance to be preferentially distributed, as well as adaptively change (as with the AdaBoost algorithm).

The weighted Gini impurity $i_w(\eta)$ in node η is then

$$i(\eta) = 1 - \sum_{k=1}^{C} \left(\frac{Q(k|\eta)}{W(\eta)} \right)^2 \tag{5.14}$$

while the weighted cross-entropy $i_w(\eta)$ is

$$i(\eta) = -\sum_{k=1}^{C} \left(\frac{Q(k|\eta)}{W(\eta)} \right) \log \left(\frac{Q(k|\eta)}{W(\eta)} \right) \tag{5.15}$$

where $Q(k|\eta)$ represents the sum of weights for those samples in node η labelled as class k:

$$Q(k|\eta) = \sum_{x_i \in \eta} w_i I(y_i = k) \tag{5.16}$$

and $W(\eta)$ the sum of all sample weights present in node η:

$$W(\eta) = \sum_{x_i \in \eta} w_i \tag{5.17}$$

The weighted impurity change for splitting a parent node into left η_L and right η_R children nodes is then

$$\Delta i(\varsigma, \eta) = i(\eta) - \frac{W(\eta_R)}{W(\eta)} i(\eta_R) - \frac{W(\eta_L)}{W(\eta)} i(\eta_L) \tag{5.18}$$

5.7.1 *Example: Rotogravure Printing*

The tree stump model as well as bagging, random forest and AdaBoost ensemble methods given above was applied to a classification problem: discerning between banding (a type of printing cylinder defect) and no banding of conditions in a rotogravure printing application (Evans and Fisher 1994). The data set consists of

Table 5.1 Tree ensemble performance on rotogravure printing data set

Training error	Random set 1	Random set 2	Test error	Random set 1	Random set 2
Tree stump	0.389	0.329	*Tree stump*	0.444	0.370
Bagging	0.370	0.329	*Bagging*	0.370	0.370
Random forest	0.319	0.324	*Random forest*	0.278	0.370
AdaBoost	0.232	0.588	*AdaBoost*	0.296	0.593

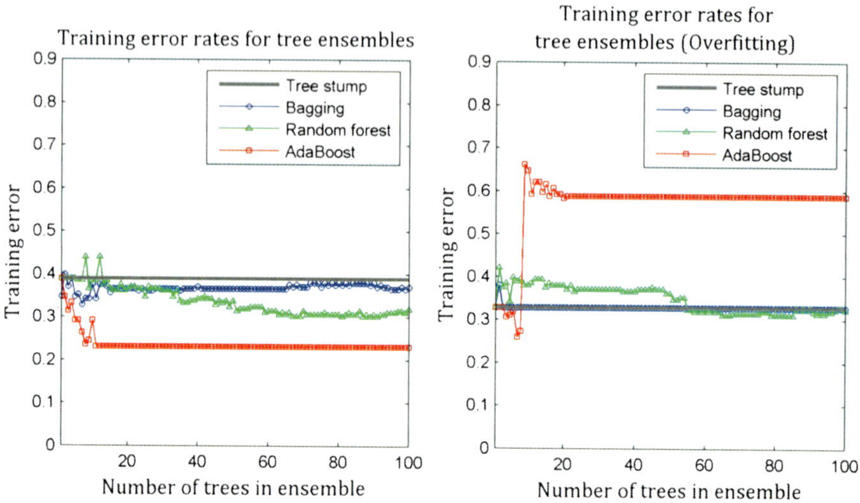

Fig. 5.12 Tree ensemble performance on rotogravure printing data set, based on two different random subsets of the training data

270 samples of 19 continuous input variables, characterizing various operating conditions: ink viscosity and temperature, humidity, blade pressure, solvent percentage and more. For more details, refer to the UCI Machine Learning repository (Frank and Asuncion 2010). Tree stump models and tree ensembles were trained on training data consisting of random subsets of 80 % of the data set, with 20 % retained as test sets. Where applicable, a bootstrap fraction of 0.5 was used, and the number of random split variables for random forests was selected as discussed earlier. The results for two random sets are reported in Table 5.1 and Fig. 5.12.

From Table 5.1, the ensemble approaches improved on the single tree stump model for random set 1, but (generally, excluding random forest) not for random set 2. The random forest and AdaBoost methods did well for random set 1. The improvement of error with random forest, compared to bagging, suggests that the added diversity introduced by random split selection strengthened the ensemble.

For random set 2, bagging and random forest could not improve significantly on the single tree stump (for both training and testing error), while AdaBoost showed decreased performance compared to the single tree stump. As can be seen in Fig. 5.12, the AdaBoost method initially showed low training error with less than 10 ensemble members, but the error dramatically increases thereafter. This could indicate that random subset 2 may have a soft margin, with the emphasis on noisy/mislabelled samples by AdaBoost reweighting causing dramatic decrease in performance.

5.7.2 *Example: Identification of Defects in Hot Rolled Steel Plate by the Use of Random Forests*

As an example of the application of tree-based models, the identification of defects in hot rolled steel plates on an industrial plant is considered. In this case, historic plant data are used to construct a model that can aid in the identification of process operating conditions that contribute to defects in the steel plate, such as inclusions and delamination of the steel.

A data set consisting of 26 operating variables and 3,017 sample was used to calibrate an ensemble of boosted trees. Each sample represented a rolled steel plate that was inspected and identified as normal (i.e. no defects) or defective (containing a defect of one type or another). The process conditions that were considered in the production of the steel plates are summarized in Table 5.2.

The random forest model was constructed with the parameters shown in Table 5.3. The misclassification rates during training and testing of the models based on the Gini information criterion are shown in Fig. 5.13. The random forest could predict the quality of the steel plates with an overall accuracy of approximately 76.6 %. Defect-free plates could be identified with an accuracy of 89.9 %, while defective plates could be identified in 51.7 % of the cases. Although not highly reliable, the performance of the random forest model is significant, given that 34.3 % of the samples in the data were defective.

Figure 5.14 shows the relative contributions of the operating variables. In this figure, the horizontal red line represents the 99 % confidence limit that was generated by Monte Carlo simulation of the random forest model, as described by Auret and Aldrich (2012). The mass loss of the steel plates during grinding (x_3) was the most important variable, while other significant variables included the final product thickness (x_{22}) and the rinse end temperature (x_{16}). With the exception of x_3, the sequence of the variables in Fig. 5.14 should not be considered as particularly significant, since these variables are highly correlated and changes in the model could result in changes in the specific sequence or even the significance of the variables close to the confidence limit.

Finally, as the projection of the data in Fig. 5.15 by the use of a linear discriminant analysis model indicates (using the significant variables shown in Fig. 5.14 only), the operating variables considered in the model were not strongly discriminative with regard to the quality of the steel plates manufactured during the hot roll process. The samples, with red dots indicating defect-free plates and black open circles defective plates, show strong overlap, which would probably make it difficult to achieve a highly reliable model with the given set of variables.

Table 5.2 Process operating variables used in prediction of quality of hot rolled steel plate

Variable	Comment	Predictor type
x_1	Level of sulphur in the steel (%)	Continuous
x_2	Level of phosphorus in the steel (%)	Continuous
x_3	Mass loss in steel slab owing to grinding (%)	Continuous
x_4	Time spent in reheat furnace (h)	Continuous
x_5	Type of powder used in mould (−)	Categorical
x_6	Width (mm)	Continuous
x_7	Casting speed of steel (m/min)	Continuous
x_8	Mould level fluctuation	Continuous
x_9	Stopper movement	Continuous
x_{10}	Tundish steel mass (ton)	Continuous
x_{11}	Superheat of steel (°C)	Continuous
x_{12}		Continuous
x_{13}	Mould heat removal variables (−)	Continuous
x_{14}		Continuous
x_{15}	Silicon level of steel in ladle (%)	Continuous
x_{16}	Rinse end temperature (°C)	Continuous
x_{17}	Time between end rinse and start of cast (h)	Continuous
x_{18}	Rinse station operational variable (h)	Continuous
x_{19}	Rinse station stir parameter	Continuous
x_{20}	Calculated Ti_3O_5 solubility product	Continuous
x_{21}	Calculated TiN solubility product	Continuous
x_{22}	Final product thickness (mm)	Continuous
x_{23}	Ladle sequence during casting	Categorical
x_{24}		Continuous
x_{25}	Temperature variables related to heat removal (°C)	Continuous
x_{26}		Continuous

Table 5.3 Parameters for construction of random forest model

No. of trees	No. of predictors	Random test set proportion (%)	Minimum samples per node
200	5	30	15

5.8 Fault Diagnosis with Tree-Based Models

Although tree-based models can be used in different ways to set up reverse models mapping features to outputs to reconstruct the process data, they are comparatively limited as far forward mapping or feature extraction from the data themselves are concerned, unless used in conjunction with other feature extraction algorithms, such as Sammon maps that require additional models to be useful in diagnostic schemes.

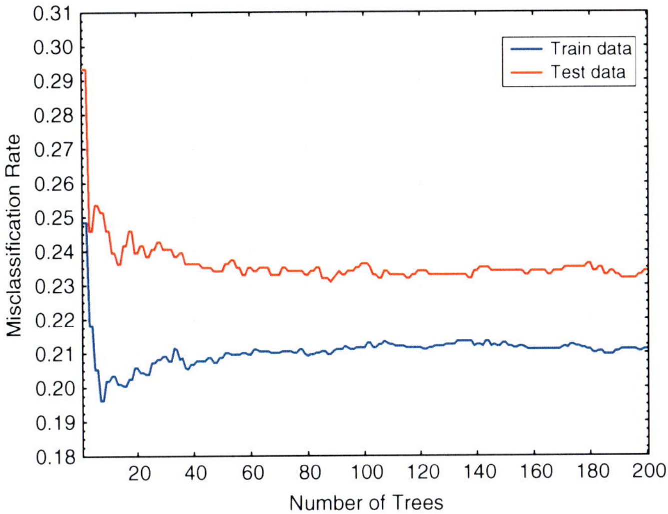

Fig. 5.13 Performance of random forests as a function of ensemble size

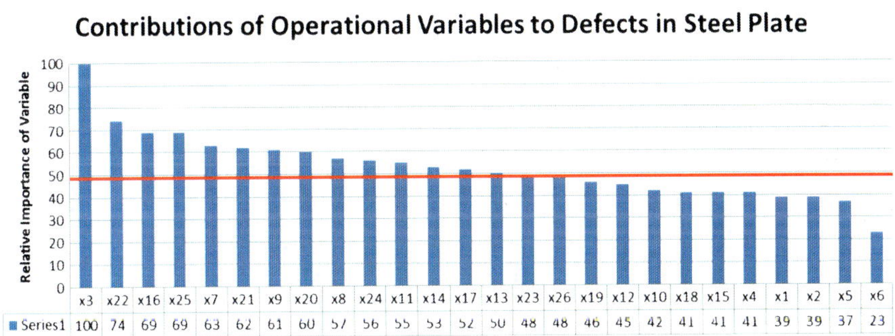

Fig. 5.14 Relative importance of variables contributing to defects in the steel plate. The *horizontal red line* represents the 99 % confidence limit

As was discussed in Sect. 5.4.4, unsupervised learning or feature extraction with tree ensemble models is based on setting up synthetic contrasts in the data. This approach could also be used with classification trees and some other machine learning methods, such as neural networks, and further investigation of this approach to unsupervised learning is required. Figure 5.16 gives a conceptual summary of the generalized framework for constructing fault diagnostic systems with tree-based methods.

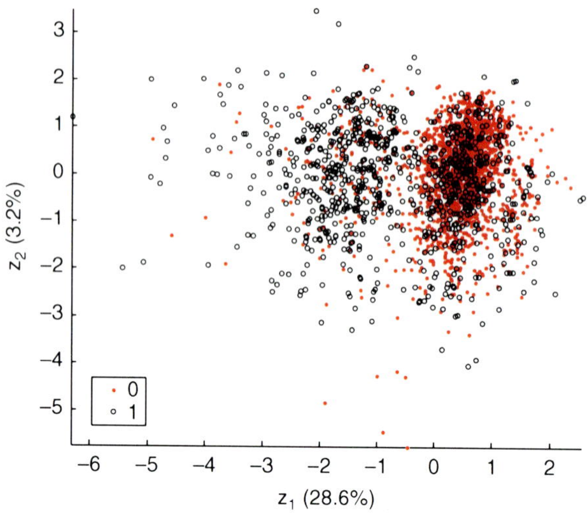

Fig. 5.15 Linear projection of the steel plant data by the use of linear discriminant analysis of the significant variables shown in Fig. 5.14

Fig. 5.16 Fault diagnosis with tree-based models, e.g. unsupervised random forests (*un-RF*), unsupervised boosted trees (*un-BT*), regression trees (*RT*), random forests (*RF*) and boosted trees (*BT*)

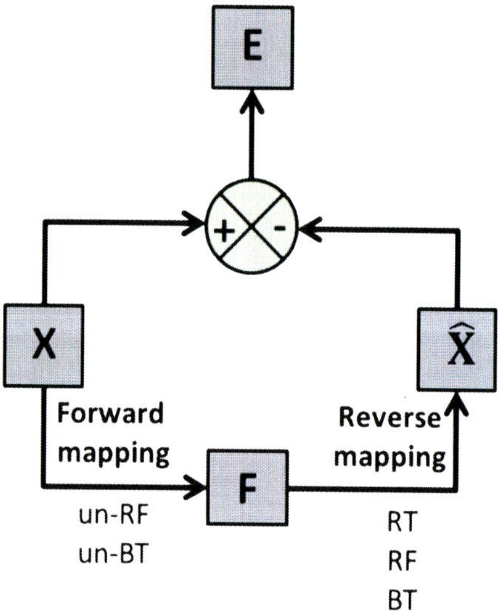

References

Amit, Y., & Geman, D. (1997). Shape quantization and recognition with randomized trees. *Neural Computation, 9*(7), 1545–1588.

Archer, K. J., & Kimes, R. V. (2008). Empirical characterization of random forest variable importance measures. *Computational Statistics & Data Analysis, 52*(4), 2249–2260.

Auret, L., & Aldrich, C. (2012). Interpretation of nonlinear relationships between process variables by use of random forests. *Minerals Engineering, 35*, 27–42.

Belson, W. A. (1959). Matching and prediction on the principle of biological classification. *Journal of the Royal Statistical Society Series C (Applied Statistics), 8*(2), 65–75.

Breiman, L. (1996). Bagging predictors. *Machine Learning, 24*(2), 123–140.

Breiman, L. (2001). Random forests. *Machine Learning, 45*(1), 5–32.

Breiman, L., & Cutler, A. (2003). *Manual on setting up, using, and understanding random forests v4.0*. Available at: http://oz.berkeley.edu/users/breiman/Using_random_forests_v4.0.pdf. Accessed 30 May 2008.

Breiman, L., Friedman, J. H., Olshen, R., & Stone, C. J. (1984). *Classification and regression trees*. Belmont: Wadsworth.

Cox, T. F., & Cox, M. A. A. (2001). *Multidimensional scaling*. Boca Raton: Chapman & Hall.

Cutler, A. (2009). Random forests. In *useR! The R User Conference 2009*. Available at: http://www.r-project.org/conferences/useR-2009/slides/Cutler.pdf

Cutler, A., & Stevens, J. R. (2006). Random forests for microarrays. In A. Kimmel & B. Oliver (Eds.), *Methods in enzymology; DNA microarrays, Part B: Databases and statistics* (vol. 441, pp. 422–432). San Diego: Academic Press. ISSN 0076-6879, ISBN 9780121828165, 10.1016/S0076-6879(06)11023-X, http://www.sciencedirect.com/science/article/pii/S007668790611023X

Dietterich, T. G. (2000a). An experimental comparison of three methods for constructing ensembles of decision trees: Bagging, boosting, and randomization. *Machine Learning, 40*(2), 139–157.

Dietterich, T. (2000b). Ensemble methods in machine learning. In *Multiple classifier systems* (Lecture notes in computer science, pp. 1–15). Berlin/Heidelberg: Springer. Available at: http://dx.doi.org/10.1007/3-540-45014-9_1.

Evans, B., & Fisher, D. (1994). Overcoming process delays with decision tree induction. *IEEE Expert, 9*(1), 60–66.

Fayyad, U. M., & Irani, K. B. (1992). On the handling of continuous-valued attributes in decision tree generation. *Machine Learning, 8*(1), 87–102.

Frank, A., & Asuncion, A. (2010). *UCI machine learning repository*. University of California, Irvine, School of Information and Computer Sciences. Available at: http://archive.ics.uci.edu/ml

Freund, Y., & Schapire, R. E. (1996). Experiments with a new boosting algorithm. In *Machine Learning. Proceedings of the Thirteenth International Conference (ICML'96)*| (pp. 148–156|558).

Freund, Y., & Schapire, R. E. (1997). A decision-theoretic generalization of on-line learning and an application to boosting. *Journal of Computer and System Sciences, 55*(1), 119–139.

Friedman, J. H. (2001). Greedy function approximation: A gradient boosting machine. *The Annals of Statistics, 29*(5), 1189–1232.

Friedman, J. H. (2002). Stochastic gradient boosting. *Computational Statistics & Data Analysis, 38*(4), 367–378.

Friedman, J., Hastie, T., & Tibshirani, R. (2000). Additive logistic regression: A statistical view of boosting. *The Annals of Statistics, 28*(2), 337–374.

Gillo, M. W., & Shelly, M. W. (1974). Predictive modeling of multivariable and multivariate data. *Journal of the American Statistical Association, 69*(347), 646–653.

Hansen, L., & Salamon, P. (1990). Neural network ensembles. *IEEE Transactions on Pattern Analysis and Machine Intelligence, 12*(10), 993–1001.

Hastie, T., Tibshirani, R., & Friedman, J. (2009). *The elements of statistical learning – Data mining, inference and prediction*. New York: Springer.

Ho, T. K. (1995). Random decision forests. In *Proceedings of the Third International Conference on Document Analysis and Recognition* (pp. 278–282). ICDAR1995. Montreal: IEEE Computer Society.

Izenman, A. (2008). *Modern multivariate statistical techniques: Regression, classification, and manifold learning*. New York/London: Springer.

Kass, G. V. (1980). An exploratory technique for investigating large quantities of categorical data. *Journal of the Royal Statistical Society Series C (Applied Statistics), 29*(2), 119–127.

Liaw, A., & Wiener, M. (2002). Classification and regression by random forest. *R News, 2*(3), 18–22.
Messenger, R., & Mandell, L. (1972). A modal search technique for predictive nominal scale multivariate analysis. *Journal of the American Statistical Association, 67*(340), 768–772.
Morgan, J. N., & Sonquist, J. A. (1963). Problems in the analysis of survey data, and a proposal. *Journal of the American Statistical Association, 58*(302), 415–434.
Nicodemus, K. K., & Malley, J. D. (2009). Predictor correlation impacts machine learning algorithms: Implications for genomic studies. *Bioinformatics, 25*(15), 1884–1890.
Polikar, R. (2006). Ensemble based systems in decision making. *Circuits and Systems Magazine, IEEE, 6*(3), 21–45.
Quinlan, J. (1986). Induction of decision trees. *Machine Learning, 1*(1), 81–106.
Quinlan, R. (1993). *C4.5: Programs for machine learning*. Palo Alto: Morgan Kaufmann.
Ratsch, G., Onoda, T., & Muller, K. (2001). Soft margins for AdaBoost. *Machine Learning, 42*(3), 287–320.
RuleQuest Research. (2011). Data mining tools See5 and C5.0. *Information on See5/C5.0*. Available at: http://www.rulequest.com/see5-info.html. Accessed 10 Feb 2011.
Sammon, J. W. (1969). A nonlinear mapping for data structure analysis. *IEEE Transactions on Computers, C-18*(5), 401–409.
Schapire, R. E. (1990). The strength of weak learnability. *Machine Learning, 5*(2), 197–227.
Schapire, R., Freund, Y., Bartlett, P., & Lee, W. (1998). Boosting the margin: A new explanation for the effectiveness of voting methods. *The Annals of Statistics, 26*(5), 1651–1686.
Shi, T., & Horvath, S. (2006). Unsupervised learning with random forest predictors. *Journal of Computational and Graphical Statistics, 15*(1), 118–138.
Strobl, C., Boulesteix, A., Kneib, T., Augustin, T., & Zeileis, A. (2008). Conditional variable importance for random forests. *BMC Bioinformatics, 9*(1), 307–317.
Strobl, C., Malley, J., & Tutz, G. (2009). An introduction to recursive partitioning: rationale, application, and characteristics of classification and regression trees, bagging, and random forests. *Psychological Methods, 14*(4), 323–348.
Valiant, L. G. (1984). A theory of the learnable. *Communications of the ACM, 27*(11), 1134–1142.

Nomenclature

Symbol	Description
$i(\eta)$	Impurity measure of node η in a classification tree
$p(k\|\eta)$	Proportion of samples in class k at node η in a classification tree
C	Number of classes in a classification problem
$\Delta i(\varsigma, \eta)$	Decrease in impurity for a candidate split position ς at node η
ς	Split point index in a decision tree
$\varsigma*$	Optimal split point index in a decision tree
η	Node index variable
$\boldsymbol{x}$	Input space
N	Sample size
p_R	Proportion of samples reporting to the right descendent node after splitting in a classification tree
p_L	Proportion of samples reporting to the left descendent node after splitting in a classification tree
c_R	Prediction in right descendent node of a regression tree after splitting
c_L	Prediction in left descendent node of a regression tree after splitting

(continued)

(continued)

Symbol	Description
η_R	Index of right descendent node
η_L	Index of left descendent node
$\mathbf{X}^k$	kth bootstrap sample of learning data set **X**
T	Set of ensemble trees
t_k	kth tree in an ensemble of trees
K	Number of trees in an ensemble of classification or regression trees
$t_k(\cdot)$	Prediction of kth tree in an ensemble of classification or regression trees
m	Number of variables considered at each split point in a random forest tree
M	Total number of input variables
$\mathbf{X}^k_{\mathrm{OOB}(j)}$	Out-of-bag (OOB) input learning data for the kth tree in an ensemble of trees with variable j permuted
$\mathbf{y}^k_{\mathrm{OOB}}$	Out-of-bag (OOB) output learning data for the kth tree in an ensemble of trees
$\omega_j(t_k)$	Importance measure for jth variable in kth tree in an ensemble of trees (random forest)
ω_j	Importance measure for jth variable in an ensemble of trees (random forest)
$\mathbf{X}_S$	Subset of variables in **X**
$\mathbf{X}_C$	Subset of variables in **X** complementary to X_S
$X_{i.C}$	Values of samples in $\mathbf{X}_C$
$\bar{f}(\mathbf{X}_S)$	Partial dependence of a predicted response to the subset of variables in $\mathbf{X}_S$
b_j	jth scalar calculation point
$\mathbf{S}$	Proximity matrix
$\mathbf{D}$	Dissimilarity matrix
g	Unknown data density
g_0	Reference distribution
$\mathbf{X}^0$	Synthetic data set obtained by random sampling from the product of marginal distributions in **X**
$\mathbf{Z}$	Concatenated matrix
$\mathbf{T}$	Scaling coordinate features
$\boldsymbol{\beta}$	K × 1 weighting vector of trees in a boosted tree ensemble
$\mathbf{w}$	N × 1 weighting vector of samples in a boosted tree ensemble
ϵ_k	Ensemble error
β_k	Weight of kth tree in boosted tree ensemble
W_k	Normalizing constant
$F(\mathbf{x})$	Output of ensemble of boosted classification or regression trees
$L(y,f(\mathbf{x}))$	Loss function of a classifier or regressor
$g_k(\mathbf{x})$	Gradient at **x** after at kth iteration
ρ_k	Optimization search step size at kth iteration
$\boldsymbol{\theta}_k$	Parameters of kth model in an ensemble
$i_w(\eta)$	Weighted cross-entropy of node η
$Q(k\|\eta)$	Sum of weights of samples in node η labelled as class k
$W(\eta)$	Sum of all sample weights present in node η
X'_t	Matrix of time series column vectors with mean centred columns
$\widehat{\mathbf{X}}_i$	ith of d
$\tilde{\mathbf{X}}_i$	ith trajectory matrix
$\rho^{(w)}_{p.q}$	Weighted or **w**-correlation between time series p and q

(continued)

(continued)

Symbol	Description
$\rho_{\max}^{(L,K)}$	Maximum of the absolute value of the correlations between the rows and between the columns of a pair of trajectory matrices $\tilde{\mathbf{X}}_i$ and $\tilde{\mathbf{X}}_j$
$\mathcal{N}(a,b)$	Normal distribution with mean a and standard deviation b
$\mathbf{u}(t)$	Input vector at time t
$\mathbf{y}(t)$	Vector of measured variables at time t
$\mathbf{v}(\mathrm{t})$	Gaussian noise with variance 0.01
$\mathbf{w}(t)$	Gaussian noise with variance 0.1

Chapter 6
Fault Diagnosis in Steady-State Process Systems

Chapter 1 introduced a generalized framework for data-driven fault diagnosis with Chap. 2 presenting an overview of approaches to multivariate statistical process control in processes assumed to exhibit linear steady-state Gaussian, nonlinear steady-state (non)Gaussian and nonlinear dynamic characteristics. This chapter presents case studies illustrating the application of neural network, tree-based and kernel-based feature extraction (discussed in Chaps. 3, 4 and 5) to the problem of linear and nonlinear data-driven fault diagnosis of steady-state processes.

This chapter is not meant to be an exhaustive comparison of all design decisions when applying data-driven fault diagnosis, but gives an overview (through specific examples) of typical considerations. The general framework for data-driven process fault diagnosis is reviewed in the context of fault diagnosis in steady-state process systems. Offline training and online implementation flow diagrams for fault diagnosis are presented. Details of particular variants of fault diagnosis through feature extraction are given, as well as typical performance measures to assess the validity of these methods. Three case studies are considered to illustrate the application of data-driven fault diagnosis of steady-state processes: a simple illustrative example (Shao et al. 2009) and two benchmark problems for data-driven fault diagnosis, i.e. the simulated Tennessee Eastman process (Russell et al. 2000b) and a real-world sugar refinery process (DAMADICS RTN 2002).

6.1 Steady-State Process Systems

Processes are considered to be at steady state when the variables representing the process do not change meaningfully over time or, if it is a stochastic system, when the probabilities of various system states being repeated remain constant. In this chapter the focus is on multivariate statistical process control of such stochastic systems of variables, rather than systems that are in dynamic equilibrium, which is considered in Chap. 7.

C. Aldrich and L. Auret, *Unsupervised Process Monitoring and Fault Diagnosis with Machine Learning Methods*, Advances in Computer Vision and Pattern Recognition,
DOI 10.1007/978-1-4471-5185-2_6,

Although true steady state in industrial processes may be nigh improbable due to ever-present disturbances and control actions, the assumption of approximate steady state for appropriate systems allows the application of linear and nonlinear steady-state multivariate process control techniques. Care should be taken in assuming steady-state conditions, as an incorrect assumption may yield inaccurate fault diagnosis results. Where the steady-state assumption is decidedly invalid, nonlinear dynamic fault diagnosis (see Chap. 7) may be a better option.

The assumption of steady state may be interrogated by means of several steady-state identification techniques, including univariate linear regression to check for non-zero slope of process variable measurements versus sample time, ratios of univariate means for different sampling intervals and ratios of univariate variances for different sampling intervals (Cao and Rhinehart 1995). The extension of univariate steady-state tests to multivariate process data by means of accumulating the results of univariate tests has also been proposed (Brown and Rhinehart 2000).

6.2 Framework for Data-Driven Process Fault Diagnosis: Steady-State Process Systems

The general framework for the data-driven construction of feature extraction models for process fault diagnosis was introduced in Chap. 1 and is repeated here (Fig. 6.1):

Data-driven process fault diagnosis consists of two stages: an offline training stage and an online application stage. Section 6.2.1 considers the general structure of the offline training stage, and Sect. 6.2.2 considers the general structure of the online implementation stage. Sections 6.2.3, 6.2.4 and 6.2.5 consider the different elements ($\mathbf{X}$, $\Im$, $\mathbf{F}$, $\aleph$, $\hat{\mathbf{X}}$ and $\mathbf{E}$) of the data-driven process fault diagnosis framework in the context of fault diagnosis for steady-state processes.

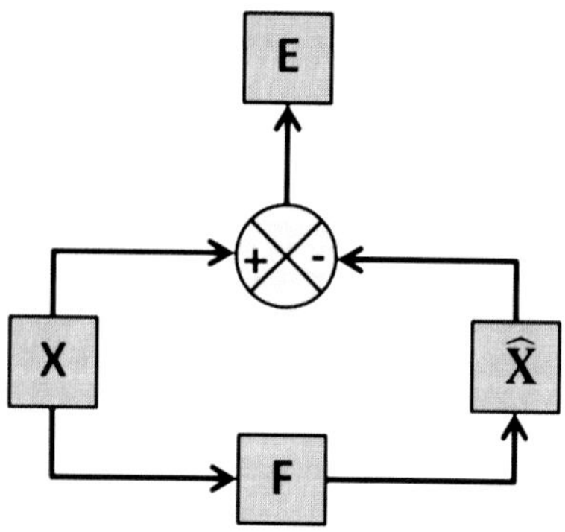

Fig. 6.1 A general framework for data-driven process fault diagnosis

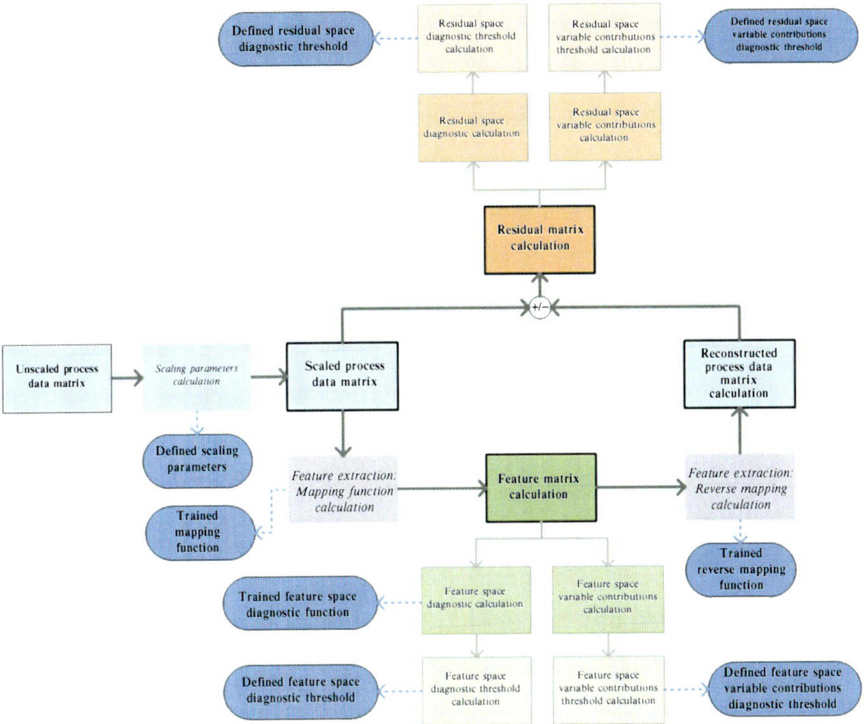

Fig. 6.2 Schematic of offline training for data-driven process fault diagnosis in steady-state process systems

6.2.1 General Offline Training Structure

An overview of the offline training stage for data-driven fault diagnosis is shown in Fig. 6.2. In the offline training stage, an unscaled process data matrix $\mathbf{X}^{(\text{unscaled})}$ undergoes scaling to standardize all process variables. Scaling parameters are calculated during the offline training stage and retained for use in the online implementation stage.

The scaled process data matrix $\mathbf{X}$ representing normal operating conditions is used in training a feature extraction process model. This feature extraction process model consists of a mapping function $\Im$ (which maps the process data matrix $\mathbf{X}$ to an information-rich feature matrix $\mathbf{F}$) and a reverse mapping function $\aleph$ (which maps the feature matrix $\mathbf{F}$ to the original process variable space, but as a reconstructed process matrix $\widehat{\mathbf{X}}$). The trained mapping and reverse mapping functions are retained for use in the online implementation stage.

The extracted feature matrix $\mathbf{F}$ determined during the offline training stage represents normal operating conditions process data projected onto a manifold which in some way captures the special cause variation in the process. The nature of the manifold (e.g. linear or nonlinear) depends on the properties of the mapping function $\Im$.

The feature space can be characterized by a feature space diagnostic T^2 (as in Hotelling's T^2) and associated diagnostic threshold T_α^2 (where α indicates the level of confidence associated with the threshold). This feature space diagnostic summarizes how central any sample is in relation to the normal operating conditions data distribution. A high diagnostic implies that a sample is further removed from the centre of the normal operating conditions data distribution in the feature space. If this diagnostic exceeds a threshold, the sample can be considered to signify a fault. The feature space diagnostic may be calculated based on data description functions, for example, one-class support vector machines (see Chap. 4). During the offline training stage, the diagnostic function and diagnostic threshold is determined and retained for use in the online implementation stage.

The difference between the original process data matrix $\mathbf{X}$ and the reconstructed process matrix $\widehat{\mathbf{X}}$ is the residual matrix $\mathbf{E}$. The residual space captures the variation of process data outside of the special cause variation accounted for by the feature space. The residual space diagnostic SPE (squared prediction error) indicates to what extent the process data cannot be adequately captured by the feature extraction process model. A large residual space diagnostic indicates that the feature extraction process model may no longer be valid, which is an indication of faulty conditions. During the offline training stage, a suitable threshold (SPE_α) for the residual diagnostic is determined and retained for use during the online implementation stage.

The contribution of individual process variables to the feature space and residual space diagnostics can be used in identifying fault conditions (e.g. a particular fault affects a particular subset of process variables and can be successfully identified, if these process variables show large contributions to either the feature space or residual space diagnostic). The contribution of the jth process variable to a fault in the feature space is represented by $C_{s,j}$ and by $C_{r,j}$ in the residual space. Thresholds for these variable contributions are determined during the offline training stage: $C_{s,j}^{\alpha}$ in the feature space and $C_{s,j}^{\alpha}$ in the residual space. These thresholds are retained and applied in the online implementation stage.

6.2.2 General Online Implementation Structure

In the online implementation stage (see Fig. 6.3), a new unscaled process data matrix is subjected to the scaling procedure and parameters derived in the offline training algorithm. The learnt mapping and reverse mapping functions ($\Im$ and $\aleph$) from the offline training stage are then applied to the scaled test process data matrix $\mathbf{X}^{(\mathrm{test})}$.

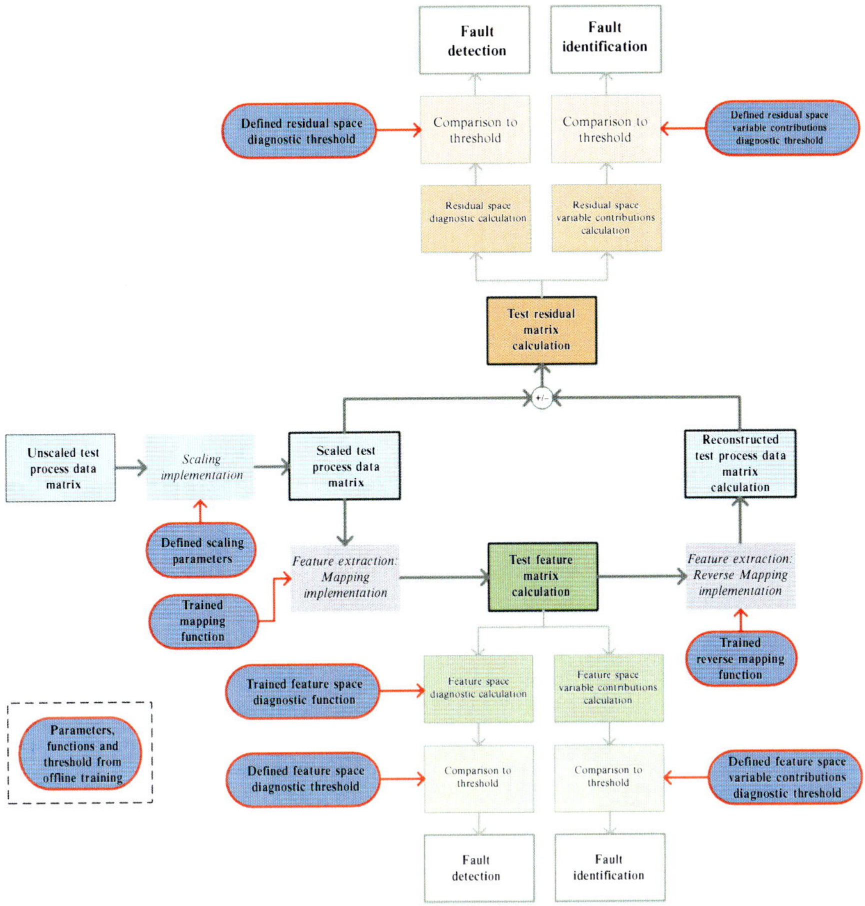

Fig. 6.3 Schematic of online implementation for data-driven process fault diagnosis in steady-state process systems

The new feature matrix $\mathbf{F}^{(\text{test})}$ and new reconstructed process data matrix $\widehat{\mathbf{X}}^{(\text{test})}$ result from these mappings. The new residual matrix $\mathbf{E}^{(\text{test})}$ is calculated from the new process data matrix and its reconstruction.

Feature space and residual space diagnostics can be calculated from the new feature matrix $\mathbf{F}^{(\text{test})}$ and residual matrix $\mathbf{E}^{(\text{test})}$ and compared to the feature space and residual space diagnostic thresholds, respectively. If a diagnostic exceeds its diagnostic threshold, an alarm is indicated. Feature space and residual space variable contributions can be calculated, and important process variables associated with faults can be identified based on whether the respective variable contribution threshold was exceeded.

6.2.3 *Process Data Matrix X*

When considering fault diagnosis of steady-state systems, the data matrix **X** consists of samples of process variables, where it is assumed that the process variables do not change meaningfully over time or that the probabilities of various system states being repeated remain constant. Steady-state process variables would exhibit near-constant means and variances under normal operating conditions. The following factors should be considered when pretreating process data to be included in the process data matrix **X**:

Scaling

Process variable measurements require autoscaling to ensure that variables with large numerical variations do not dominate the process model (e.g. PCA loading matrix). For example, one process variable may be a reactor temperature with a numerical range of 100 K, while another process variable may be product concentration with a numerical range of 0.2. If left unscaled, the reactor temperature process variable would contribute a disproportionately large load to a PCA loading matrix (for example) even if this variable did not capture more information (in terms of common cause variation) than the product concentration variable.

Autoscaling for each process variable is accomplished by subtracting from each measurement the sample mean calculated from the normal operating conditions measurements and dividing by the sample standard deviation, also calculated from the normal operating conditions measurements. The sample mean μ_i and sample standard deviation σ_i for (unscaled) process variable $\mathbf{X}_i^{(\text{unscaled})}$ (with measurements $X_{i,j}^{(\text{unscaled})}, j = 1, \ldots, N$) based on N samples in the training data set are given by

$$\mu_i = \frac{1}{N}\sum_{j=1}^{N} X_{i,j}^{(\text{unscaled})}$$

$$\sigma_i = \sqrt{\frac{1}{N-1}\sum_{j=1}^{N}\left(X_{i,j}^{(\text{unscaled})} - \mu_i\right)^2}. \tag{6.1}$$

An autoscaled process variable measurement $X_{i,j}$ for the jth sample of variable i is calculated as

$$X_{i,j} = \frac{X_{i,j}^{(\text{unscaled})} - \mu_i}{\sigma_i}. \tag{6.2}$$

During the online implementation stage, the sample means and standard deviations of process variables calculated from the offline training data are used to scale the new data of the process variables. Autoscaling ensures that process variables are standardized before feature extraction algorithms are applied.

Sufficient Sampling

The included process variables must be routinely sampled to ensure that large data sets of normal operating conditions are available and that abnormal events can be detected timeously. Large, representative data sets of normal operating conditions are essential, since data-driven fault diagnosis methods rely heavily on process data to glean the underlying process model structure.

In general, it is assumed that process variable measurements are statistically independent from measurements at previous sampling times. If the sampling frequency is high, process variable observations may not be statistically independent from observations at previous time instances. When such autocorrelation exists in the process data, a modified process data matrix $\mathbf{X}^{*(h)}$ can be created by including h previous autocorrelated measurements, where h is the number of previous measurements to include. Let $\mathbf{X}^{(N)}$ be the original process data matrix up to the most recent sample time consisting of N samples, $\mathbf{X}^{(N-1)}$ be the original process data matrix up to the previous sample consisting of $N-1$ samples, etc. The modified process data matrix which includes h previous autocorrelated measurements as additional variables is given by

$$\mathbf{X}^{*(h)} = \left[\mathbf{X}^{(N)}\mathbf{X}^{(N-1)}\ldots\mathbf{X}^{(N-h)}\right]. \tag{6.3}$$

Dynamic PCA makes use of such an augmented process data matrix to exploit autocorrelated process variable measurements (Ku et al. 1995). Chapter 7 expands further on dynamic process monitoring.

Process Variable Sensitivity to Abnormal Process Conditions

The entirety of included process variables must reflect the process state and be sensitive to abnormal events while robust to random fluctuations. Whereas traditional univariate process monitoring would require much energy to be expended on the selection of a few critical key performance indicators and process variables for individual monitoring, multivariate process monitoring is more forgiving, in that implicit process variable selection and combination is automated.

For example, if a certain measured process variable does not show large variation during normal operating conditions, this variable will have a lower contribution to the process model (e.g. weighting in a PCA loading matrix) than a process variable that exhibits higher variation during normal operating conditions. However, a process variable with a minor contribution to the process model is not discounted from fault detection: The monitoring of the residual space (represented by the residual matrix **E**) focuses on determining whether the learnt process model is still valid for process data garnered from new sampling intervals.

Fault Identification Utility of Process Variables

Process variables included in the data matrix **X** should be conducive to informative fault identification and aid root cause analysis and possible process recovery. If an abnormal event is detected (through monitoring of the feature space **F** and/or the residual space **E**), the contribution of each included process variable to the fault can be determined: either through its contribution to the process model or its residual magnitude. These process variable contributions do not discriminate between causes and effects/symptoms. For example, if an important upstream valve controlling feed flow fails, all downstream process variables may eventually show significant contributions to a detected fault. The upstream feed flow would be related to the cause of the fault and should show a significant contribution to a fault detected in the feature and/or residual space. However, certain downstream process variables would show symptoms of the fault in the form of deviations that may result in significant contributions to the detected fault.

Expert knowledge of process topology and possible fault propagation paths is required to differentiate between the cause of a fault and its symptoms. Progress has been made in the automation of root cause and fault propagation analysis from phenomenological process models, from expert knowledge and from process data (Yang and Xiao 2012).

Although root cause and fault propagation analysis is not explicitly considered in the general framework for data-driven fault diagnosis presented in Fig. 6.1, brief attention will be given to some interesting aspects. Certain principles can be exploited to capture process topology from process data.

One such principle is that propagation of excitation from a causal process variable to an effect process variable implies that there is a positive time delay between the effect and causal variables. Pairwise time delays can be estimated between all process variable pairings by means of cross-correlation analysis (Bauer and Thornhill 2008). Significant process variable cause–effect pairs can be visualized by means of a process topology map, where process variables are represented by nodes and cause–effect relations are represented by directed edges connecting nodes. An example of a process topology derived from process data (albeit simulated, in this example) is given in Fig. 6.4.

Another principle employed to extract process topology is that the propagation of excitation from a causal process variable to an effect process variable implies information transfer from the causal to the effect variable. Transfer entropy is an example of an information theoretic measure which may indicate cause–effect connections between two process variable vectors (Bauer et al. 2007). As with cross-correlation analysis, a process topology map can be constructed to represent significant transfer entropy connections between process variables.

A probability theoretic view of causality is another principle used in extraction process topology from process data. This view considers that the propagation of excitation from a causal process variable to an effect process variable would imply that the effect process variable measurement exhibits a probability conditional on the causal process variable measurement. Bayesian nets can be learnt from process

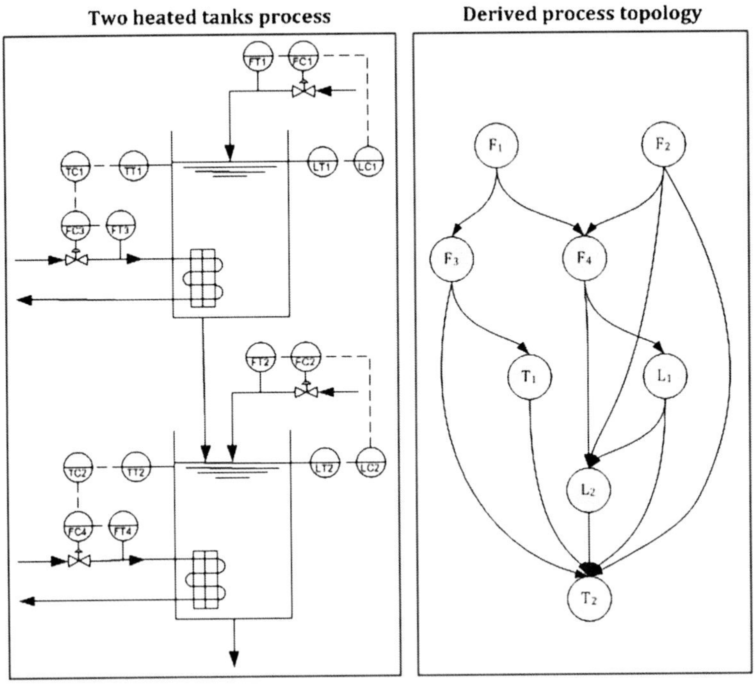

Fig. 6.4 Example of process topology derived from simulated data of a two heated tanks process by means of cross-correlation analysis (*Nodes* represent process variables, and *edges* represent estimated causal paths) (Adapted from Lindner 2012)

data, where the process topology is represented as a structure of nodes representing fault modes and process variables, while edges represent conditional probabilities (Yang and Xiao 2006).

Incorporation of Process Time Lags

The time delays between cause and effect process variables are not only of interest during fault identification and root cause analysis but may also be considered in the construction of the process data matrix $\mathbf{X}$ to be subjected to feature extraction model construction.

Consider a process of connected process units in which a feed stream undergoes reaction and separation processes, captured by 15 measured process variables (see Fig. 6.5). In general, the process data matrix for such a process would be constructed by simply concatenating process variables to ensure that process variables are synchronized in time:

$$\mathbf{X} = [\mathbf{X}_1(k)\mathbf{X}_2(k)\ldots\mathbf{X}_{15}(k)]. \tag{6.4}$$

Above, k indicates a specific sample time.

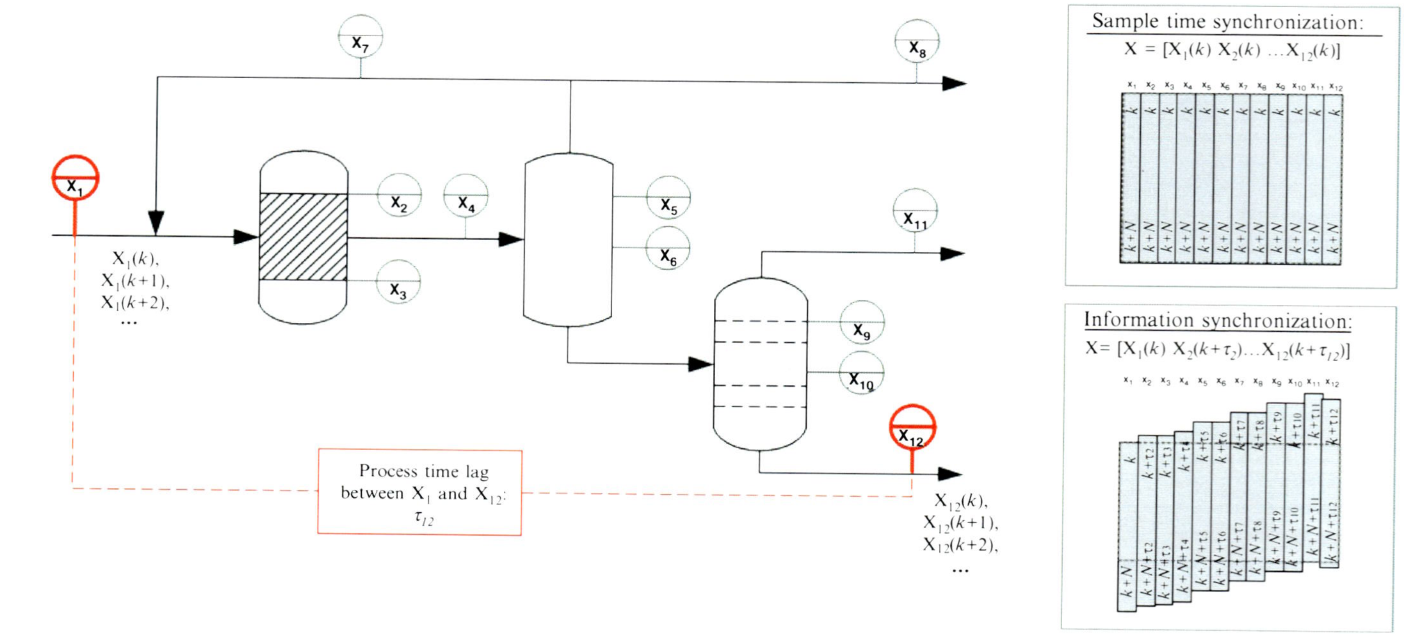

Fig. 6.5 Illustration of sample time versus information synchronization of the process data matrix for a typical process

An upstream process variable (e.g. reactant concentration in the raw feed stream, $\mathbf{X}_1$) could reflect a disturbance in the feed stream some time (e.g. a time delay of τ_{15}) before a downstream process variable (e.g. final product concentration, $\mathbf{X}_{15}$). A feed condition fault at sample time k would induce a change in the raw feed stream reactant concentration at sample time k: $\mathbf{X}_1(k)$, while only generating a change in the final product concentration at sample time $k + \tau_{15}$: $\mathbf{X}_{15}(k + \tau_{15})$. The initial change in the feed stream reactant concentration might not immediately trigger a fault alarm, since all downstream process variables would still be unaffected. For temporary disturbances, the situation might also arise that once the product concentration variable $\mathbf{x}_{15}$ exhibits a significant change, the reactant feed concentration $\mathbf{x}_1$ no longer shows abnormal behaviour.

One approach to addressing process time lag between variables involves synchronizing process variables based on event propagation, the so-called information synchronization (Lu et al. 2012). In the example above, the process variables may be shifted by their relevant process time lags before concatenation to form the process data matrix $\mathbf{X}$:

$$\mathbf{X} = [\mathbf{X}_1(k)\mathbf{X}_2(k + \tau_2)\ldots\mathbf{X}_{15}(k + \tau_{15})]. \tag{6.5}$$

The process time lags (τ_i) for each process variable need to be estimated, for example, by means of cross-correlation analysis (as mentioned earlier in the context of data-driven process topology extraction). By lagging process variables, a delay (equivalent to the maximum process time lag) is introduced in online monitoring applications. The undesirability of this delay depends on whether faults will be detected due to information synchronization that would not have been detected without it. Another factor to consider before implementing information synchronization concerns the ease of calculation and accuracy of process time lags.

6.2.4 Mapping $\Im$ and Feature Matrix F

The procedure for extracting informative features from process data is termed mapping. Mapping entails transforming process data measurements to informative features with reduced redundancy and noise with the intention to capture special cause variation more optimally than the original process data. A mapping function $\Im$ is typically learnt during offline training from a training process data matrix $\mathbf{X}$. This mapping function $\Im$ transforms an m-dimensional process data sample vector $\mathbf{x}_i = [X_{i,1}, X_{i,2}, \ldots, X_{i,m}] \in \mathcal{R}^m$ to a q-dimensional feature sample vector $\mathbf{f}_i = [F_{i,1}, F_{i,2}, \ldots, F_{i,q}] \in \mathcal{R}^q$:

$$\mathbf{f}_i = \Im(\mathrm{x}_i). \tag{6.6}$$

The number of features q does not necessarily need to be less than the number of process variables m, but this is generally the case. It is often true that $q \ll m$, indicating high levels of redundancy in the original process data. For the general case where $q < m$, the feature extraction mapping functions can be considered similar to dimension reduction techniques. Dimension reduction techniques specifically aim to reduce the dimensionality of data in order to circumvent the curse of dimensionality and extract informative data models (Carreira-Perpinan 1997). When q is equal to two or three, the extracted features can be visualized, which may assist in process understanding.

Motivation for Feature Extraction

Feature extraction aims to exploit the presence of redundancy in the process data. Redundancy is an indication of uninformative measurements in process data, resulting in an intrinsic dimensionality of informative process data being lower than the number of process measurement variables available. This lower-dimensional *information space* is defined by features or latent variables constructed as combinations and functions of measured process variables. These features aim to capture the relationship among process variables, which often arise due to fundamental aspects of physical systems (Kresta et al. 1991). The division of the original measurement space into so-called *information-rich* and *information-poor* subspaces allows higher robustness to noise (Venkatasubramanian et al. 2003). A further advantage is the fact that no explicit process system based model has to be developed.

The theoretical advantages of multivariate feature extraction fault diagnosis pertain to the compression and information-retention characteristics of the feature extraction transformation. By utilizing data compression, these methods can handle many correlated measurement variables. The presence of structure in the data is explicitly exploited by assuming that a lower intrinsic dimensionality is applicable. Where feature extraction is deterministic (e.g. principal component analysis and kernel principal component analysis; see Chap. 4), model development is straightforward and computational requirements are low. For methods where feature extraction is not deterministic, e.g. feature extraction methods based on neural networks (see Chap. 3) and random forests (see Chap. 5), model development is complicated by repeatability and parameter selection issues. Random forest feature extraction is further complicated by the fact that separate direct mapping and reverse functions must be learnt (see Chap. 5).

If two or three features can be extracted which represents the process sufficiently, these features and projected process measurements can be visualized. This allows plant managers to exploit the powerful human talent of pattern recognition in order to gain an understanding of process conditions and fault patterns. Above and beyond the visualization of process data in low-dimensional feature space, the inspection of simple derived-statistic control charts enables a quick and easy way to determine whether a process is in control and whether it shows trends towards out-of-control behaviour.

Selection of Number of Features

The number of features q to extract to sufficiently represent normal operating conditions in the feature space is an important design decision in data-driven process fault diagnosis schemes. If too few features are extracted, the feature space may not capture all special cause variation. If too many features are extracted, the feature space may contain noise and redundancy. A number of techniques have been suggested for the selection of the number of PCA features in a fault diagnostic framework, with a good survey of these methods presented by Himes et al. (1994). Depending on the nature of the feature extraction algorithm, these methods may be applicable to other feature extraction methods as well.

A challenge with evaluating the best technique for determining the number of features relates to the unsupervised nature of feature extraction: the true number of features that best summarizes the normal operating conditions is not known, since a universal criterion for "best" feature space does not exist. Considering the case where feature extraction is employed in a fault detection scheme, the performance of a specific feature space dimension selection technique may be assessed based on its detection performance on benchmark process data sets.

Percentage of Variance Explained

The most widespread technique for selecting the number of PCA features to retain is by inspection of the percentage of variance explained by each feature and retaining the number of features which account for a defined cumulative percentage of the variance, e.g. 90 %. This technique may not be readily extended to other feature extraction techniques where feature space variance is not additive, such as feature extraction methods based on neural networks (see Chap. 3).

Scree Test

Another popular technique for PCA feature selection is the scree test (Cattell 1966). The motivation for the scree test comes from the observation that PCA eigenvalues (which represent the variance explained by each feature) generally show an initial rapid decrease for the first number of features, followed by a longer tail of slowly decreasing eigenvalues (see Fig. 6.6). The break point between the initial rapid decrease and subsequent slower decrease is said to represent a separation of features capturing process variation from features capturing noise and redundancy. The scree test can be applied to any feature extraction method which involves eigendecomposition of a covariance or similarity matrix, including kernel principal component analysis (see Chap. 4) and random forest feature extraction (see Chap. 5).

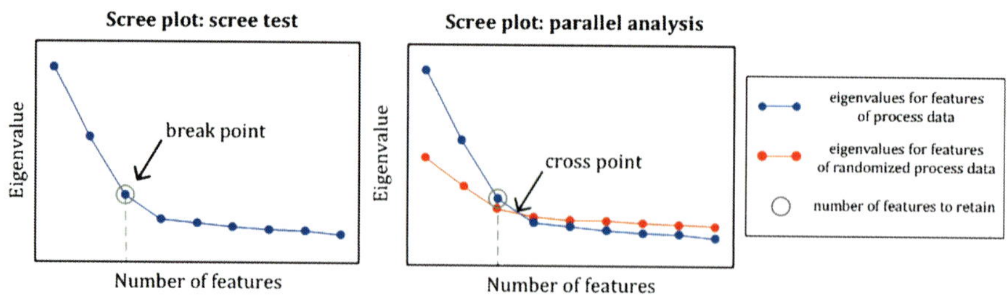

Fig. 6.6 Illustration of scree test and parallel analysis for the selection of the number of features to retain

Parallel Analysis

The parallel analysis technique (Horn 1965) is an extension of the scree test. In a parallel analysis, the scree plot based on features extracted from process data is compared with the scree plot(s) of features extracted from a resampled, randomized process data set(s). This resampled, randomized process data set is created by randomly sampling process variables *independently* in order to destroy the correlation between process variables. (This technique of random resampling to create a synthetic data set with reduced intervariable correlation is similar to the technique for creating a synthetic contrast in random forest feature extraction; see Chap. 5).

The scree plot of the features of the random process data represents spurious covariance that may be present in the process data. It is expected that the scree plot of the actual process data will lie above the scree plot of the random process data for the first q number of features (i.e. the first q eigenvalues for the actual process data features will be larger than the first q eigenvalues for the random process data features), where q is the number of features accounting for non-spurious, special cause process variation (see Fig. 6.6). For numbers of features larger than q, the scree plot of the actual process data will lie on or below the scree plot of the random process data. These actual process data features are assumed to capture only random variation and can be excluded from the feature space.

As with the scree test, parallel analysis can be applied to any feature extraction method which involves eigendecomposition of a covariance or similarity matrix, including kernel principal component analysis (see Chap. 4) and random forest feature extraction (see Chap. 5).

Prediction Sum of Squares (PRESS)

The prediction sum of squares, or PRESS, technique (Wold 1978; Eastment and Krzanowski 1982) compares the normal operating conditions process data $\mathbf{X}$ with its reconstruction $\widehat{\mathbf{X}}$ after feature extraction. The average sum of squared errors between the actual data and reconstructed data is calculated by means of cross-validation for a selection of number of extracted features. Cross-validation ensures that the reconstruction integrity of the feature extraction technique can be verified on unseen data. The number of features that results in the lowest average sum of squared errors is selected as the optimal number of features. A feature extraction technique which performs well on reconstructing unseen process data suggests that sufficient process variation has been preserved in the feature space to represent normal operating conditions but that noise and redundancy in the feature space have been minimized. The PRESS technique can be extended to any feature extraction method, since only the residual matrix **E** is required. For example, this technique can be used with feature extraction methods based on neural networks (see Chap. 3).

Feature Space Diagnostics and Confidence Bounds

Control limits for feature extraction-based fault diagnosis can be determined parametrically or non-parametrically. As shown with the modified Hotelling's T^2 approach (see PCA score distance in Chap. 2), by assuming an underlying parametric distribution of the measurements, the distribution properties can be retained through the respective mappings and parametric distribution-based confidence limits calculated.

Another approach is to not make any parametric distribution assumptions about the process measurement or feature variables, but to rather use empirical techniques to establish the structure of the normal operating conditions data from the data itself. Such an approach was first developed by Martin and Morris (1996), who exploited kernel density estimation in the construction of so-called likelihood-based confidence regions. This approach was motivated by the fact that the authors found that for many industrial process, multivariate normality tests confirmed that the scores in the feature space rarely followed a multivariate normal distribution. It was also shown that modified Hotelling's T^2 confidence limits for industrial data sometimes resulted in very conservative confidence regions, with large areas devoid of any NOC data.

Another non-parametric approach for determining confidence limits in the feature space is one-class support vector machines (see Chap. 4). One-class support vector machines (1-SVM), also known as support vector data description (SVDD), have been used to derive so-called "kernel-distance" diagnostics in the input space (Sun and Tsung 2003; Camci et al. 2008) as well as in the feature space (Ypma et al. 1999; Jemwa and Aldrich 2006; Liu et al. 2008).

Process Variable Contributions in the Feature Space

Once a fault has been detected in the feature space (i.e. the calculated feature space diagnostic for a new process data sample was greater than a previously defined feature space diagnostic threshold), further information can be gathered in order to identify the fault. Fault identification can be aided by inspecting the contributions of process variables to the feature space diagnostic. For example, the contributions of process variables to the modified Hotelling's T^2-statistic (a PCA feature space diagnostic) can be calculated (Nomikos 1996; Russell et al. 2000a).

The identification of process variables responsible for abnormal conditions is not always a simple matter in the case of other feature extraction approaches to fault diagnosis. Complexities arise when the mapping from the measurement space to the feature space is not explicit, thus providing no direct weightings, coefficients or "fingerprints" of process variables in the calculated features. When such limitations are encountered, fault identification in the residual space is often still an option.

6.2.5 Reverse Mapping ℵ and Residual Matrix E

Reverse mapping entails the reconstruction of process variables from the feature space. Since it is intended that the feature space only retains special cause variation, the reconstructed process matrix $\widehat{\mathbf{X}}$ should, in the ideal case, be free from noise. The reverse mapping function ℵ outputs an m-dimensional reconstructed process data sample $\hat{\mathbf{x}}_i = \left[\widehat{X}_{i,1}, \widehat{X}_{i,2}, \ldots \widehat{X}_{i,m}\right] \in \mathcal{R}^m$ from a q-dimensional feature space sample $\mathbf{f}_i = [F_{i,1}, F_{i,2}, \ldots, F_{i,q}] \in \mathcal{R}^q$:

$$\hat{\mathbf{x}}_i = \aleph\left(\mathbf{f}_i\right). \tag{6.7}$$

For certain feature extraction algorithms, reverse mapping is straightforward. For example, PCA reverse mapping simply requires the transform of the projection matrix used in mapping. In comparison, reconstruction from kernel principal component analysis features is not straightforward (the so-called pre-image problem; see Chap. 4). As mentioned in Chap. 4, one of the approaches for solving the reconstruction problem in KPCA is through the learning/training of a multivariate regression function which can map features back to the original input space, e.g. KPCA reconstruction with kernel ridge regression (Bakır et al. 2004; Bakır 2005). Other feature extraction methods might even require learning/training of a multivariate regression function for both the mapping and reverse mapping procedures, for example, random forest feature extraction which uses random forest regression functions for mapping and reverse mapping of new data (Auret and Aldrich 2010; Auret 2010).

Once the process variables have been reconstructed as $\widehat{\mathbf{X}}$, the residual matrix $\mathbf{E}$ can be calculated as the difference between the original process data and its reconstruction:

$$\mathbf{E} = \mathbf{X} - \widehat{\mathbf{X}}. \tag{6.8}$$

The residual matrix $\mathbf{E}$ captures the mismatch between the actual process data and its reconstruction from the process model encapsulated by the feature space.

Optimal feature extraction mapping and reverse mapping would result in a residual matrix **E** for normal operating conditions process data consisting only of random noise.

Residual Space Diagnostics and Confidence Bounds

The residual space diagnostic SPE is the sum of squared prediction errors (residuals) for all process variables:

$$\mathrm{SPE}_i = \sum_{j}^{m} \left(E_{i,j}\right)^2 = \sum_{j}^{m} \left(X_{i,j} - \widehat{X}_{i,j}\right)^2. \tag{6.9}$$

Above, SPE_i is the residual space diagnostic for sample i.

The threshold SPE_α for the residual space diagnostic above has been stated based on an approximate distribution determined by Jackson and Mudholkar in 1979 (Russell et al. 2000a).

Process Variable Contributions in the Residual Space

When a fault has been detected in the residual space (i.e. the calculated residual space diagnostic exceeds the residual space diagnostic threshold), the contribution $\{C_{r,j}\}_i$ of each process variable j to the residual diagnostic r for sample i can be expressed as

$$\left\{C_{r,j}\right\}_i = \left(X_{i,j} - \widehat{X}_{i,j}\right)^2. \tag{6.10}$$

The average contribution $C_{r,j}$ of process variable j can be determined by averaging the per-sample contributions for samples that have been identified as being faulty.

6.3 Details of Fault Diagnosis Algorithms Applied to Case Studies

Four algorithms are considered here, viz. fault detection and identification based on the use of principal component analysis (see Chap. 2), nonlinear principal component analysis with an inverse autoassociative neural network (see Chap. 3), kernel principal component analysis (see Chap. 4) as well as random forests (see Chap. 5). These algorithms stem from the machine learning paradigms considered in this book, while principal component analysis is retained as a benchmark in all cases. The implementation of the different approaches is discussed below.

The general offline training and online implementation frameworks shown in Figs. 6.2 and 6.3 are adhered to by all methods. The following subsections highlight specific design choices for the different feature extraction methods for the following aspects: mapping training and implementation, selection of number of variables, feature space diagnostic calculation, feature space contribution calculation, reverse mapping training and implementation, residual space diagnostic calculation and residual space contribution calculation (Tables 6.1, 6.2, 6.3, 6.4, 6.5, 6.6 and 6.7).

6.4 Performance Metrics for Fault Detection

When knowledge is available on the samples associated with NOC and fault conditions (as is the case in the above-mentioned case studies), the performance of monitoring algorithms can be compared based on the number of faults that are correctly and incorrectly identified, run lengths to first alarms and receiver operating curve characteristics.

In order to ensure a fair comparison of fault detection between different feature extraction methods, the diagnostic thresholds for all algorithms are aligned to correspond to a 1 % false alarm rate on a validation normal operating conditions data using the percentile approach for limits correction.

6.4.1 Alarm Rates, Alarm Run Lengths and Detection Delays

The false alarm rate (FAR) is the number of detections that are logged for validation normal operating conditions data set, i.e. a data set that was unseen during training, but is known a priori, consists only of normal operating condition data. The alarm run length (ARL) for false alarms is the number of consecutive samples before an alarm is made for validation data; the larger ARL (false), the better.

The missing alarm rate (MAR) is the fraction of known fault samples that are not detected for a given data set. The alarm run length (ARL) for true alarms is the number of consecutive samples before an alarm is made for fault data; the smaller ARL (true), the better.

The detection delay (DD) is the number of consecutive faulty samples missed before a detection is logged, with the overall detection delay, the minimum of the feature space diagnostic detection delay DD_s and the residual distance detection delay DD_r. For detection delay calculation purposes, a detection is defined as three consecutive alarms.

Table 6.1 Details for mapping training and implementation for data-driven fault diagnosis techniques applied in case studies

Feature extraction algorithm	Mapping offline training	Mapping online implementation
Principal component analysis (PCA)	Eigendecomposition of covariance matrix of process data to determine loading matrix Projection of process data to q principal components	Projection of process data to q principal components
Nonlinear principal component analysis (NLPCA)	Hierarchic, inverse autoassociative neural network (Scholz et al. 2005, 2008; Scholz and Vigário 2002), using MATLAB NLPCA toolbox (Scholz 2011)	Implementation of trained hierarchic, inverse autoassociative neural network
Kernel principal component analysis (KPCA)	Radial basis function/Gaussian kernel Kernel parameter selection: σ selected as 50th percentile of all pairwise distances between input training samples	Calculation of test kernel matrix relative to training data, centring of test kernel matrix, projection onto retained eigenvectors with test kernel matrix vectors
Random forest feature extraction (RF)	Unsupervised random forest feature extraction (Shi and Horvath 2006) Contrast construction: random sampling (without replacement) from the product of marginal distributions of the training data Forest construction: –1,000 trees, number of random split selection variables is equal to the floored square root of the number of variables –Minimum leaf size of 1 Average dissimilarities of five forests used –Using randomForest package (Liaw and Wiener 2002) in R (R Development Core Team 2010) –Features calculated by classical multidimensional scaling Training of regression random forests for online implementation, one forest for each feature to be calculated –Using randomForest package (Liaw and Wiener 2002) in R (R Development Core Team 2010) –100 trees per forest, number of random split selection variables: $m/3$ floored –Minimum leaf size of 5	Application of trained regression forests

Table 6.2 Details for selection of number of features for data-driven fault diagnosis techniques applied in case studies

Feature extraction algorithm	Selection of number of features (q)
Principal component analysis (PCA)	Parallel analysis (average of 5 randomly resampled data sets)
Nonlinear principal component analysis (NLPCA)	PRESS (fivefold cross-validation; maximum of m features, where m is number of process variables)
Kernel principal component analysis (KPCA)	Parallel analysis (average of 5 randomly resampled data sets)
Random forest feature extraction (RF)	Parallel analysis (average of 5 randomly resampled data sets)

Table 6.3 Details for feature space diagnostic and threshold calculation for data-driven fault diagnosis techniques applied in case studies

Feature extraction algorithm	Feature space diagnostic and threshold calculation
Principal component analysis (PCA)	Modified Hotelling's T^2 (see Chap. 2) Threshold set at 99th percentile of diagnostic values for training set
Nonlinear principal component analysis (NLPCA)	One-class support vector machines (1-SVM) were used to characterize the feature space Selection of ν-parameter (which controls the complexity of the support subset): $\nu = \alpha$, where $(1 - \alpha)\%$ is the confidence level, 99 % in this case
Kernel principal component analysis (KPCA)	A Gaussian kernel was used with the kernel width parameter σ determined by fivefold cross-validation (Chang and Lin 2011). σ-values considered during cross-validation were ten equispaced values between the first and 50th percentile of pairwise distances in the training input data. The loss function for the cross-validation optimization was the error rate, i.e. the number of test samples falling outside the 1-SVM support
Random forest feature extraction (RF)	To prevent excessively disjointed or overfitted support boundaries, an additional heuristic constraint is placed on the fraction of support vectors: –The largest kernel width which delivers a 1-SVM model with less than 10 % of the training samples used as support vectors while still delivering a test error rate (fraction of samples outside 1-SVM support) of less than 10 % The LIBSVM toolbox was used (Chang and Lin 2011) Threshold set at 99th percentile of diagnostic values for training set

6.4.2 Receiver Operating Characteristic Curves

A receiver operating characteristic curve plots true alarm rates versus false alarm rates, given the full extent of possible threshold values. By considering the full extent of possible threshold values, different classifiers (alarm systems, in this case) can be compared without being limited by the choice of threshold values. A specific

Table 6.4 Details for feature space contribution calculation for data-driven fault diagnosis techniques applied in case studies

Feature extraction algorithm	Feature space contribution calculation
Principal component analysis (PCA)	The contribution of each process variable j to the modified Hotelling's T^2-statistics can be calculated as (Nomikos 1996): $C_{s,j} = \mathbf{F}^1 \mathbf{\Lambda}^{-1} \mathbf{X}_j \mathbf{P}_j$
Nonlinear principal component analysis (NLPCA) Kernel principal component analysis (KPCA) Random forest feature extraction (RF)	No process variable contributions calculated for feature space

Table 6.5 Details for reverse mapping offline training and online implementation for data-driven fault diagnosis techniques applied in case studies

Feature extraction algorithm	Reverse mapping offline training	Reverse mapping online implementation
Principal component analysis (PCA)	Reconstruction of by means of transposed q principal components matrix	Reconstruction of by means of transposed q principal components matrix
Nonlinear principal component analysis (NLPCA)	Hierarchic, inverse autoassociative neural network (Scholz et al. 2005; 2008; Scholz and Vigário 2002), using MATLAB NLPCA toolbox (Scholz 2011)	Implementation of trained hierarchic, inverse autoassociative neural network
Kernel principal component analysis (KPCA)	Reconstruction by learning: Training of multiple kernel ridge regression model (Bakır et al. 2004; Bakır 2005) with training features as input and training process data as output Selection of hyperparameters (kernel width for kernel ridge regression and ridge parameter) with efficient leave-one-out cross-validation for multiple kernel ridge regression (An et al. 2007) Kernel width range: 10th, 20th, 50th and 100th percentile of training feature space distances Ridge parameter range: 1×10^{-6}, 1×10^{-5}, 1×10^{-4}, 1×10^{-3}, 1×10^{-2}, 1×10^{-1}, 1×10^{0}, 1×10^{1}, 1×10^{2}, 1×10^{3}	Application of trained kernel ridge regression model
Random forest feature extraction (RF)	Training of regression random forests, one forest for each reconstructed variable to be calculated –Using randomForest package (Liaw and Wiener 2002) in R (R Development Core Team 2010) –100 trees per forest, number of random split selection variables: $m/3$ floored –Minimum leaf size of 5	Application of trained regression forests

Table 6.6 Details for residual space diagnostic and threshold calculation for data-driven fault diagnosis techniques applied in case studies

Feature extraction algorithm	Residual space diagnostic and threshold calculation
Principal component analysis (PCA)	
Nonlinear principal component analysis (NLPCA)	The residual space diagnostic SPE_i for sample i is the sum of squared prediction errors (residuals) for all process variables: $\mathrm{SPE}_i = \sum_j^m (E_{i,j})^2 = \sum_j^m \left(X_{i,j} - \widehat{X}_{i,j}\right)^2$
Kernel principal component analysis (KPCA)	Threshold set at 99th percentile of diagnostic values for training set
Random forest feature extraction (RF)	

Table 6.7 Details for residual space diagnostic and threshold calculation for data-driven fault diagnosis techniques applied in case studies

Feature extraction algorithm	Feature space contribution calculation
Principal component analysis (PCA)	Average contribution of each process variable to squared prediction error for samples that have identified as being faulty
Nonlinear principal component analysis (NLPCA)	
Kernel principal component analysis (KPCA)	
Random forest feature extraction (RF)	

classifier may be generally powerful but restricted by a poor choice of threshold. ROC curves level the playing field in comparing alarm systems, by removing the effect of threshold selection. To construct an ROC curve, the true classification must be known, i.e. which samples are indeed faulty and which are not.

The ROC curve is a parameterized curve, where a one-sided detection threshold is the curve parameter. The ROC curve succinctly visualizes the trade-off between true alarms and false alarms for a classification system. Given a specific application of a diagnostic statistic with a continuous (or pseudo-continuous) distribution, the true alarm rate (TAR) and false alarm rate (FAR) are solely dependent on the selection of the detection threshold, θ:

$$\mathrm{TAR} = f_{\mathrm{TAR}}(\theta) \tag{6.11}$$

$$\mathrm{FAR} = f_{\mathrm{FAR}}(\theta). \tag{6.12}$$

The parametric equations above are used to construct the ROC curve. The system of two equations (6.11 and 6.12) with three unknowns (TAR, FAR and θ) results in a one-dimensional parametric curve.

Figure 6.7 shows a simulated example of a diagnostic statistic distribution and its associated ROC curve. Diagnostic statistics representing NOC and fault conditions show different, but overlapping, distributions. The ROC curve is constructed by calculating true alarm rates and false alarm rates for different (upper limit) threshold

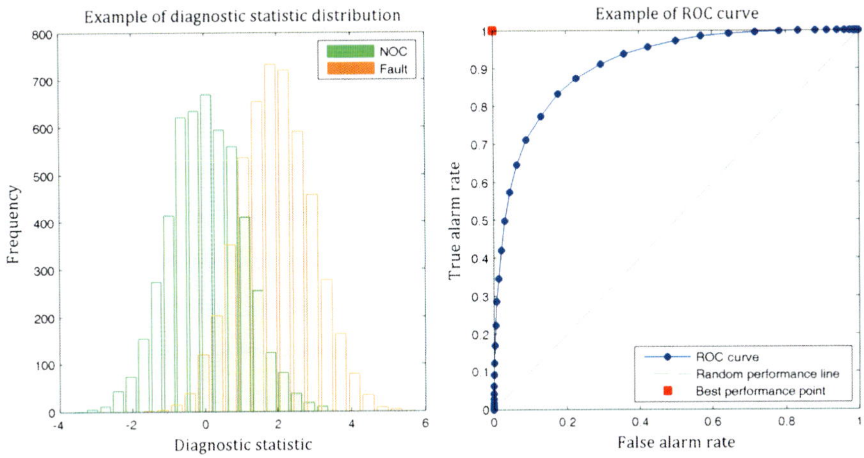

Fig. 6.7 Example of the distribution of a diagnostic statistic and its associated ROC curve

values. The resulting true alarm rate and false alarm rate for a specific choice of threshold value is indicated with a circular marker in the ROC space.

For this alarm system, when the threshold value is low,[1] there is a higher probability of correctly classifying the fault data (high true alarm rate) but also a higher probability of incorrectly classifying the NOC data (high false alarm rate). This corresponds to the top right corner of the ROC space: a liberal classifier which errs in the direction of alarming. When the threshold value is high, there is a lower probability of false alarms but also of true alarms. This corresponds to the bottom left corner of the ROC space: a conservative classifier which errs in the direction of not alarming.

An ideal classifier would have a true alarm rate of 1 and a false alarm rate of 0 (the top left corner of the ROC space). The closer a classifier is to the top left corner, the better its performance. The presence of overlap in the NOC and fault distributions of the example case in Fig. 6.7 restricts the optimal performance of this diagnostic statistic: A true alarm rate of 1 and a false alarm rate of 0 cannot be achieved.

A classifier that lies on the line of identity (diagonal) in the ROC space does no better than random guessing; such a classifier is known as a random classifier. A classifier with a ROC curve (or point) in the lower right triangle of the ROC space performs worse than random guessing. Although such a classifier contains some useful information about the NOC and fault diagnostic statistics, the information is applied incorrectly. Reversing the alarms of such a classifier would give improved performance: by considering the threshold as a lower limit, instead of as an upper limit.

[1]Note: threshold values are not indicated on the ROC curve.

The area under the ROC curve (AUC) is a statistic that summarizes the ROC curve for a specific classifier. The AUC is interpreted as the probability that a classifier will rank a randomly chosen true alarm above a randomly chosen false alarm. The higher the AUC, the better the performance of the classifier. Where AUC is equal to 0.5, this translates to a random classifier, while AUC values below 0.5 translate to classifiers that can be improved by reversing the classification result. The AUC is a convenient scalar summary for comparing ROC curves of different alarm systems.

6.5 Case Study: Simple Nonlinear System

As a first case study to demonstrate the efficacy of the process diagnostic approaches discussed above, a simulated nonlinear system is considered. With only a few variables to consider, direct visualization of the data is possible.

6.5.1 *Description*

The simple simulated nonlinear system (Shao et al. 2009) consists of three variables and two degrees of freedom. Normal operating conditions are defined by the following set of equations:

$$\begin{aligned} x_1 &= t_1 + e_1 \\ x_2 &= t_1^3 - 4.5t_2^2 + 6t_1 + t_2 + e_2 \\ x_3 &= 3t_1^2 - t_2^3 + 3t_2^2 + e_3. \end{aligned} \quad (6.13)$$

Above, t_1 and t_2 are uniform distributed over the range [0.01, 3] and e_1, e_2 and e_3 are independent Gaussian noise variables with mean of 0 and variance of 0.01. Two faults are introduced by changing certain parameters, with 200 samples generated for each fault (the normal and fault data are shown in Fig. 6.8). Of the 200 samples representing normal operating condition (NOC), the first 100 were used for model calibration, while the 2nd 100 samples were used for model validation (the validation NOC data set).

Two faults were simulated:

- Fault 1: A linear increase of 0.1(*i*-50) is added to x_3 from sample 51 to sample 150, where *i* is the sample number.
- Fault 2: The coefficient of the t_1^2 term in the expression for x_3 is linearly increased from 3 at sample 50 to 4 at sample 150. The coefficient then remains 4 up to sample 200.

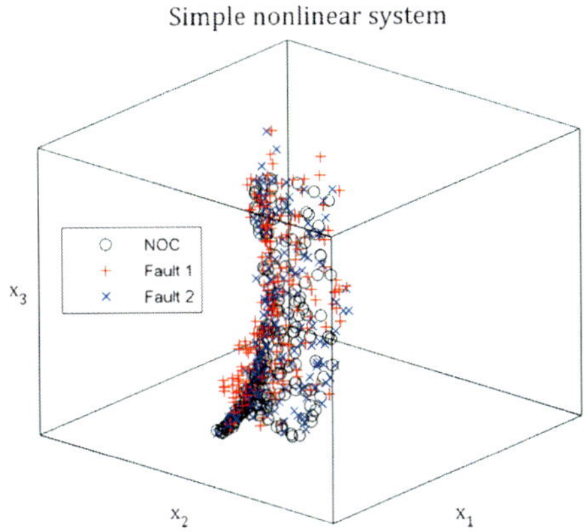

Fig. 6.8 Normal operating condition and fault data for the simple nonlinear system case study

6.5.2 *Results of Fault Diagnosis*

The results obtained with the process monitoring schemes based on PCA, KPCA, NLPCA and RF are discussed below.

Selecting the Number of Features

Selection of the number of features to retain with each algorithm is considered based on the scree plots or scree plot equivalents generated by the mappings of the variables to the feature space. In this case, there are only three variables to consider in a two-dimensional subspace, but it is still instructive the view the automated selection of the number of features to retain.

As indicated in the summary description of the algorithms, feature selection was based on parallel analysis for PCA, KPCA and RF, while PRESS cross-validation was used to select the number of features for NLPCA. The results of these algorithms are shown in Fig. 6.9. The red curves in the scree plots are the mean values for five runs with random data and the retention of two features is suggested when using PCA and KPCA. The PRESS cross-validation procedure in the NLPCA approach suggested the retention of three features, while parallel analysis with the random forests suggested the retention of 13 components. Although this seems excessive compared to the results obtained with the other algorithms (and the true dimensionality of the feature space), the rightmost bottom panel of Fig. 6.9 suggests that a more conservative estimate of the number of features to be retained would probably be justifiable, given that the red curve representing the random data runs very close to the actual scree plot.

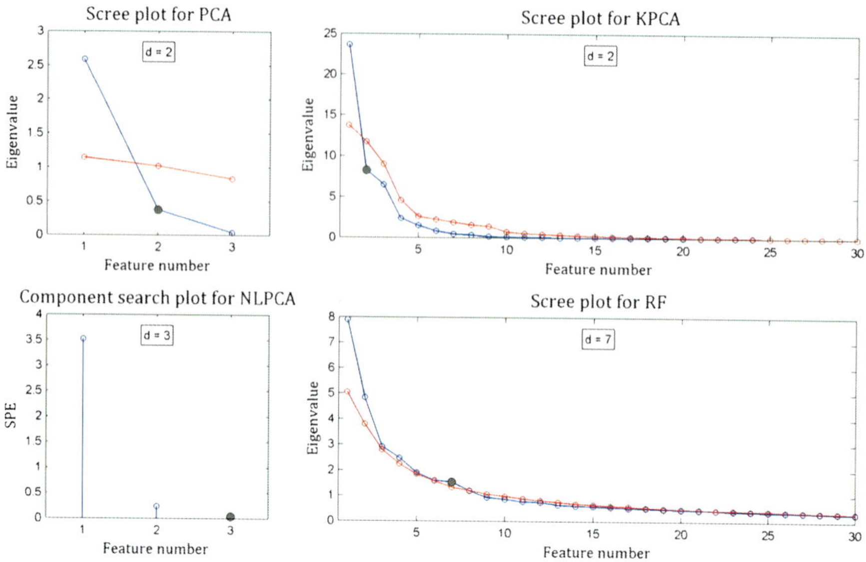

Fig. 6.9 Scree plots for the simple nonlinear system case study. KPCA and RF scree plots are truncated at 30 features; eigenvalues exist up to 100 features. *Red shows* average scree plot derived from five randomized data sets. *Grey circles* show selected number of features

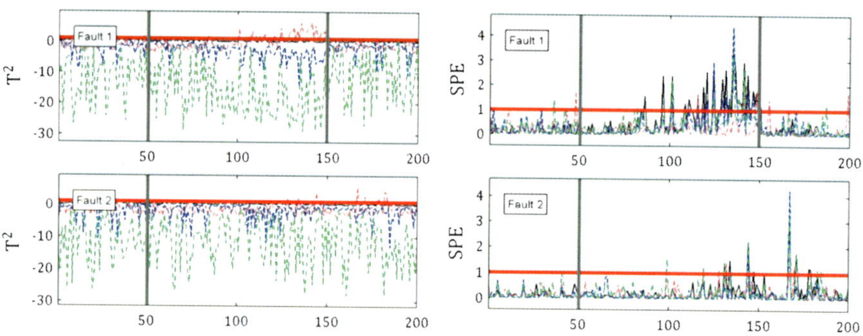

Fig. 6.10 Diagnostic sequence plots for the simple nonlinear system. PCA = *black*, NLPCA = *pink*, KPCA = *blue*, RF = *green* (Diagnostics scaled to have threshold of 1, indicated in *red*)

Performance of the Different Models

The T^2 and squared prediction error (SPE) control charts for the two different faults are shown in Fig. 6.10. In these figures, the control limits are indicated by bold horizontal red lines, while the diagnostic variables, i.e. the T^2- and SPE-statistics derived from the features for PCA, NLPCA, KPCA and RF, are indicated by black, pink, blue and green curves, respectively.

Table 6.8 False alarm rates on validation NOC data for the simple nonlinear system, before and after adjustments of the model thresholds for more reliable comparison

FAR (T^2/SPE)	PCA	NLPCA	KPCA	RF
Before adjustment	0/0.02	0.07/0	0.08/0.01	0/0.02
After adjustment	0.01/0.01	0.01/0.01	0.01/0.01	0.01/0.01

Table 6.9 False alarm run length on the validation data of the simple nonlinear system with before and after adjustments of the model thresholds for more reliable comparison

ARL (T^2/SPE)	PCA	NLPCA	KPCA	RF
Before adjustment	–/8	15/–	7/46	–/6
After adjustment	44/8	67/58	12/46	62/6

Table 6.10 False alarm rates on the fault condition data for the simple nonlinear system

FAR (T^2/SPE)	PCA	NLPCA	KPCA	RF
Fault 1	0.03/0	0.05/0.06	0.01/0.03	0.02/0.04
Fault 2	0/0	0.02/0	0/0	0/0

Validation Results

Table 6.8 shows the false alarm rates (FAR) of the different models for the T^2- and SPE-statistics on the NOC validation data, before and after adjustment on these data to ensure fair comparison. It consisted of alignment of the detection thresholds of all the algorithms to correspond to a 1 % false alarm rate on the validation data for the normal operating conditions by using the percentile approach for limits correction.

Considering the T^2- and SPE-statistics collectively, PCA and RF had the lowest false alarm rates of 0.02 before adjustment. Obviously, after adjustment, all the algorithms had false alarm rates of 0.01.

The false alarm run lengths of the different models on the validation data are shown in Table 6.9, i.e. the run lengths on 100 samples of NOC data not used in calibration of the models. Here the autoassociative neural network (NLPCA) performed consistently better than the other models, showing the largest minimum of the two statistics in each case.

The false alarm rates of the models on the fault condition data are shown in Table 6.10. None of the models outperformed the other strongly.

Unlike the other algorithms, KPCA had a zero false alarm average run length on Fault 1, as indicated in Table 6.11.

The missing alarm rates (MAR) of the models for the fault condition data are shown in Table 6.12. As indicated, the neural network model had the lowest missing alarm rates when the two statistics are considered collectively, as would be the case in practice.

Table 6.11 False alarm run length for the simple nonlinear system on the fault condition data

ARL (false) (T^2/SPE)	PCA	NLPCA	KPCA	RF
Fault 1	45/–	2/10	0/2	52/35
Fault 2	–/–	30/–	–/–	–/–

Table 6.12 Missing alarm rates for the simple nonlinear system on the fault condition data

MAR (T^2/SPE)	PCA	NLPCA	KPCA	RF
Fault 1	0.93/0.71	0.66/0.96	1/0.86	1/0.93
Fault 2	0.953/0.94	0.90/0.967	0.993/0.987	1/0.94

Table 6.13 True alarm run lengths for the simple nonlinear system

ARL (true) (T^2/SPE)	PCA	NLPCA	KPCA	RF
Fault 1	13/33	33/64	–/33	–/31
Fault 2	9/80	68/68	98/93	–/48

Table 6.14 Detection delay for three consecutive samples (First of three samples indicated) for the simple nonlinear system

DD (T^2/SPE)	PCA	NLPCA	KPCA	RF
Fault 1	–/77	64/–	–/83	–/–
Fault 2	–/–	–/–	–/–	–/–

Table 6.15 Area under receiver operator characteristic curve for the simple nonlinear system

AUC (T^2/SPE)	PCA	NLPCA	KPCA	RF
Fault 1	0.388/0.745	0.680/0.620	0.317/0.722	0.319/0.693
Fault 2	0.555/0.546	0.480/0.550	0.463/0.559	0.487/0.587

The true alarm run lengths for the different models are shown in Table 6.13. Interestingly, the PCA model performed significantly better than the other models, even after adjustment of the detection thresholds on the validation NOC data.

The detection delay results are shown in Table 6.14. The NLPCA model performed best on the Fault 1 data, but none of the models could flag Fault 2 at three consecutive points, as can also be seen in Fig. 6.10.

Finally, the areas under the receiver operator characteristic (ROC) curves for all the models and fault conditions are shown in Table 6.15. After adjustment of the thresholds of the models, the random forest model performed the best with the largest ROC area, although this was not markedly larger than the areas generated by the other models. The SPE-statistics generally showed higher AUC values than the T^2-statistics for the NLPCA, KPCA and RF models. This indicates that the SPE-statistic is more sensitive to the presence of faulty conditions of the type of Fault 1 and Fault 2. The NLPCA, KPCA and RF models have the 1-SVM approach

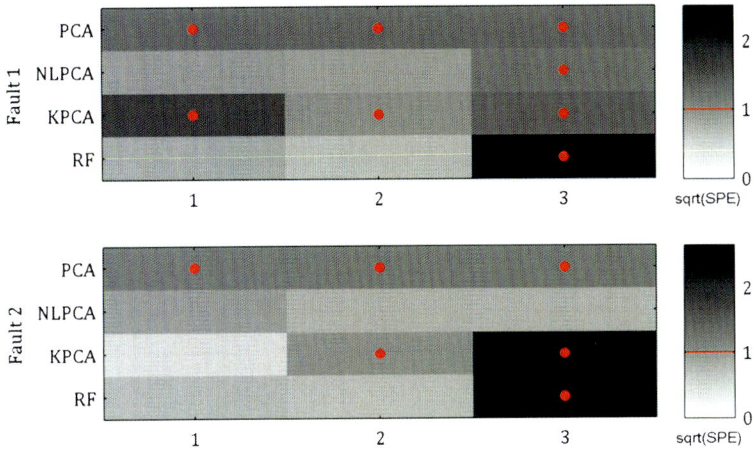

Fig. 6.11 Contribution plots based on the squared prediction error of reconstructed variables of the simple nonlinear system (Square roots of relative contributions shown: contributions divided by 99 % confidence limit and relative contributions exceeding 1 indicated with *red markers*)

to feature space characterization in common. This characterization approach may be less robust than the SPE-statistic due to complexity of optimizing the 1-SVM parameters, given the unsupervised nature of the problem.

Contribution Results

The results of the contribution plots based on the squared prediction errors of the reconstructed variables are shown in Fig. 6.11. NLPCA and RF correctly identified implication of variable x_3 only in the first fault condition, while RF was the only algorithm that could identify the implicated variable (x_3) in the second fault.

Effect of Number of Retained Features

Figures 6.12 and 6.13 show the effect of the number of retained features on the false alarm rates, missing alarm rates and areas under the ROC curves for Fault 1 and 2 data, respectively.

For Fault 1, PCA and NLPCA show very similar trends: Feature space detection is most optimal when three features are retained (which negates the utility of feature extraction, since the measured process data is three-dimensional). However, the residual space fault detection is most optimal for two retained features (and with better detection performance than the feature space diagnostics), reflecting the true underlying dimensionality of the system. For most number of retained features, KPCA and RF perform worse than optimal PCA and NLPCA.

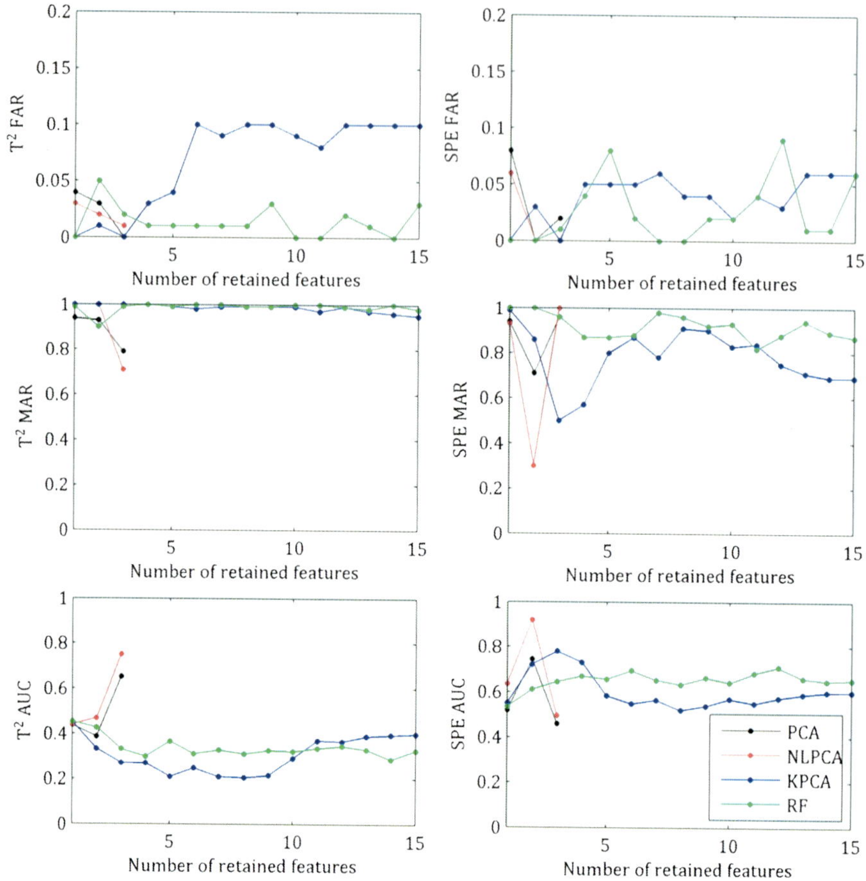

Fig. 6.12 Effect of varying the number of retained features on the false alarm rates, missing alarm rates and area under ROC curve for the simple nonlinear system, for Fault 1 data

For Fault 2, the detection performance of the different methods is relatively insensitive to the number of retained features, bar the feature space false alarm rate for RF.

6.6 Case Study: Tennessee Eastman Problem

The Tennessee Eastman process is a well-known simulated chemical process developed to provide a realistic industrial process on which to evaluate the performance of process control and monitoring methods (Russell et al. 2000b). The Eastman Chemical Company created a simulation of an actual chemical process with five major units and eight components (Downs and Vogel 1993). Simulation data of

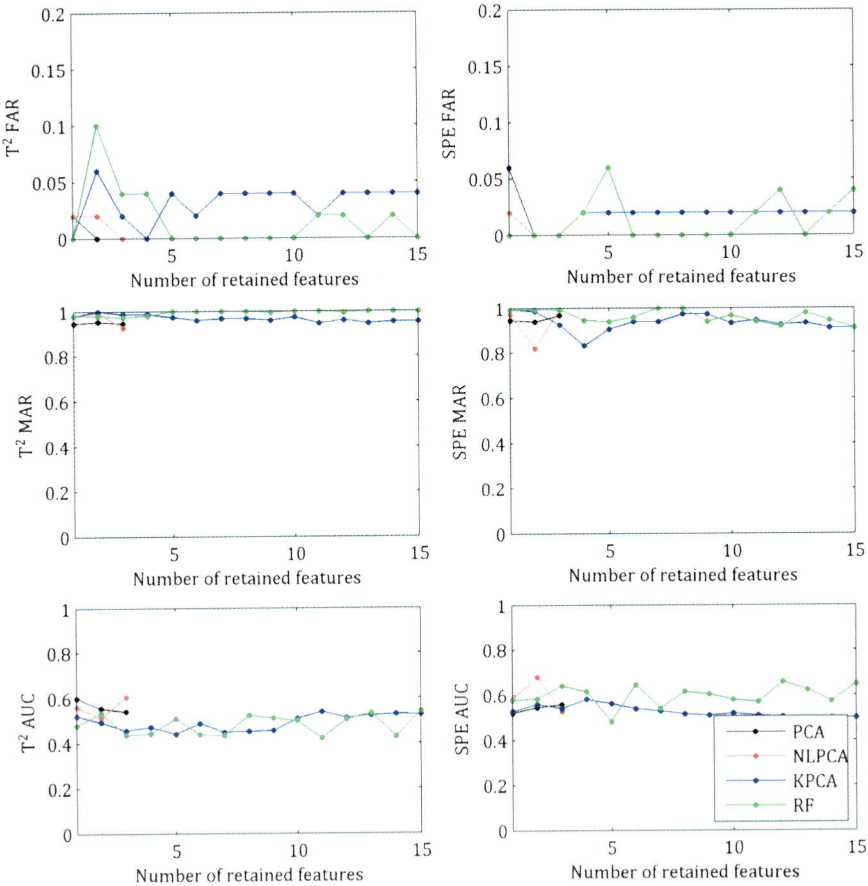

Fig. 6.13 Effect of varying the number of retained features on the false alarm rates, missing alarm rates and area under ROC curve for the simple nonlinear system, for Fault 2 data

the Tennessee Eastman process, with plant-wide control based on proportional and proportional–integral control as suggested by (Lyman and Georgakis 1995), are available for normal operating conditions, as well as 21 fault conditions.

6.6.1 Process Description

The flowsheet for the process is given in Fig. 6.14. Gaseous reactants A, C, D and E, with inert B, are fed to a water-cooled reactor, where liquid products G and H and liquid by-product F are formed through irreversible exothermic reactions. The reactor product is cooled in a condenser and then separated in a vapour liquid separator. The vapour phase from the separator is recycled via a compressor to the

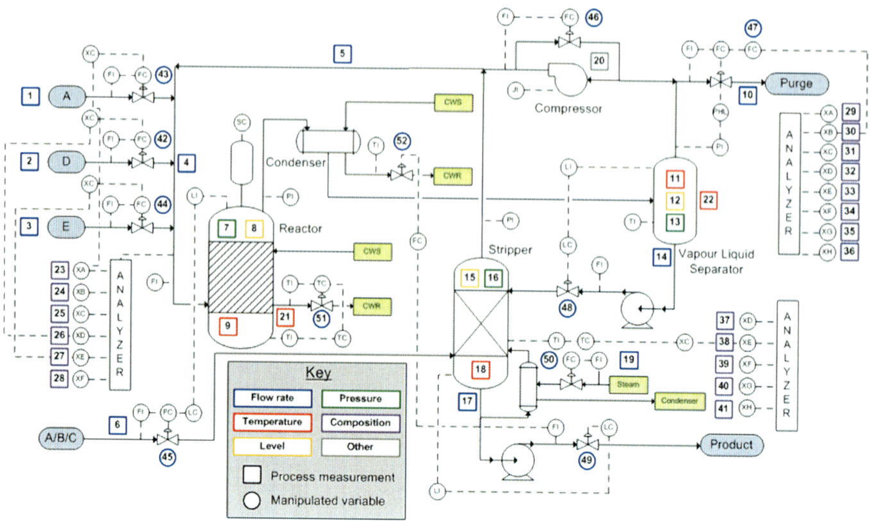

Fig. 6.14 Process flow diagram of Tennessee Eastman process (After Russell et al. 2000b)

reactor, with a portion purged to avoid accumulation of inert B and by-product F in the system. The liquid from the separator is pumped to a stripper to remove remaining reactants in the stream for recycle. Liquid products G and H report to the product stream.

6.6.2 Control Structure

The control structure implemented by Lyman and Georgakis (1995) uses single-input, single-output proportional–integral–derivative (PID) controllers, with the plant control scheme designed in a tiered fashion.

The primary control objective is the product production rate; the suggested scheme controls the product production rate by manipulating the condenser cooling water valve. Since the reactor products G and H are condensed to liquid form, a reduction in condenser cooling water flow will reduce the amount of product reporting to the final product stream.

Another level of control concerns the reactor, separator and stripper levels. When the product production rate is decreased, the increased amount of G and H reporting back to the reactor (through the recycle stream) will increase the reactor level. The reactor level is controlled by a control loop that manipulates the feed flow of reactants A and C. The separator and stripper levels are controlled by control loops manipulating the liquid flow out of these vessels.

Table 6.16 Input variables for the Tennessee Eastman process

Variable	Description	Variable	Description
1	*A* feed (PM)	27	Reactor feed component *E* (CM)
2	*D* feed (PM)	28	Reactor feed component *F* (CM)
3	*E* feed (PM)	29	Purge component *A* (CM)
4	Total feed (PM)	30	Purge component *B* (CM)
5	Recycle flow (PM)	31	Purge component *C* (CM)
6	Reactor feed rate (PM)	32	Purge component *D* (CM)
7	Reactor pressure (PM)	33	Purge component *E* (CM)
8	Reactor level (PM)	34	Purge component *F* (CM)
9	Reactor temperature (PM)	35	Purge component *G* (CM)
10	Purge rate (PM)	36	Purge component *H* (CM)
11	Separator temperature (PM)	37	Product component *D* (CM)
12	Separator level (PM)	38	Product component *E* (CM)
13	Separator pressure (PM)	39	Product component *F* (CM)
14	Separator underflow (PM)	40	Product component *G* (CM)
15	Stripper level (PM)	41	Product component *H* (CM)
16	Stripper pressure (PM)	42	D feed flow (MV)
17	Stripper underflow (PM)	43	E feed flow (MV)
18	Stripper temperature (PM)	44	A feed flow (MV)
19	Stripper steam flow (PM)	45	Total feed flow (MV)
20	Compressor work	46	Compressor recycle valve (MV)
21	Reactor cooling water outlet temp. (PM)	47	Purge valve (MV)
22	Separator cooling water outlet temp. (PM)	48	Separator product liquid flow (MV)
23	Reactor feed component *A* (CM)	49	Stripper product liquid flow (MV)
24	Reactor feed component *B* (CM)	50	Stripper steam valve (MV)
25	Reactor feed component *C* (CM)	51	Reactor cooling water flow (MV)
26	Reactor feed component *D* (CM)	52	Condenser cooling water flow (MV)

PM process measurement, *CM* concentration measurement, *MV* manipulated variable

Another aspect of control concerns the pressure build-up in the circuit. Pressure control is achieved by controlling the gas content in the recycle stream through manipulation of the flow rate of the purge stream. Finally, control loops for reactor temperature, component feed flow rate and final product composition are included.

6.6.3 *Process Measurements*

The Tennessee Eastman process monitoring data sets, as described by Russell et al. (2000a), were used in this case study. Process data consist of 52 variables: 41 measured variables (process and composition measurements) and 11 manipulated variables (see Table 6.16), sampled at 3 min intervals over a 25 h window.

Composition measurements for the total feed and purge streams have a delay of 6 min (2 samples), while the product stream composition measurements have a delay of 15 min (5 samples). Simulations for the normal operating conditions generated 500 samples for a training set that was used for model calibration, as well as 960 samples for a validation data set associated with normal operating conditions.

6.6.4 *Process Faults*

The Tennessee Eastman process monitoring data sets include 21 fault conditions. Simulations for the fault conditions generated 960 samples for each fault condition, with each fault condition introduced after 160 samples, giving 800 faulty samples in total. A summary of the 21 fault conditions is given in Table 6.17. Table 6.17 also gives an indication of variables considered to be directly related to specific faults, based on careful consideration of the process flowsheet and control structures (Russell et al. 2000a).

6.6.5 *Performance of the Different Models*

The performance of the four different algorithms is discussed below, starting with the automated selection of model features. As explained above, the models were evaluated on 22 data sets in total, each containing 960 samples, i.e. one data set containing NOC data only and 21 data sets each reflecting a particular fault condition.

Selecting the Number of Features

As before, the scree plots for PCA, KPCA and RF are shown, as well as the cross-validation search plot for NLPCA. The RF model contained the lowest number of features, i.e. 9. As before, the angle between the curves representing the permuted (red) and actual (blue) variables is rather small, so that the number of features selected may not be particularly reliable for the RF model and to a lesser extent also the KPCA model (Fig. 6.15).

Diagnostic Sequence Plots

Diagnostic plots for all the fault conditions are shown in Figs. 6.16, 6.17 and 6.18. As before, in these figures, the control limits are indicated by bold horizontal red lines, while the diagnostic variables for PCA, NLPCA, KPCA and RF are indicated by black, pink, blue and green curves, respectively.

Table 6.17 Process faults for Tennessee Eastman process, with indication of affected variables, where known

Fault	Description	Type	Variables directly involved	Additional variables considerably affected
1	Stripper feed *A/C* feed ratio, *B* composition constant	Step change	1, 6, 23, **43**, 45	Most
2	Stripper feed *B* composition, *A/C* feed ratio constant	Step change	47	Most
3	*D* feed temperature	Step change	–	–
4	Reactor cooling water inlet temperature	Step change	9, 21, **51**	None
5	Condenser cooling water inlet temperature	Step change	11, **22**, 52	Most
6	Reactor feed *A* loss	Step change	44	Most
7	Stripper feed *C* header pressure loss, reduced availability	Step change	–	Most
8	*A*, *B* and *C* stripper feed composition	Random variation	–	Most
9	Reactor feed *D* temperature	Random variation	–	–
10	Stripper feed *C* temperature	Random variation	–	–
11	Reactor cooling water inlet temperature	Random variation	9, 21, **51**	None
12	Condenser cooling water inlet temperature	Random variation	22	–
13	Reaction kinetics	Slow drift	–	–
14	Reactor cooling water valve	Sticking	51	Most
15	Condenser cooling water valve	Sticking	–	–
16–20	Unknown		–	–
21	Stripper feed valve fixed at steady-state position	Constant position	45	–

Bold face indicates single most important variable where more than one variable is involved

Validation Results

The performance of the models with regard to the false alarm rate (FAR) on the NOC data is shown in Table 6.18. PCA appeared to perform the best on the NOC data prior to adjustment of the thresholds of the different models, after which all of them performed similarly at a level of approximately 0.01, as could be expected.

The performance of the models with regard to the false alarm run length on the NOC data is shown in Table 6.19. PCA appeared to perform the best on the NOC data prior to adjustment of the thresholds of the different models, after which the kernel principal component (KPCA) model outperformed the others.

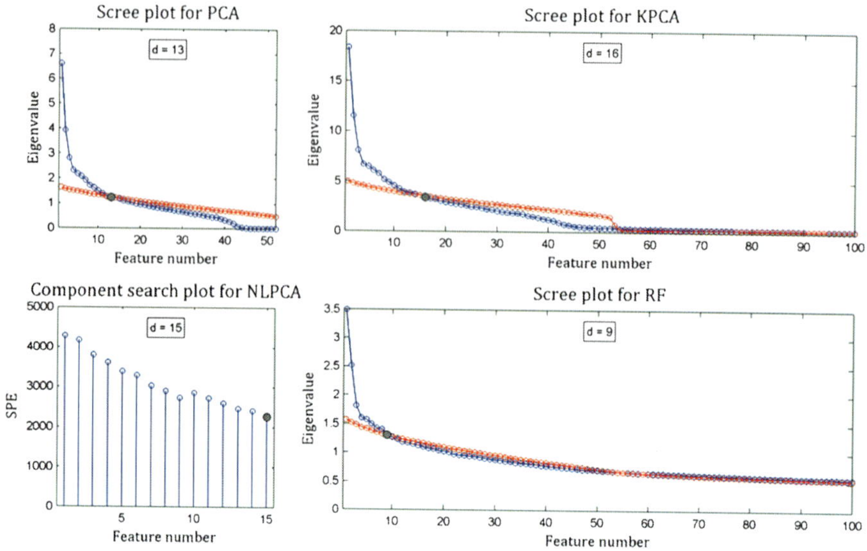

Fig. 6.15 Scree plots for the Tennessee Eastman process case study. KPCA and RF scree plots are truncated at 100 features; eigenvalues exist up to 500 features. *Red shows* average scree plot derived from five randomized data sets. *Grey circles* show selected number of features

The false alarm rates (FAR) of the models on the 21 fault conditions are shown in Table 6.20. On average, PCA and KPCA performed best, with the lowest overall false alarm rates (0.007), which were approximately 70 % of that of NLPCA, which had the highest overall mean false alarm rate (0.010), higher than that of the RF model (0.009).

The false alarm run lengths of the models on the 21 fault conditions are shown in Table 6.21. Here KPCA emerged as the best performer overall, with the largest run length in 16 of the 21 cases, followed by RF, with a score of 12, PCA with 10 and NLPCA with 9 (where more than one model had the same longest run length, such as infinite run lengths at different faults, each model was credited equally).

The missing alarm rates of the models on the 21 fault conditions are shown in Table 6.22. When average overall missing alarm rates on all the faults are considered, then the models performed as follows (from best to worst): KPCA (0.3645), PCA (0.3709), RF (0.3917) and NLPCA (0.3953). The five most difficult faults to detect were Faults 3 (0.981), 9 (0.975), 15 (0.941), 19 (0.871) and 5 (0.748), with missing alarm rates of the best models shown in parentheses. Of these five faults, PCA and NLPCA could each detect two with the lowest MAR, while the RF model could detect one.

It is noticeable that the RF and KPCA feature space diagnostic (T^2) exhibits very high missing alarm rates for all faults (with the lowest RF T^2 MAR being 0.904 and KPCA T^2 MAR being 0.981). This cannot be due solely to the feature space characterization method, as the same feature space characterization method is used for NLPCA, which shows T^2 MAR as low as 0.013. A probable explanation for the

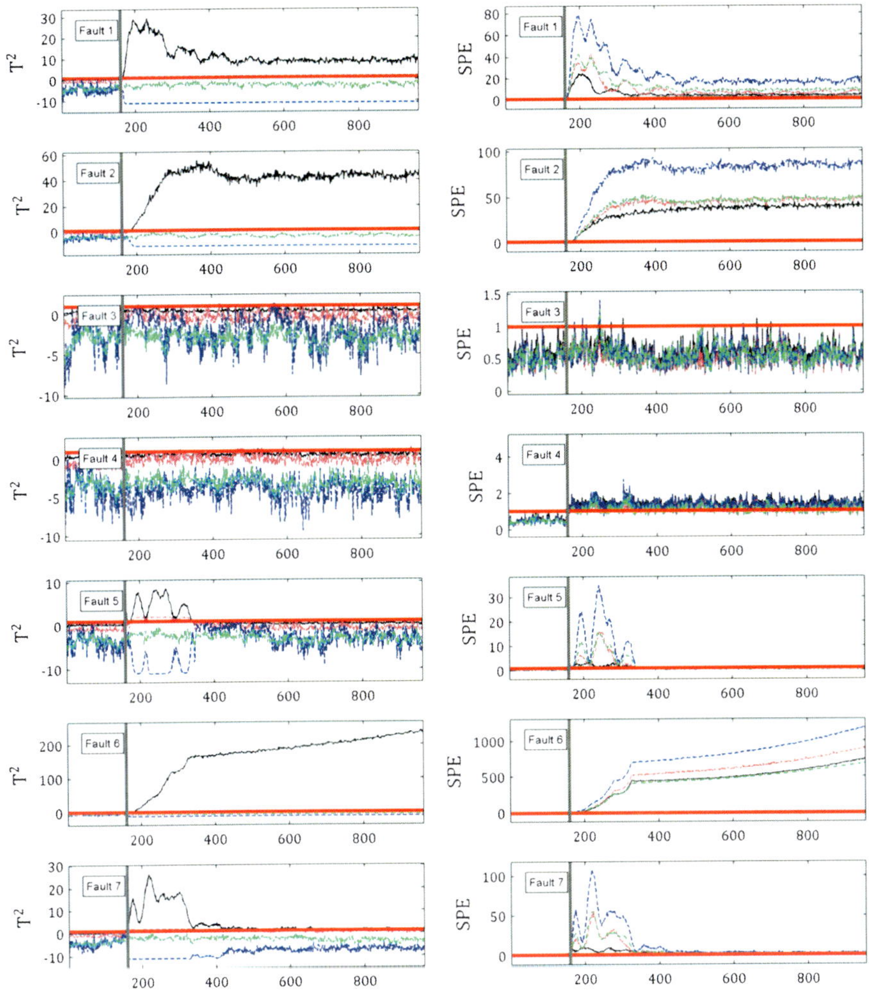

Fig. 6.16 Diagnostic sequence plots for Faults 1–7 for the Tennessee Eastman process. PCA = *black*, NLPCA = *pink*, KPCA = *blue*, RF = *green* (Diagnostics scaled to have threshold of 1, indicated in *red*)

poor RF T^2 MAR performance relates to the constrained response characteristic (see Chap. 5). The constrained response characteristic may lead to high missing alarm rates in the feature space but is also responsible for low missing alarm rates in the residual space.

Table 6.23 shows the performance of the models with regard to true run length on the fault conditions. When scored by assigning a 1 to the best performer at each fault condition (co-winners each getting 1), as before, then interestingly, PCA performed best on this criterion, with a score of 16, followed by KPCA with 12 and RF and NLPCA jointly with 11 each.

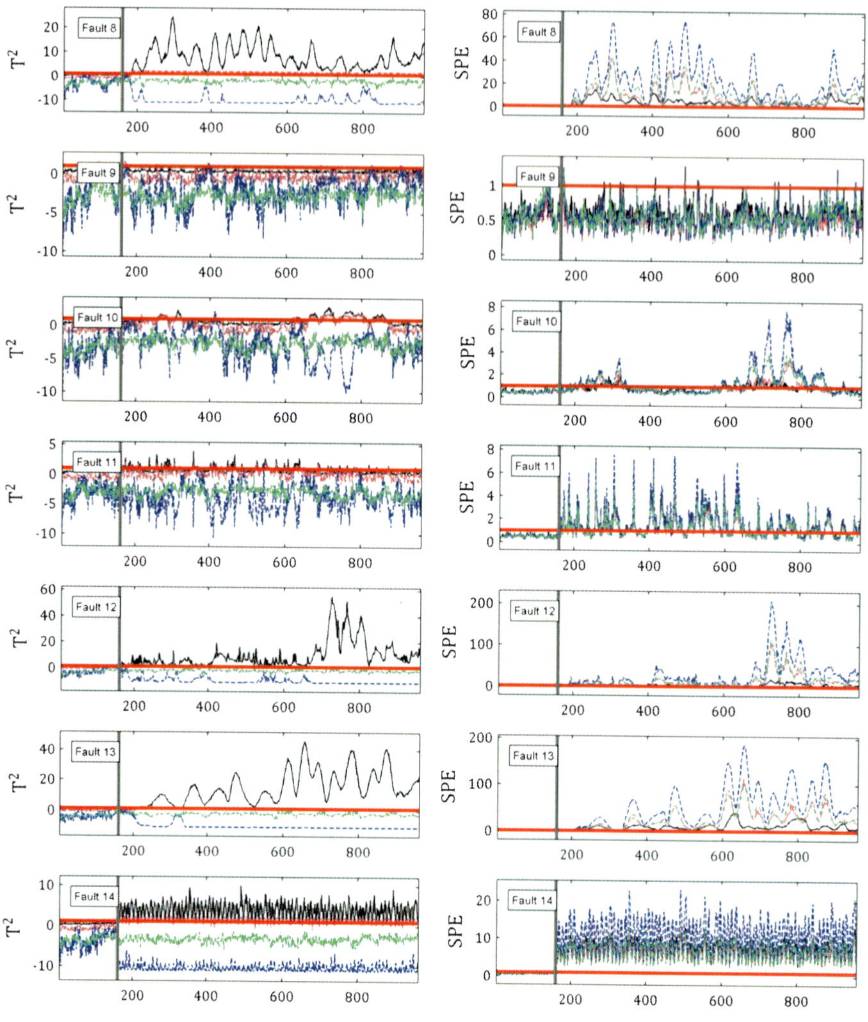

Fig. 6.17 Diagnostic sequence plots for Faults 8–14 for the Tennessee Eastman process. PCA = *black*, NLPCA = *pink*, KPCA = *blue*, RF = *green* (Diagnostics scaled to have threshold of 1, indicated in *red*)

The detection delay times of the models on the different fault conditions are summarized in Table 6.24. When scoring the winners as before by assigning a 1 to the best performer on each fault (co-winners each getting 1), then the results are as follows: PCA (15), KPCA (14), RF (11) and NLPCA (6).

Finally, the ROC area values of the models are shown in Table 6.25. The different models showed average values as follows: PCA (0.883), NLPCA (0.871), RF (0.868) and KPCA (0.866). Discrimination between the models on the basis of these similar results is probably not possible, given the relatively large standard deviations in the values.

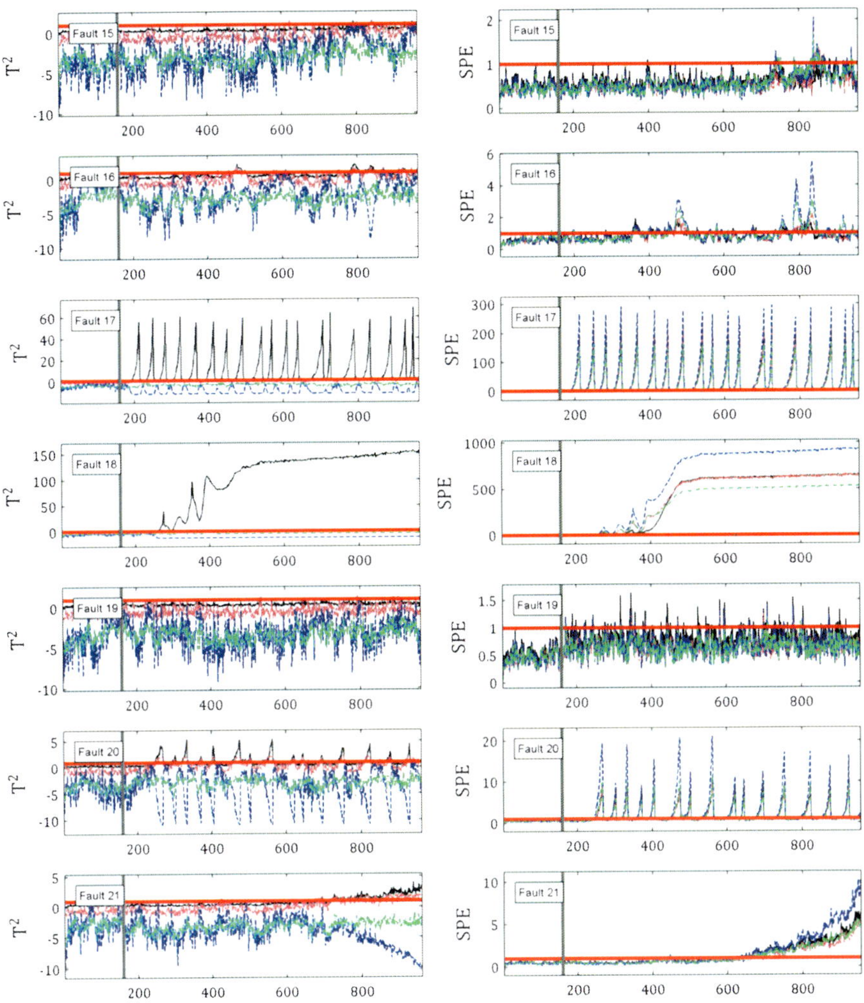

Fig. 6.18 Diagnostic sequence plots for Faults 15–21 for the Tennessee Eastman process. PCA = *black*, NLPCA = *pink*, KPCA = *blue*, RF = *green* (Diagnostics scaled to have threshold of 1, indicated in *red*)

Table 6.18 False alarm rates on the NOC validation data for the Tennessee Eastman process, before and after adjustments of the model thresholds for more reliable comparison

FAR (T^2/SPE)	PCA	NLPCA	KPCA	RF
Before adjustment	0.0156/0.0917	0.2490/0.1750	0.0583/0.3948	**0**/0.6375
After adjustment	**0.0104/0.0104**	**0.0104/0.0104**	**0.0104/0.0104**	**0.0104/0.0104**

Bold values show the overall best for the T^2- and SPE-statistics

Table 6.19 False alarm run length on the NOC validation data for the Tennessee Eastman process, before and after adjustments of the model thresholds for more reliable comparison

ARL (T^2/SPE)	PCA	NLPCA	KPCA	RF
Before adjustment	30/16	6/16	30/8	960/8
After adjustment	30/256	189/807	578/256	63/807

Table 6.20 False alarm rates on the fault conditions in the Tennessee Eastman process

FAR (T^2/SPE)	PCA	NLPCA	KPCA	RF
Fault 1	**0**/0.006	0.006/**0**	**0/0**	0.006/**0**
Fault 2	**0**/0.006	0.013/**0**	**0/0**	**0/0**
Fault 3	0.006/0.013	0.013/**0**	0.025/**0**	0.013/**0**
Fault 4	0.006/0.006	0.006/**0**	**0/0**	0.013/**0**
Fault 5	0.006/0.006	0.013/**0**	**0/0**	0.013/**0**
Fault 6	**0/0**	0.006/**0**	**0/0**	**0/0**
Fault 7	**0/0**	**0/0**	**0/0**	**0/0**
Fault 8	**0/0**	0.006/**0**	0.006/**0**	0.006/**0**
Fault 9	0.013/0.006	0.031/**0**	0.019/0.019	0.006/0.05
Fault 10	**0/0**	**0/0**	**0/0**	**0/0**
Fault 11	**0**/0.006	**0/0**	**0/0**	0.019/**0**
Fault 12	**0/0**	0.006/**0**	0.006/**0**	0.006/**0**
Fault 13	**0/0**	**0/0**	**0/0**	**0/0**
Fault 14	**0/0**	0.006/**0**	**0/0**	**0/0**
Fault 15	**0**/0.006	**0/0**	**0/0**	**0/0**
Fault 16	0.044/**0.006**	0.088/0.019	0.1/0.019	0.025/0.025
Fault 17	**0**/0.013	**0/0**	**0/0**	**0/0**
Fault 18	**0**/0.006	0.006/**0**	**0/0**	0.013/**0**
Fault 19	**0/0**	**0/0**	**0/0**	**0/0**
Fault 20	**0/0**	**0/0**	**0/0**	**0/0**
Fault 21	**0**/0.013	0.006/**0**	**0/0**	0.013/**0**

Bold values show the overall best for the T^2- and SPE-statistics

The contribution plots for the different fault conditions and algorithms are shown in Figs. 6.19, 6.20 and 6.21. Some of the faults could not be identified, for example, none of the algorithms could identify variable no 45 associated with Fault 21, but otherwise reasonable results were obtained with most of the models.

6.7 Case Study: Sugar Refinery Benchmark

A fault detection and identification benchmark was developed by the Development and Application Methods for Diagnosis of Actuators in Industrial Controls Systems (DAMADICS) research training network (DAMADICS RTN 2002).

Table 6.21 False alarm run lengths on the fault conditions in the Tennessee Eastman process

ARL (false) (T^2/SPE)	PCA	NLPCA	KPCA	RF
Fault 1	–/51	54/–	–/–	54/–
Fault 2	–/82	75/–	–/–	–/–
Fault 3	44/135	41/–	44/–	66/–
Fault 4	74/72	74/–	–/–	74/–
Fault 5	74/72	70/–	–/–	74/–
Fault 6	–/–	26/–	–/–	–/–
Fault 7	–/–	–/–	–/–	–/–
Fault 8	–/–	142/–	55/–	151/–
Fault 9	111/118	98/–	98/118	83/111
Fault 10	–/–	–/–	–/–	–/–
Fault 11	–/158	–/–	–/–	41/–
Fault 12	–/–	26/–	149/–	155/–
Fault 13	–/–	–/–	–/–	–/–
Fault 14	–/–	144/–	–/–	–/–
Fault 15	–/99	–/–	–/–	–/–
Fault 16	60/145	56/138	63/139	121/138
Fault 17	–/58	–/–	–/–	–/–
Fault 18	–/17	145/–	–/–	100/–
Fault 19	–/–	–/–	–/–	–/–
Fault 20	–/–	–/–	–/–	–/–
Fault 21	–/71	145/–	–/–	26/–

– indicates no detection occurred

This benchmark is based on a real sugar refinery (Cukrownia Lublin SA) in Poland, where three actuators have been modified to allow the introduction of mechanically and electrically induced faults (Bartyś et al. 2006). Faults were introduced under supervised conditions to prevent the sugar factory from operating outside acceptable quality limits. Both simulated models and real data are available as benchmark data for fault detection and identification. Detailed information, simulators and process data of this benchmark is available from the DAMADICS research group website (DAMADICS RTN 2002).

6.7.1 Process Description

A brief overview of the production of crystallized sugar from sugar beets at the Lublin sugar refinery is given here. The raw material (whole sugar beet) is prepared through washing and size reduction (slicing to obtain thin strips of sugar beet). The sugar component of the beet strips is then extracted by leaching with hot water, producing the so-called raw juice. This raw juice undergoes chemical refining to remove impurities. The chemical refining involves the treatment of the raw juice

Table 6.22 Missing alarm rates for the fault conditions in the Tennessee Eastman process

MAR (T^2/SPE)	PCA	NLPCA	KPCA	RF
Fault 1	0.008/0.003	0.008/0.004	1/0.003	0.969/0.005
Fault 2	0.02/0.014	0.029/0.015	1/0.015	0.983/0.015
Fault 3	0.998/0.989	0.981/0.999	0.994/0.988	0.989/0.99
Fault 4	0.934/0.046	0.958/0.186	1/0.084	0.989/0.335
Fault 5	0.775/0.748	0.779/0.773	0.991/0.753	0.973/0.761
Fault 6	0.011/0	0.013/0	1/0	1/0
Fault 7	0.076/0	0.409/0	1/0	0.96/0
Fault 8	0.033/0.026	0.046/0.023	1/0.024	0.904/0.024
Fault 9	0.996/0.979	0.975/0.998	0.981/0.989	0.991/0.993
Fault 10	0.66/0.639	0.724/0.641	0.989/0.503	0.978/0.541
Fault 11	0.734/0.358	0.821/0.468	0.999/0.366	0.99/0.439
Fault 12	0.019/0.026	0.05/0.024	1/0.01	0.929/0.011
Fault 13	0.06/0.045	0.076/0.045	1/0.048	0.929/0.054
Fault 14	0.048/0	0.149/0	1/0	1/0
Fault 15	0.976/0.974	0.979/0.968	0.988/0.948	0.988/0.941
Fault 16	0.835/0.731	0.845/0.803	0.986/0.648	0.981/0.735
Fault 17	0.244/0.104	0.363/0.136	1/0.106	0.988/0.131
Fault 18	0.113/0.099	0.121/0.105	1/0.103	0.989/0.108
Fault 19	1/0.871	0.985/0.97	1/0.95	0.998/0.99
Fault 20	0.701/0.548	0.756/0.569	0.999/0.521	0.995/0.536
Fault 21	0.695/0.58	0.733/0.616	1/0.596	0.998/0.611

leachate with calcium hydroxide and carbon dioxide, which creates precipitates of impurities. These precipitated impurities are removed by mechanical filtering, producing thin juice.

The thin juice is thickened through a series of evaporation steps. This stepwise boiling procedure is situated in the evaporator station. The product of the evaporator station is thick juice, which has a higher sugar concentration than the thin juice. The thick juice undergoes crystallization operations, in which crystallization nuclei are added to the thick juice to induce crystal formation. A final product separation step involves centrifuges to separate the sugar crystal product from the liquid caramel by-product.

6.7.2 *Benchmark Actuators Description*

Three actuators of the sugar refinery were selected for creating the benchmark. The first actuator controls the flow of thin juice to the first evaporator in the evaporator section (Fig. 6.22), the second actuator controls the flow of thick juice from the last

Table 6.23 True alarm run lengths for the fault conditions in the Tennessee Eastman process

ARL (true) (T^2/SPE)	PCA	NLPCA	KPCA	RF
Fault 1	6/2	4/3	−/2	88/4
Fault 2	16/11	19/12	−/12	77/12
Fault 3	83/42	35/20	33/20	6/85
Fault 4	0/0	0/0	−/0	158/0
Fault 5	0/1	1/1	206/0	80/0
Fault 6	9/0	10/0	−/0	−1/0
Fault 7	0/0	0/0	−/0	4/0
Fault 8	22/19	24/14	−/15	41/15
Fault 9	8/112	0/327	8/5	45/5
Fault 10	22/26	28/35	29/26	79/26
Fault 11	6/6	11/6	382/6	173/6
Fault 12	2/2	2/3	−/2	116/2
Fault 13	48/36	36/36	−/37	265/38
Fault 14	1/0	1/0	−/0	−/0
Fault 15	572/132	174/578	615/311	470/575
Fault 16	288/14	1/123	1/14	597/35
Fault 17	28/24	30/24	−/24	105/24
Fault 18	88/17	82/84	−/17	160/86
Fault 19	−1/10	85/83	−/10	716/10
Fault 20	86/83	85/86	777/82	37/81
Fault 21	508/13	262/285	−/285	46/284

− indicates no detection occurred

evaporator in the evaporator section (Fig. 6.23), while the third actuator controls the flow of water to one of the boilers in the boiler section (Fig. 6.24), which delivers steam to the sugar refinery processes.

Each benchmark actuator consists of a control valve, a pneumatic servomotor and a positioner. The valve controls the flow of fluid (thin juice, thick juice or water in the case of the benchmark actuators) that passes through a pipe. The servomotor controls the rate of flow through the control valve by setting the position of the valve plug. The positioner corrects for mispositioning of the servomotor shaft.

6.7.3 Actuator and Process Measurements

For the benchmark data set, 32 measured variables are available for the benchmark actuators and some process variables. The actuator measurements include fluid pressures at the valve inlet and outlet, fluid temperature at the valve outlet, the control value output of the actuator control loop, the process value of the actuator control loop and the servomotor rod displacement.

Table 6.24 Detection delay for three consecutive samples (first of three samples indicated). Tennessee Eastman process

DD (T^2/SPE)	PCA	NLPCA	KPCA	RF
Fault 1	7/3	8/4	–/3	–/5
Fault 2	17/12	28/13	–/13	80/13
Fault 3	–/–	–/–	–/86	–/86
Fault 4	75/5	429/5	–/3	–/11
Fault 5	1/2	2/2	–/1	94/1
Fault 6	10/1	11/1	–/1	–/1
Fault 7	1/1	1/1	–/1	–/1
Fault 8	23/22	25/22	–/22	42/21
Fault 9	–/361	–/–	–/–	–/–
Fault 10	79/49	102/51	251/39	–/39
Fault 11	11/11	12/11	–/11	–/11
Fault 12	3/3	8/8	–/3	180/3
Fault 13	49/37	50/37	–/41	464/45
Fault 14	4/1	6/1	–/1	–/1
Fault 15	677/741	678/685	–/578	–/583
Fault 16	312/36	265/200	–/36	758/36
Fault 17	29/25	31/25	–/25	–/25
Fault 18	93/84	99/85	–/85	165/87
Fault 19	–/185	–/223	–/617	–/–
Fault 20	87/87	89/87	–/83	–/86
Fault 21	557/285	521/497	–/286	–/488

– indicates no detection occurred

Table 6.25 Areas under receiver operator characteristic curves for the different models tested on the Tennessee Eastman process

AUC (T^2/SPE)	PCA	NLPCA	KPCA	RF
Fault 1	0.992/0.993	0.993/0.992	0.004/0.993	0.836/0.993
Fault 2	0.986/0.989	0.989/0.987	0.012/0.986	0.604/0.987
Fault 3	0.577/0.545	0.607/0.557	0.54/0.577	0.525/0.593
Fault 4	0.839/0.992	0.808/0.993	0.452/0.993	0.628/0.992
Fault 5	0.747/0.7	0.731/0.717	0.553/0.724	0.67/0.739
Fault 6	0.987/0.986	0.992/0.985	0/0.986	0.173/0.985
Fault 7	0.993/0.994	0.98/0.994	0.053/0.994	0.797/0.994
Fault 8	0.987/0.983	0.985/0.988	0.033/0.983	0.789/0.987
Fault 9	0.435/0.484	0.498/0.451	0.475/0.46	0.513/0.445
Fault 10	0.832/0.876	0.834/0.893	0.627/0.897	0.725/0.892
Fault 11	0.803/0.908	0.776/0.916	0.414/0.916	0.542/0.917
Fault 12	0.989/0.99	0.989/0.978	0.046/0.98	0.833/0.986
Fault 13	0.973/0.974	0.979/0.966	0.066/0.965	0.848/0.966
Fault 14	0.992/0.994	0.987/0.994	0.004/0.994	0.558/0.994
Fault 15	0.654/0.605	0.627/0.638	0.684/0.644	0.604/0.689
Fault 16	0.567/0.787	0.594/0.747	0.371/0.796	0.45/0.727
Fault 17	0.9/0.861	0.922/0.841	0.121/0.848	0.643/0.865
Fault 18	0.935/0.928	0.951/0.93	0.074/0.934	0.719/0.933
Fault 19	0.615/0.844	0.622/0.834	0.584/0.824	0.559/0.822
Fault 20	0.878/0.871	0.855/0.891	0.565/0.888	0.714/0.906
Fault 21	0.693/0.76	0.747/0.747	0.353/0.75	0.42/0.754

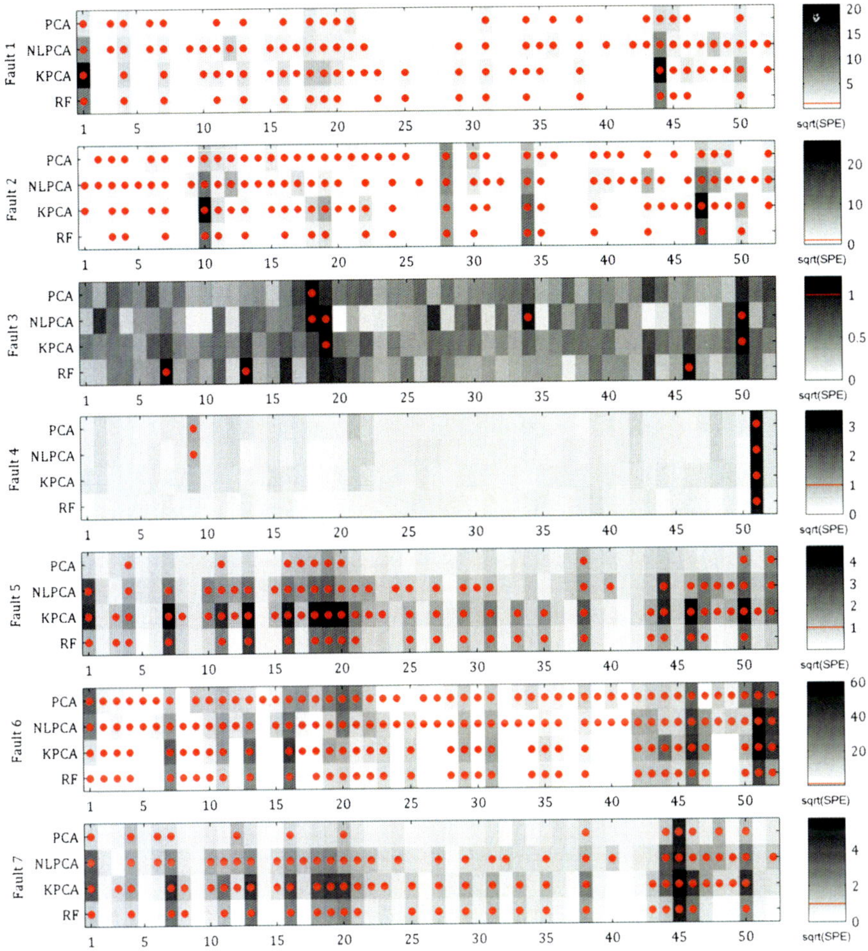

Fig. 6.19 Contribution plots (Based on squared prediction error of reconstructed variables). Tennessee Eastman process (Square root of relative contributions shown: contributions divided by 99 % confidence limit; relative contributions exceeding 1 indicated with *red markers*)

Samples are available at one second intervals. Two days' worth of normal operating conditions data are available (172,800 samples). For this study, a random sample of 1,000 samples was drawn from one day, to represent training normal operating conditions data. The normal operating conditions validation data consisted of the last 1,000 samples of the next day, representing continuously sampled normal operating conditions. Table 6.26 gives a summary of the input variables for the sugar refinery case study, as well as which benchmark actuator these variables are associated with.

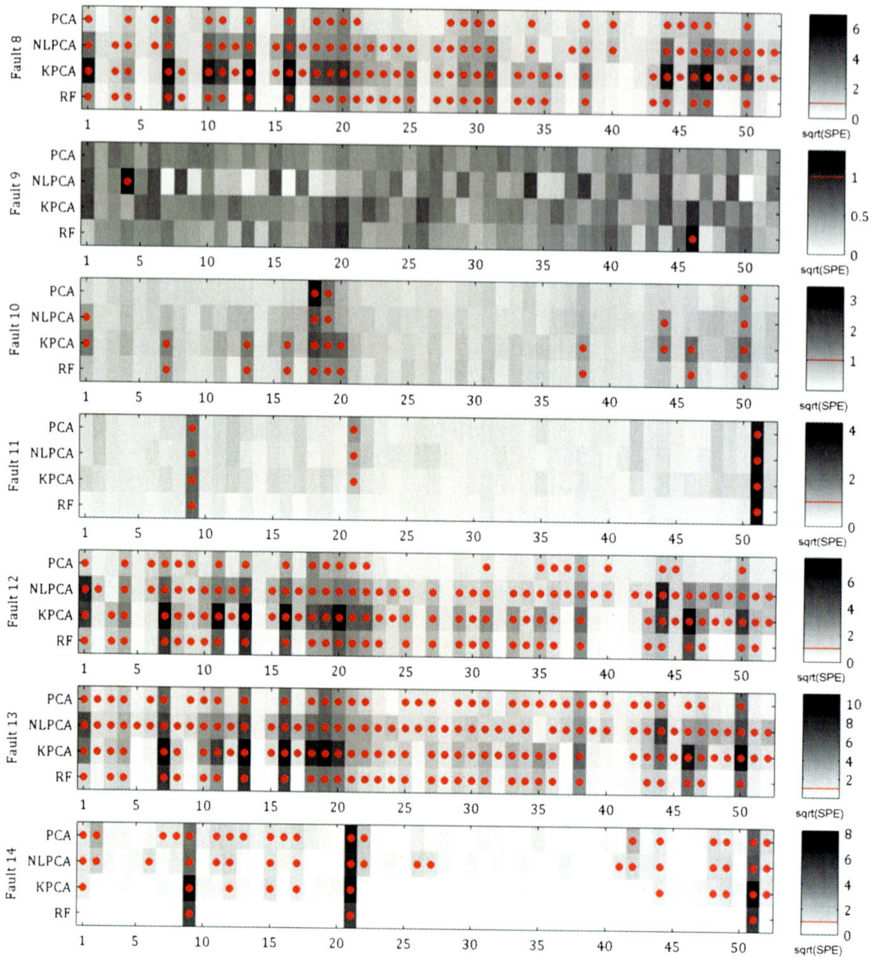

Fig. 6.20 Contribution plots (Based on squared prediction error of reconstructed variables). Tennessee Eastman process (Square root of relative contributions shown: contributions divided by 99 % confidence limit; relative contributions exceeding 1 indicated with *red markers*)

6.7.4 *Process Faults*

Artificial faults were introduced to the benchmark actuators, while data were captured before, during and after these faults. A summary of the actuator faults[2] is given in Table 6.27. For this case study, fault data sets were constructed by considering the 200 samples (i.e. 200 s) before the start of the fault, as well as the samples during the fault condition (Bartyś and Syfert 2002).

[2]The faults of the original benchmark considered in this work are Faults 5, 6, 7, 10, 11, 12, 17 and 19 (Bartyś and Syfert 2002).

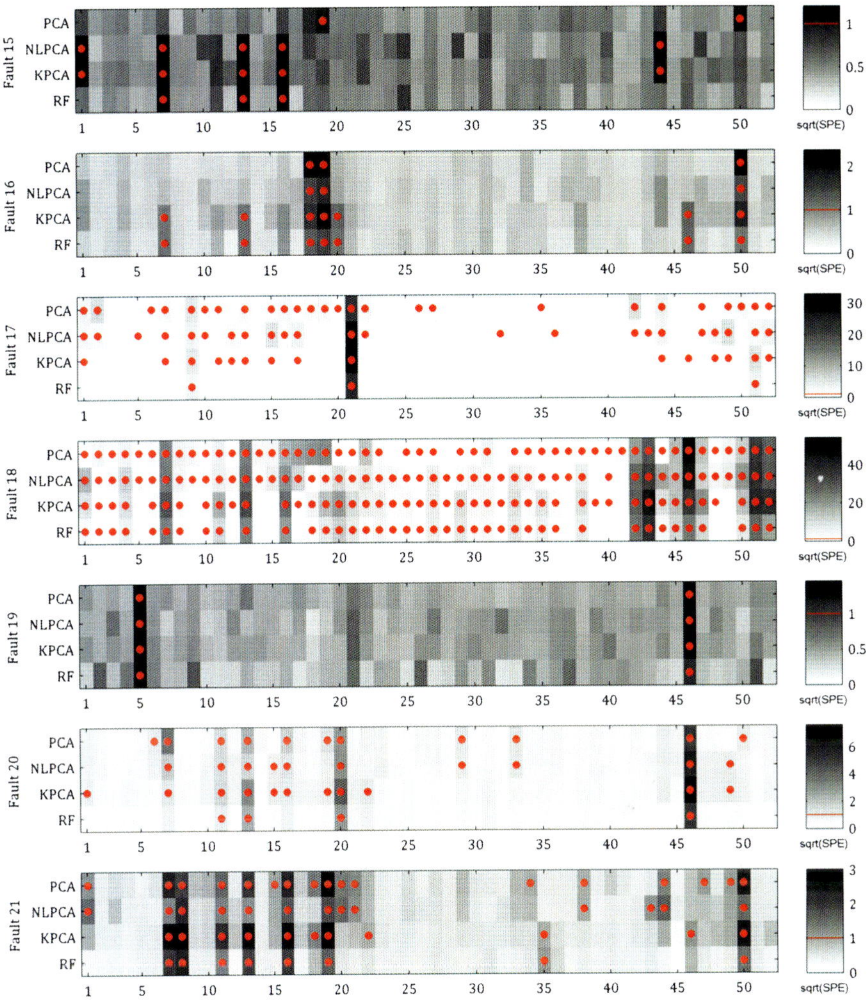

Fig. 6.21 Contribution plots (Based on squared prediction error of reconstructed variables). Tennessee Eastman process (Square root of relative contributions shown: contributions divided by 99 % confidence limit; relative contributions exceeding 1 indicated with *red markers*)

6.7.5 *Results of Fault Diagnosis*

Selecting the Number of Features

As before, the automated selection of the number of model components to retain, based on parallel plots and a component search plot for the NLPCA model, is shown in Fig. 6.25. The PCA model has the fewest components (7), while the random forest model had the largest number (30).

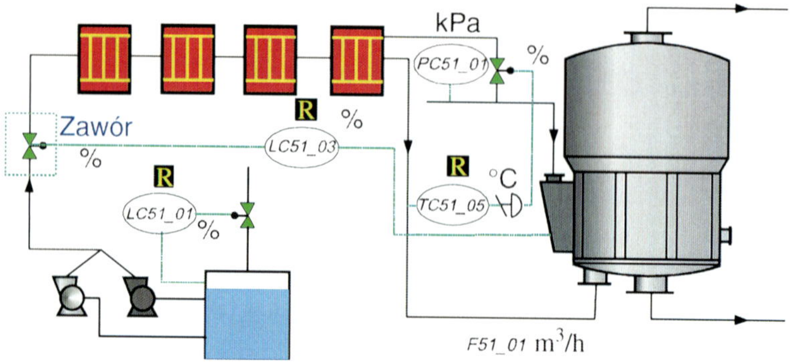

Fig. 6.22 Schematic showing the position of actuator 1 of the sugar refinery case study. Actuator 1 controls the flow of thin juice into the first evaporator (Image reprinted from Bartyś et al. (2006), Copyright (2006). With permission from Elsevier)

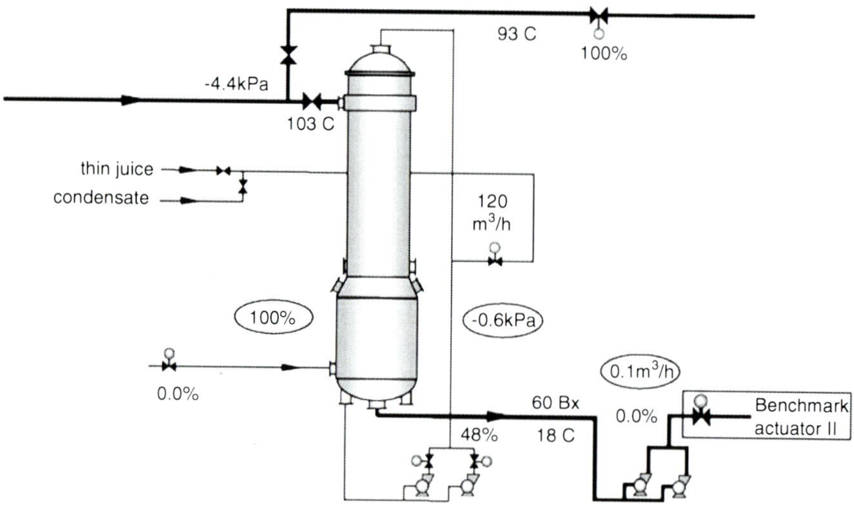

Fig. 6.23 Schematic showing the position of actuator 2 of the sugar refinery case study. Actuator 2 controls the flow of thick juice from the last evaporator (Image reprinted from Bartyś et al. (2006), Copyright (2006). With permission from Elsevier)

Performance of the Different Models

As before, the Hotelling's T^2 and squared prediction error (SPE) control charts for the two different faults are shown in Fig. 6.26. In these figures, the control limits are indicated by bold horizontal red lines, while the diagnostic variables, i.e. the T^2- and Q-statistics derived from the features for PCA, NLPCA, KPCA and RF, are indicated by black, pink, blue and green curves, respectively.

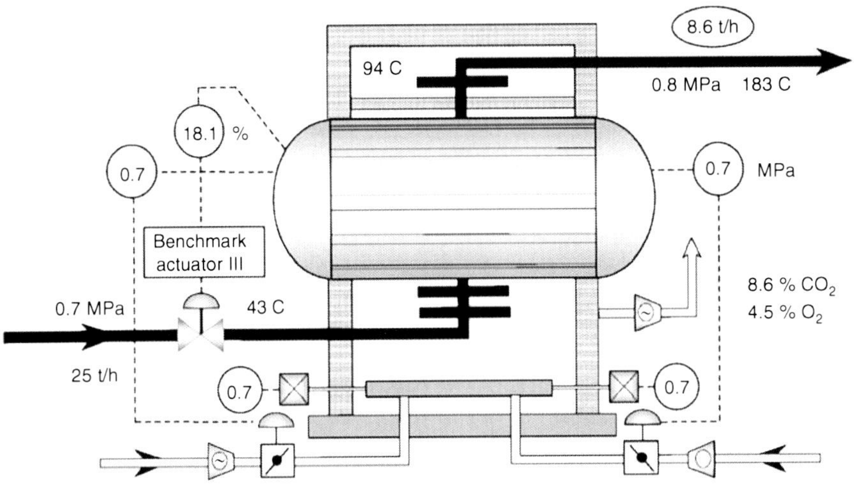

Fig. 6.24 Schematic showing the position of actuator 3 of the sugar refinery case study. Actuator 3 controls the flow of water into one of the steam boilers, which generates steam for the refinery (Image reprinted from Bartyś et al. (2006), Copyright (2006). With permission from Elsevier)

Validation Results

The performance of the different models is displayed based on the same criteria as for the previous two case studies. Table 6.28 shows the results for the false alarm rates for the NOC validation data of the sugar refinery, before and after adjustments of the model thresholds. The RF model performed best prior to adjustment of the model thresholds, after which all the models showed false alarm rates of approximately 0.01.

In Table 6.29, the false alarm run lengths for the NOC validation data are shown, again before and after adjustment of the model thresholds. The RF and KPCA models showed the best performance before and after the adjustments, respectively.

Table 6.30 shows the false alarm rates of the different models on all the fault conditions on the plant. The overall average false alarm rates of the models were as follows, with FAR values shown in parentheses: NLPCA (0.241), KPCA (0.264), PCA (0.276) and RF (0.479).

Table 6.31 shows the false alarm run lengths of the different models on all the fault conditions on the plant. The models could be ranked as follows: PCA (4), KPCA (4), RF (3) and NLPCA (2) in terms of their performance on each fault. As before, tied models were each awarded a score of 1.

The missing alarm rates for the models are summarized in Table 6.32. The models can be ranked as follows, based on their overall average missing alarm rates (shown in parentheses): RF (0.054), PCA (0.061), NLPCA (0.076) and KPCA (0.100).

Table 6.26 Input variables for the sugar refinery case study

Variable number	Actuator	Variable description
1	1	Juice pressure (valve inlet) (AM)
2		Juice pressure (valve outlet) (AM)
3		Juice temperature (valve outlet) (AM)
4		Juice flow (first evaporator inlet) (AM)
5		Control value (controller output) (AM)
6		Servomotor rod displacement (AM)
7		Process value (juice level in first evaporator) (AM)
8		Juice temperature (first evaporator inlet) (PM)
9		Juice temperature (first evaporator outlet) (PM)
10		Juice density (first evaporator inlet) (PM)
11		Juice density (first evaporator outlet) (PM)
12		Steam flow (PM)
13		Steam pressure[a] (PM)
14		Steam temperature (PM)
15		Vapour pressure (PM)
16		Vapour temperature (PM)
17	2	Juice pressure (valve inlet) (AM)
18		Juice pressure (valve outlet) (AM)
19		Juice temperature (valve inlet) (AM)
20		Process value (juice flow, last evaporator outlet) (AM)
21		Control value (controller output) (AM)
22		Servomotor rod displacement (AM)
23	3	Water pressure (valve inlet) (AM)
24		Water pressure (valve outlet) (AM)
25		Water temperature (valve outlet) (AM)
26		Water flow (steam boiler inlet) (AM)
27		Control value (controller output) (AM)
28		Servomotor rod displacement (AM)
29		Process value (water level in steam boiler)
30		Steam flow (steam boiler outlet) (PM)
31		Steam pressure (steam boiler outlet) (PM)
32		Steam temperature (steam boiler outlet) (PM)

[a]Here, steam refers to heated water vapour from the boiler house, while vapour refers to heated water vapour recycled from the evaporators

AM indicates actuator measurement, *PM* indicates process measurement

The true alarm run lengths for the fault conditions on the sugar refinery are shown in Table 6.33. By awarding a score of one to the best model on each fault (co-winners each getting one as before), the models can be ranked as follows: RF (6), KPCA (5), NLPCA (4) and PCA (4).

The performance of the different models with regard to the delay in fault detection is shown in Table 6.34. The models were scored on a similar basis as that used to assess performance on average run lengths, resulting in the RF model performing best with a score of 7 and the other models tying with 5 wins each.

Table 6.27 Actuator faults for the sugar refinery case study, indicating the actuator involved a general fault description and the number of samples in the fault data set (including 200 samples preceding fault inception)

Fault	Actuator	Fault description	Number of samples
1	1	Partly opened bypass valve	301
2		Positioner supply pressure drop	301
3		Unexpected pressure drop across the valve	821
4	2	Flow rate sensor fault	236
5		Flow rate sensor fault	239
6		Flow rate sensor fault	243
7	3	Positioner supply pressure drop	256
8		Flow rate sensor fault	376

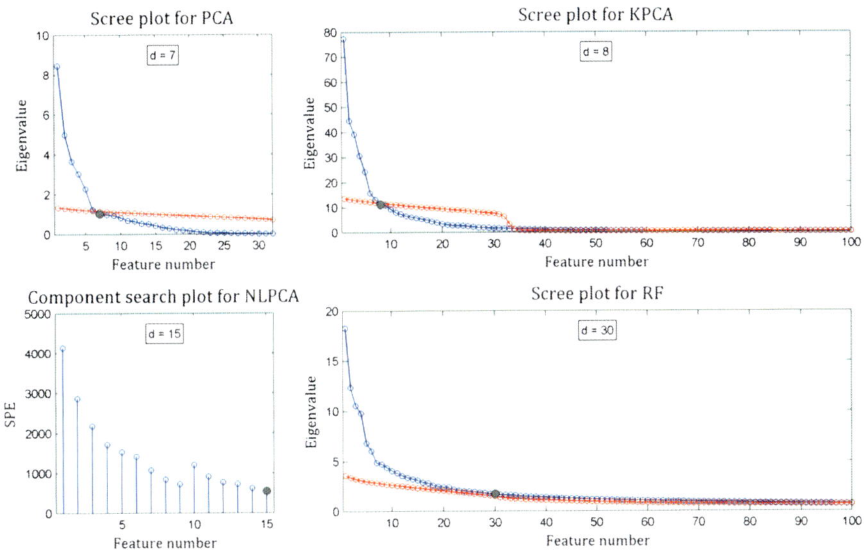

Fig. 6.25 Scree plots for the sugar refinery case study. Note that only the first 100 of the 1,000 eigenvalues calculated for the KPCA and RF feature space are shown. *Red shows* random scree plots. *Grey circle* shows selected number of features

Finally, as far as the areas under the ROC curves are concerned, as shown in Table 6.35, the models could be ranked nominally as in the other two case studies based on their overall average maximum ROC areas as follows: PCA (0.966), NLPCA (0.963), RF (0.959) and KPC (0.954). These values are very close in all cases and it is likely that differentiation between the models is not possible, based on these results. The contribution plots for the 8 fault conditions are shown in Fig. 6.27.

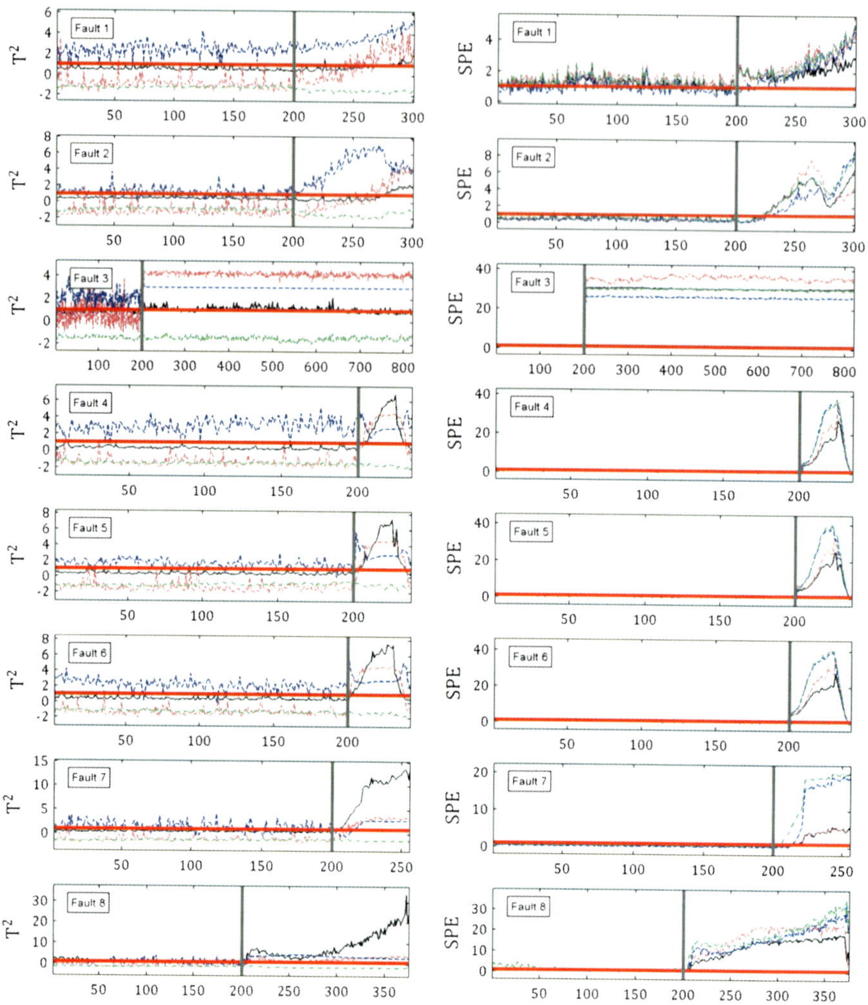

Fig. 6.26 Diagnostic sequence plots for the sugar refinery case study. PCA = *black*, NLPCA = *pink*, KPCA = *blue*, RF = *green* (Diagnostics scaled to have threshold of 1, indicated in *red*)

6.7.6 Discussion

In the first case study, when the results are considered for the models with adjusted thresholds where applicable on the NOC data, no single model outperformed the others. This includes the linear PCA model, the two-dimensional representation of the nonlinear system of which was sufficient to eventually capture the simulated deviations from the process. When the different criteria are considered individually,

Table 6.28 False alarm rates for the NOC validation data of the sugar refinery, before and after adjustments of the model thresholds for more reliable comparison

FAR (T^2/SPE)	PCA	NLPCA	KPCA	RF
Before adjustment	0/0.005	0.034/0.002	0/0.847	0/0
After adjustment	0.01/0.01	0.01/0.01	0.01/0.01	0.01/0.01

Table 6.29 False alarm run length for the NOC validation data from the sugar refinery, before and after adjustments of the model thresholds for more reliable comparison

ARL (T^2/SPE)	PCA	NLPCA	KPCA	RF
Before adjustment	–/21	23/371	–/0	–/–
After adjustment	64/21	23/21	65/83	7/64

Table 6.30 False alarm rates for the fault condition on the sugar refinery

FAR (T^2/SPE)	PCA	NLPCA	KPCA	RF
Fault 1	0.015/0.74	0.02/0.89	0.01/0.37	0.015/0.885
Fault 2	0/0.035	0.03/0	0.305/0.005	0.785/0.03
Fault 3	0.19/0.485	0.275/0.525	0.02/0.405	0/0.99
Fault 4	0.01/0.06	0.005/0.31	0.005/0.01	0/0.145
Fault 5	0.005/0.005	0.02/0.025	0.055/0	0.39/0.045
Fault 6	0.035/0.065	0.005/0.095	0.045/0.01	0.12/0.065
Fault 7	0.055/0.025	0.015/0.01	0.36/0	0.025/0.04
Fault 8	0.49/0.035	0.035/0.1	0.56/0.02	0/0.48

Table 6.31 False alarm run lengths for the sugar refinery case study

ARL (false) (T^2/SPE)	PCA	NLPCA	KPCA	RF
Fault 1	65/0	3/0	12/3	170/0
Fault 2	–/47	24/–	9/47	4/37
Fault 3	6/1	1/0	7/2	–/0
Fault 4	8/35	29/2	8/35	–/8
Fault 5	175/150	23/17	23/–	50/20
Fault 6	5/7	34/7	61/5	6/5
Fault 7	46/109	10/60	2/–	149/60
Fault 8	0/0	0/0	0/65	–/0

– indicates no detection occurred

the picture changes. For example, when the area under the receiver operator curves are considered as the sole performance criterion, then it can be concluded that the random forest model performed best. Likewise, when false alarm run length on the NOC validation data is considered in isolation, then the NLPCA model appears to be the best, as another example.

Table 6.32 Missing alarm rates on the different fault conditions in the sugar refinery

MAR (T^2/SPE)	PCA	NLPCA	KPCA	RF
Fault 1	0.683/0	0.525/0	1/0.099	1/0
Fault 2	0.743/0.198	0.624/0.248	1/0.267	0.941/0.198
Fault 3	0.343/0	0/0	1/0	0.997/0
Fault 4	0.194/0.083	0.222/0.028	1/0.056	1/0.028
Fault 5	0.154/0.077	0.179/0.077	1/0.077	0.615/0.077
Fault 6	0.186/0.047	0.14/0.047	1/0.047	1/0.047
Fault 7	0.071/0.232	0.179/0.304	0.732/0.232	0.911/0.071
Fault 8	0.011/0.028	0.068/0.028	0.983/0.023	0.994/0.011

Table 6.33 True alarm run lengths for the fault conditions on the sugar refinery

ARL (true) (T^2/SPE)	PCA	NLPCA	KPCA	RF
Fault 1	25/0	30/0	–/5	–/0
Fault 2	75/20	50/25	–/27	0/20
Fault 3	2/0	0/0	–/0	256/0
Fault 4	1/0	3/0	–/0	–/0
Fault 5	2/1	3/1	–/1	0/1
Fault 6	2/0	1/0	–/0	–/0
Fault 7	4/13	3/17	0/13	½
Fault 8	2/5	5/5	1/4	15/2

– indicates no detection occurred

Table 6.34 Detection delay for three consecutive samples (First of three samples indicated) for the sugar refinery

DD (T^2/SPE)	PCA	NLPCA	KPCA	RF
Fault 1	70/1	60/1	–/11	–/1
Fault 2	76/21	66/26	–/28	1/21
Fault 3	3/1	1/1	–/1	–/1
Fault 4	5/3	4/1	–/1	–/1
Fault 5	3/2	4/2	–/2	1/2
Fault 6	3/1	2/1	–/1	–/1
Fault 7	5/14	12/18	1/14	–/6
Fault 8	3/6	6/6	–/5	–/3

– indicates no detection occurred

Table 6.35 Area under receiver operator characteristic curve for the sugar refinery

AUC (T^2/SPE)	PCA	NLPCA	KPCA	RF
Fault 1	0.778/0.959	0.917/0.928	0.124/0.901	0.05/0.952
Fault 2	0.525/0.884	0.847/0.895	0.045/0.882	0.06/0.877
Fault 3	0.812/0.995	0.995/0.995	0.082/0.995	0.498/0.995
Fault 4	0.963/0.982	0.975/0.965	0.515/0.989	0.227/0.966
Fault 5	0.962/0.965	0.976/0.971	0.055/0.978	0.41/0.968
Fault 6	0.954/0.974	0.98/0.965	0.158/0.968	0.184/0.962
Fault 7	0.982/0.933	0.975/0.866	0.343/0.928	0.617/0.972
Fault 8	0.978/0.987	0.976/0.983	0.032/0.987	0.247/0.982

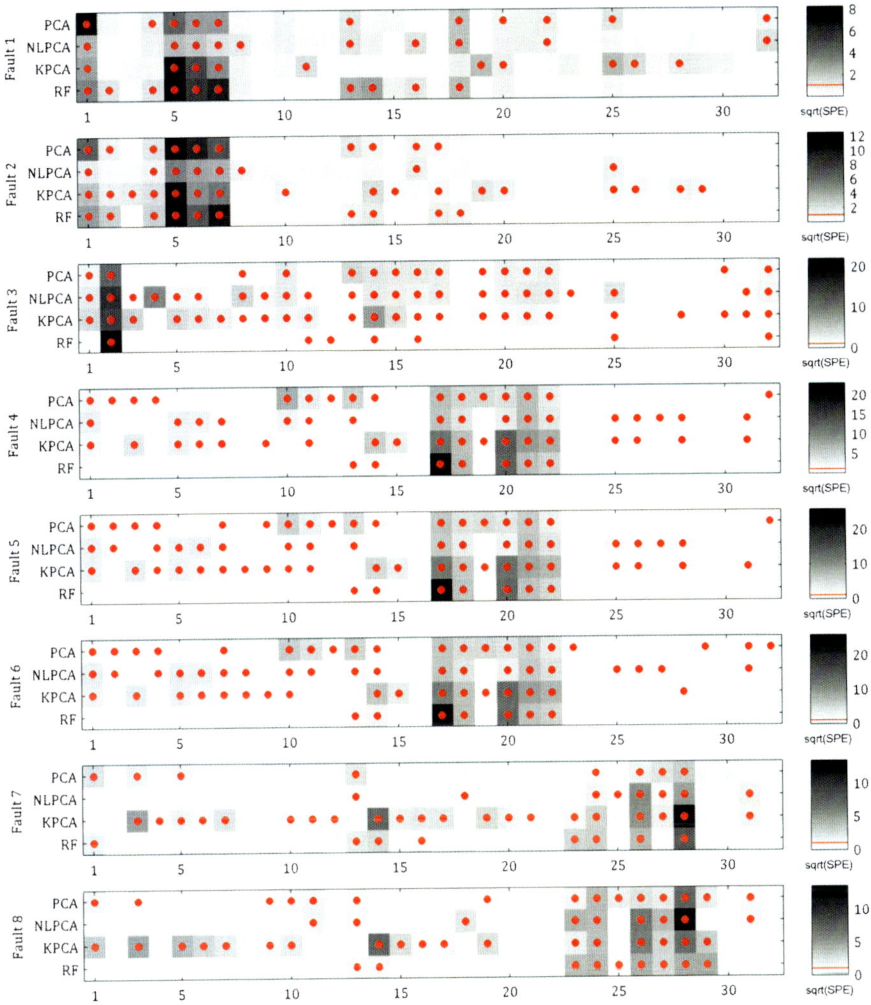

Fig. 6.27 Contribution plots (Based on squared prediction error of reconstructed variables). Sugar refinery case study (Square root of relative contributions shown: contributions divided by 99 % confidence limit; relative contributions exceeding 1 indicated with *red markers*)

In the second and third case studies on the Tennessee Eastman plant and sugar refinery, no single model stood out as consistently better than its competitors. These problems were more complicated in that they comprised larger sets of variables of which the true dimensionality is unknown. In addition, the variables are highly correlated and the relationships between different variables are not easily characterized. It is therefore more difficult to evaluate the models against an actual benchmark, as was the case with the first case study, even though the second and third case studies are based on simulated data.

6.8 Concluding Remarks

The performance of four fault diagnostic models representative of the major classes of machine learning models, as well as principal component analysis, was considered in three simulated case studies involving steady-state processes. A variety of performance measures were considered and on the whole, no model stood out as the best performer. This is not to say that the different models do not have their comparative strengths and weaknesses, and in practice logistic issues related to implementation and maintenance of the models would also be a serious consideration. This could have an impact on the cost of the machine learning models in particular, where model optimization and automated selection of the model features could incur relative large computational costs.

References

An, S., Liu, W., & Venkatesh, S. (2007). Face recognition using kernel ridge regression. In *IEEE Conference on Computer Vision and Pattern Recognition, 2007. CVPR'07* (pp. 1–7), Minneapolis

Auret, L. (2010). *Process monitoring and fault diagnosis using random forests*. Doctoral dissertation. Stellenbosch: Stellenbosch University. Available at: http://hdl.handle.net/10019.1/5360

Auret, L., & Aldrich, C. (2010). Unsupervised process fault detection with random forests. *Industrial and Engineering Chemistry Research*. doi:10.1021/ie901975c.

Bakır, G. H. (2005). *Extension to kernel dependency estimation with applications to robotics*. Ph.D. dissertation. Berlin: Technical University of Berlin.

Bakır, G. H., Weston, J., & Schölkopf, B. (2004). Learning to find pre-images. In *Advances in neural information processing systems* (pp. 449–456). MIT Press, Cambridge, MA, USA.

Bartyś, M., & Syfert, M. (2002). *Lublin sugar factory data description file*. Institute of Automatic Control and Robotics – Warsaw University of Technology. Available at: diag.mchtr.pw.edu.pl/damadics/download/Lublin/damadics-lublin-data-description-v02March2002.zip

Bartyś, M., Patton, R., Syfert, M., de las Heras, S., & Quevedo, J. (2006). Introduction to the DAMADICS actuator FDI benchmark study. *Control Engineering Practice*, *14*(6), 577–596.

Bauer, M., & Thornhill, N. F. (2008). A practical method for identifying the propagation path of plant-wide disturbances. *Journal of Process Control*, *18*(7–8), 707–719.

Bauer, M., Cox, J. W., Caveness, M. H., Downs, J. J., & Thornhill, N. F. (2007). Finding the direction of disturbance propagation in a chemical process using transfer entropy. *IEEE Transactions on Control Systems Technology*, *15*(1), 12–21.

Brown, P. R., & Rhinehart, R. R. (2000). Automated steady-state identification in multivariable systems. *Hydrocarbon Processing*, *79*(9), 79.

Camci, F., Chinnam, R. B., & Ellis, R. D. (2008). Robust kernel distance multivariate control chart using support vector principles. *International Journal of Production Research*, *46*(18), 5075.

Cao, S., & Rhinehart, R. R. (1995). An efficient method for on-line identification of steady state. *Journal of Process Control*, *5*(6), 363–374.

Carreira-Perpinan, M. A. (1997). *A review of dimension reduction techniques*, Department of Computer Science, University of Sheffield.

Cattell, R. B. (1966). The scree test for the number of factors. *Multivariate Behavioral Research*, *1*, 245–276.

Chang, C.-C., & Lin, C.-J. (2011). LIBSVM: A library for support vector machines. *ACM Transactions on Intelligent Systems and Technology, 2*(3), 27:1–27:27.

DAMADICS RTN. (2002). DAMADICS RTN information web site. *Institute of Automatic Control and Robotics – Warsaw University of Technology*. Available at: http://diag.mchtr.pw.edu.pl/damadics/. Accessed 26 Jan 2012.

Downs, J. J., & Vogel, E. F. (1993). A plant-wide industrial process control problem. *Computers and Chemical Engineering, 17*(3), 245–255.

Eastment, H. T., & Krzanowski, W. J. (1982). Cross-validatory choice of the number of components from a principal component analysis. *Technometrics, 24*(1), 73–77.

Himes, D. M., Storer, R. H., & Georgakis, C. (1994, June 29–July 1). Determination of the number of principal components for disturbance detection and isolation. In *American control conference* (vol. 2, pp. 1279–1283). doi: 10.1109/ACC.1994.752265, URL: http://ieeexplore.ieee.org/stamp/stamp.jsp?tp=&arnumber=752265&isnumber=16245

Horn, J. L. (1965). A rationale and test for the number of factors in factor analysis. *Psychometrika, 30*, 179–185.

Jemwa, G. T., & Aldrich, C. (2006). Kernel-based fault diagnosis on mineral processing plants. *Minerals Engineering, 19*(11), 1149–1162.

Kresta, J. V., MacGregor, J. F., & Marlin, T. E. (1991). Multivariate statistical monitoring of process operating performance. *Canadian Journal of Chemical Engineering, 69*(1), 35–47.

Ku, W., Storer, R. H., & Georgakis, C. (1995). Disturbance detection and isolation by dynamic principal component analysis. *Chemometrics and Intelligent Laboratory Systems, 30*(1), 179–196.

Liaw, A., & Wiener, M. (2002). Classification and regression by randomForest. *R News, 2*(3), 18–22.

Lindner, B. S. (2012). *Demonstration of root cause analysis with process and data connectivity methods for data-based monitoring methods*. Process Engineering, Stellenbosch University.

Liu, X., Xie, L., Kruger, U., Littler, T., & Wang, S. (2008). Statistical-based monitoring of multivariate non-Gaussian systems. *AICHE Journal, 54*(9), 2379–2391.

Lu, N., Jiang, B., Wang, L., Lu, J., & Chen, X. (2012). A fault prognosis strategy based on Time-Delayed Digraph Model and Principal Component Analysis. *Mathematical Problems in Engineering, 2012*, 1–17. Available at: Accessed 10 Jan 2013.

Lyman, P. R., & Georgakis, C. (1995). Plant-wide control of the Tennessee Eastman problem. *Computers and Chemical Engineering, 19*(3), 321–331.

Martin, E. B., & Morris, A. J. (1996). Non-parametric confidence bounds for process performance monitoring charts. *Journal of Process Control, 6*(6), 349–358.

Nomikos, P. (1996, May). Detection and diagnosis of abnormal batch operations based on multi-way principal component analysis World Batch Forum, Toronto. *ISA Transactions, 35*(3), 259–266.

R Development Core Team. (2010). *R: A language and environment for statistical computing*. Vienna: R Foundation for Statistical Computing. Available at: http://www.R-project.org

Russell, E. L., Chiang, L. H., & Braatz, R. D. (2000a). *Data-driven techniques for fault detection and diagnosis in chemical processes*. London/New York: Springer

Russell, E. L., Chiang, L. H., & Braatz, R. D. (2000b). Fault detection in industrial processes using canonical variate analysis and dynamic principal component analysis. *Chemometrics and Intelligent Laboratory Systems, 51*(1), 81–93.

Scholz, M. (2011). Nonlinear PCA toolbox for Matlab – Matthias Scholz. *Nonlinear PCA*. Available at: http://www.nlpca.de/matlab.html. Accessed 22 June 2011.

Scholz, M., & Vigário, R. (2002). Nonlinear PCA: A new hierarchical approach. In *ESANN 2002 proceedings*. European Symposium on Artificial Neural Networks (pp. 439–444). Bruges: d-side publi.

Scholz, M., Kaplan, F., Guy, C. L., Kopka, J., & Selbig, J. (2005). Non-linear PCA: a missing data approach. *Bioinformatics, 21*(20), 3887–3895. Available at: Accessed 23 June 2011.

Scholz, M., Fraunholz, M., & Selbig, J. (2008). Nonlinear principal component analysis: Neural network models and applications. In A. N. Gorban, B. Kégl, D. C. Wunsch, & A. Y. Zinovyev (Eds.), *Principal manifolds for data visualization and dimension reduction* (pp. 44–67). Berlin/Heidelberg: Springer. Available at: http://www.springerlink.com/index/10.1007/978-3-540-73750-6_2. Accessed 22 June 2011.

Shao, J.-D., Rong, G., & Lee, J. M. (2009). Generalized orthogonal locality preserving projections for nonlinear fault detection and diagnosis. *Chemometrics and Intelligent Laboratory Systems, 96*(1), 75–83.

Shi, T., & Horvath, S. (2006). Unsupervised learning with random forest predictors. *Journal of Computational and Graphical Statistics, 15*(1), 118–138.

Sun, R., & Tsung, F. (2003). A kernel-distance-based multivariate control chart using support vector methods. *International Journal of Production Research, 41*(13), 2975.

Venkatasubramanian, V., Rengaswamy, R., Kavuri, S. N., & Yin, K. (2003). A review of process fault detection and diagnosis Part III: Process history based methods. *Computers and Chemical Engineering, 27*(3), 327–346.

Wold, S. (1978). Cross-validatory estimation of the number of components in factor and principal components models. *Technometrics, 20*(4), 397–405.

Yang, F., & Xiao, D. (2006). Model and fault inference with the framework of probabilistic SDG. In *IEEE* (pp. 1–6). Available at: http://ieeexplore.ieee.org/lpdocs/epic03/wrapper.htm?arnumber=4150163. Accessed 10 Jan 2013.

Yang, F., & Xiao, D. (2012). Progress in root cause and fault propagation analysis of large-scale industrial processes. *Journal of Control Science and Engineering, 2012*, 1–10.

Ypma, A., Tax, D. M. J., & Duin, R. P. W. (1999). Robust machine fault detection with independent component analysis and support vector data description. In *Neural Networks for Signal Processing IX. Proceedings of the 1999 IEEE Signal Processing Society Workshop* (pp. 67–76). Madison: IEEE. Available at: http://ieeexplore.ieee.org/lpdocs/epic03/wrapper.htm?arnumber=788124. Accessed 17 Jan 2013.

Nomenclature

Symbol	Description
s_x	Standard deviation of variable x
$\bar{\mathbf{x}}$	Mean of variable x
z_i	Zero mean unit variance scaling of x_i
C_i^+	Cumulative sum (CUSUM) control statistic of the ith sample x_i
δ	Magnitude of shift in process variable
E_i	Exponentially weighted moving average (EWMA) control statistic of the ith sample x_i
β	Desired missing alarm rate
α	Desired false alarm rate
λ	Smoothing parameter
$\sum$	Sample covariance matrix
$s_{\mathrm{H.i}}$	Score distance based on Hotelling's T^2-statistic
s	Score distance
s_α	Score distance detection threshold
$\mathbf{X}'$	Reconstructed variables $\mathbf{X}$

(continued)

(continued)

Symbol	Description
r	Residual distance
r_α	Residual distance detection threshold
C_s	Feature space contributions
C_r	Residual space contributions
Q	Squared reconstruction error
$\mathbf{R}$	Correlation matrix
θ	Detection threshold
$f_{FAR}(\theta)$	Functional relationship between false alarm rate (FAR) and detection threshold θ
$f_{TAR}(\theta)$	Functional relationship between true alarm rate (TAR) and detection threshold θ
$C_{s,j}$	Score distance contribution of jth variable
$C_{r,j}$	Residual distance contribution of jth variable
$\mathbb{F}$	Feature space
ρ	Threshold or significance level
$\boldsymbol{\xi}$	Margin error vector
$\mathbf{w}$	Vector defining separating hyperplane
ν	A decision surface complexity parameter

Chapter 7
Dynamic Process Monitoring

The emergence of a range of new sensors and equipment for data acquisition has enabled the collection of data from chemical and manufacturing process or equipment at high frequencies. The application of principal component analysis to monitor inherently nonlinear dynamic systems may lead to inefficient and unreliable process monitoring. This has led to the development of extended versions of principal component analysis and other multivariate methods for dynamic process monitoring, as reviewed earlier in Chap. 2. A large class of these methods rely on a phase space embedding of the data. This leads to a static, but often highly nonlinear, representation of the monitored system. In this chapter, some of the state-of-the-art approaches to dynamic process monitoring are illustrated by means of selected case studies.

Three feature extraction and reconstruction approaches (singular spectrum analysis, random forest feature extraction and inverse nonlinear principal component analysis) are compared to a phase space distribution estimation approach (one-class support vector machines) and another phase space characterization approach, recurrence quantification analysis. Three case studies are considered to illustrate the application of data-driven fault diagnosis of dynamic processes: the Lotka–Volterra predator–prey system (specifically the system considered by Lindfield and Penny 2000), the Belousov–Zhabotinsky reaction system (Zhang et al. 1993) and the autocatalytic reaction system (Lee and Chang 1996).

7.1 Monitoring Dynamic Process Systems

Whereas steady-state process variables do not change meaningfully over time, continuous dynamic systems exhibit stable dynamic behaviour or (quasi)periodic behaviour within a bounded region in the variable space. Steady-state process monitoring methods do not account for such behaviour and may be unsuitable for monitoring such systems for abnormal behaviour.

C. Aldrich and L. Auret, *Unsupervised Process Monitoring and Fault Diagnosis with Machine Learning Methods*, Advances in Computer Vision and Pattern Recognition, DOI 10.1007/978-1-4471-5185-2_7, © Springer-Verlag London 2013

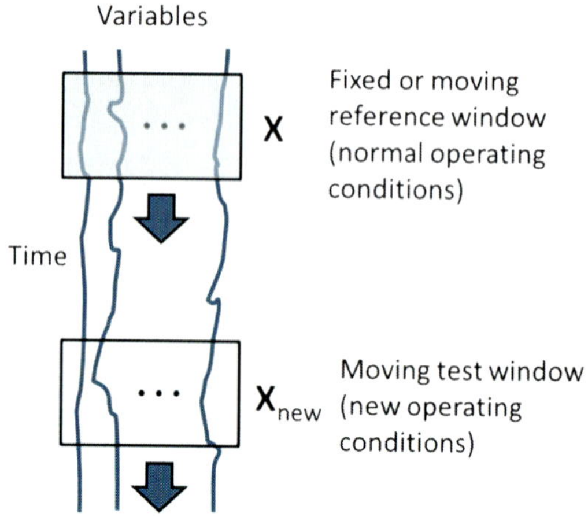

Fig. 7.1 Monitoring of continuous dynamic systems based on the generation of an augmented data matrix obtained by sliding a window along the time series variables

The periodic behaviour of a stable/(quasi)periodic dynamic system can be described by its attractor in phase space. The attractor represents every possible state of a particular dynamic system. If the dynamic system should change, for example, due to an abnormal event, this abnormal behaviour would manifest as states outside of the normal behaviour attractor. Monitoring the phase space of a dynamic system for changes in its normal operating conditions attractor is similar to monitoring the input space of a steady-state system.

The phase space of a dynamic system can be approximated by the construction of an augmented data matrix that contains lagged copies of the variables in the original data matrix, as mentioned previously in Chap. 2. A reference window (fixed or moving) of lagged process data represents normal operating conditions, against which a moving testing window of lagged process data is compared (see Fig. 7.1). This chapter considers the case of a fixed reference window.

Feature extraction methods can be used to characterize the phase space of the reference window such that an information-rich feature space captures the structure of the normal operating conditions attractor. The feature space for new test data can be directly monitored (e.g. with one-class support vector machines or recurrence quantification analysis), or the accuracy of reconstruction to the original phase space can be monitored. If the reconstruction is inaccurate, it may indicate that the feature extraction process model is no longer valid due to an abnormal event.

7.2 Framework for Data-Driven Process Fault Diagnosis: Dynamic Process Monitoring

The general framework for the data-driven construction of feature extraction models for process fault diagnosis was introduced in Chap. 1 and is repeated here (Fig. 7.2).

Data-driven process fault diagnosis consists of two stages: an offline training stage and an online implementation stage.

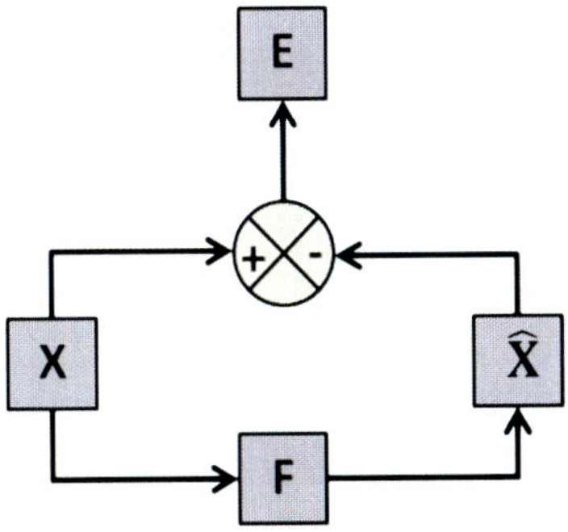

Fig. 7.2 A general framework for data-driven process fault diagnosis

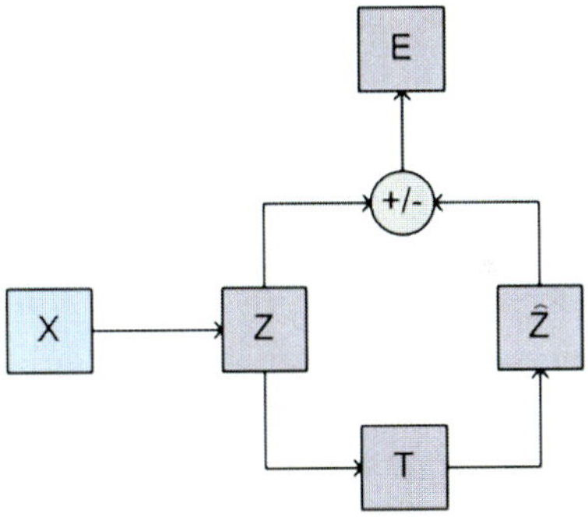

Fig. 7.3 A framework for data-driven process fault diagnosis for dynamic process systems

In steady-state process monitoring, the mapping function $\Im$ from the process data matrix $\mathbf{X}$ to the feature matrix $\mathbf{F}$ is learnt during the offline training stage. A feature space diagnostic function to characterize the feature matrix $\mathbf{F}$, as well as an appropriate diagnostic threshold, is also learnt during the offline stage. Furthermore, the demapping/reverse mapping function $\aleph$, which calculates the reconstructed process data matrix $\widehat{\mathbf{X}}$ from the feature matrix $\mathbf{F}$, and an appropriate diagnostic threshold for the residual space are learnt during the offline training stage.

During the online implementation stage of steady-state process monitoring, a new process data matrix $\mathbf{X}^{(\text{test})}$ is subjected to the mapping and demapping functions ($\Im$ and $\aleph$) learnt during the offline training space, yielding the new feature matrix $\mathbf{F}^{(\text{test})}$ and (after subtraction) the new residual matrix $\mathbf{E}^{(\text{test})}$. Feature space and residual space diagnostics are then calculated and compared to the appropriate thresholds in order to determine whether a fault has occurred.

The general framework can be adapted for dynamic process monitoring by substituting the process data matrix $\mathbf{X}$ by the lagged trajectory matrix $\mathbf{Z}$: The mapping algorithm then maps the lagged trajectory matrix $\mathbf{Z}$ to an informative feature matrix $\mathbf{T}$ which captures the structure of the normal operating conditions attractor. Reverse mapping from the feature matrix $\mathbf{T}$ yields the reconstructed lagged trajectory matrix $\widehat{\mathbf{Z}}$, and by subtraction, the residual matrix $\mathbf{E}$ (see Fig. 7.3).

In this chapter, two monitoring approaches will be investigated which focus on the feature space for diagnostic purposes, one-class support vector machines (1-SVM) and recurrence quantification analysis (RQA), and three monitoring approaches will be investigated which focus on the residual space: singular spectrum analysis (SSA), random forest feature extraction (RF) and inverse nonlinear principal component analysis (NLPCA). These methods all have a comparative quantification of the phase space in common, where a phase space associated with normal operating conditions is compared to a phase space or part thereof associated with some new test data.

7.2.1 *Offline Training Stage*

An initial extent of a time series $\mathbf{x}$ is assumed to represent normal operating conditions (NOC): $\mathbf{x}^{(\mathrm{NOC})}$. The extent of the time series used in model building is known as the window size, N_{w}. Scaling of this time series ($\mathbf{x}^{(\mathrm{NOC})}$ scaled to $\mathbf{z}^{(\mathrm{NOC})}$) is done to ensure practical, comparable variable ranges. To capture the dynamic structure of the NOC data, the time series data are embedded in a lag-trajectory matrix $\mathbf{Z}^{(\mathrm{NOC})}$, representing the phase space of the system. Embedding parameters are determined with average mutual information and false nearest neighbour approaches. Where applicable, quantification of the phase space is done, depending on the nature of the technique, giving rise to a diagnostic statistic.

To determine a confidence threshold for the diagnostic statistic, validation data for normal operating conditions are employed. The phase space quantification models built during the NOC training steps are applied to N_{w}-sized windows of the validation portion of the time series, $\mathbf{x}^{(\mathrm{valid})}$. The distribution of the validation diagnostic statistic allows for the definition of a diagnostic statistic limit, with a desired expected false alarm rate.

7.2.2 *Online Application Stage*

Given the phase space quantification models and their associated diagnostic statistic limit, new time series test data $\mathbf{x}^{(\mathrm{test})}$ can be monitored for changes. Moving windows of size N_{w} are considered at each time step. Scaling, embedding and phase space quantification models (learnt from the NOC stage) are applied to each window. If the diagnostic statistic associated with a specific test data window falls outside the limits (calculated during the validation stage), an alarm is made to signify a possible change in the time series.

In this framework, the training of NOC models is done once and not updated again: The reference window is fixed. A variant of dynamic monitoring, the so-called change point detection, continually updates the NOC models along a moving window. Auret and Aldrich (2010) investigated change point detection with SSA and RF models. Although this continual updating is beneficial in avoiding outdated and irrelevant NOC data, the computational costs of updating may be prohibitive. In the dynamic monitoring framework considered here, the judicious selection of NOC data is assumed.

7.3 Feature Extraction and Reconstruction Approaches: Framework

In the case of SSA, RF and NLPCA, quantification is achieved in the form of a reconstruction error for an embedded (lag-trajectory) matrix. In the mapping modelling step, a feature extraction model is learnt to extract a number of significant features from the phase space of data representing normal operating conditions. In a demapping modelling step, a reconstruction model is learnt to reconstruct the phase space from the significant features. The success of the reconstruction is quantified in terms of a reconstruction distance, which is essentially a scaled sum of squared errors between the original phase space and its reconstruction. The mapping and demapping models can then be applied to test data, and the test reconstruction distances or errors compared to the reconstruction distances or errors obtained from normal operating conditions. The general form of the feature extraction and reconstruction approach is illustrated in Fig. 7.4.

The three stages of dynamic monitoring with feature extraction and reconstruction methods are discussed here:

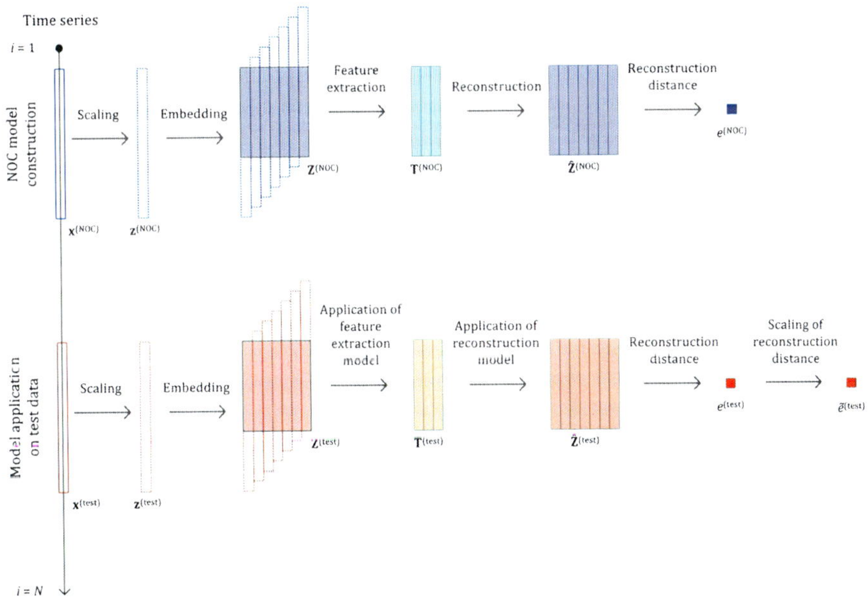

Fig. 7.4 Illustration of dynamic monitoring with feature extraction and reconstruction methods. The scaled reconstruction distance serves as the monitored statistic

7.3.1 Training Stage with NOC Data

For a time series $\mathbf{x}$ with N samples, define a monitoring window size N_w. The first N_w samples of the time series ($\mathbf{x}^{(NOC)}$) are considered to construct the feature extraction and reconstruction models. First, the NOC data $\mathbf{x}^{(NOC)}$ are scaled to $\mathbf{z}^{(NOC)}$, to ensure unit variance and zero mean, according to the scaling parameters $\mu^{(NOC)}$ and $\sigma^{(NOC)}$:

$$\mu^{(NOC)} = \frac{1}{N_w}\sum_{i=1}^{N_w} x_i^{(NOC)}; \quad \sigma^{(NOC)} = \sqrt{\frac{1}{N_w - 1}\sum_{i=1}^{N_w}\left(x_i^{(NOC)} - \mu^{(NOC)}\right)^2} \tag{7.1}$$

$$z_i^{(NOC)} = \frac{x_i^{(NOC)} - \mu^{(NOC)}}{\sigma^{(NOC)}}; \quad i = 1, \ldots, N_w. \tag{7.2}$$

The scaled NOC time series data $\mathbf{z}^{(NOC)}$ are embedded to give the lag-trajectory matrix $\mathbf{Z}^{(NOC)}$ with embedding parameters k (lag) and m (embedding dimension). The jth column of $\mathbf{Z}^{(NOC)}$ is given by

$$\mathbf{Z}_j^{(NOC)} = \left(z_j^{(NOC)}, z_{j+k}^{(NOC)}, z_{j+2k}^{(NOC)}, \ldots, z_{j+N_w-k}^{(NOC)}\right); \quad j = 1, \ldots, m. \tag{7.3}$$

Feature extraction, or mapping modelling, is then applied to the lag-trajectory matrix, and d significant features retained in the feature matrix $\mathbf{T}^*$. The mapping models $\mathcal{F}_{\mathbf{Z}\to\mathbf{T}^*}$ are retained for feature extraction application to test data. Reconstruction, or demapping modelling, is then applied to the feature matrix to obtain a reconstruction of the embedded data, $\hat{\mathbf{Z}}^{(NOC)}$. The demapping models $\mathcal{F}_{\mathbf{T}^*\to\hat{\mathbf{Z}}}$ are retained for reconstruction of test features. The nature of feature extraction and reconstruction, as well as the choice of number of features to retain, depends on the technique applied.

The reconstruction distance $e_i^{(NOC)}$ from an actual entry of the lag-trajectory matrix $\mathbf{Z}_i^{(NOC)}$ to its reconstruction $\hat{\mathbf{Z}}_i^{(NOC)}$ is given by

$$e_i^{(NOC)} = \mathbf{Z}_i^{(NOC)}\left(\mathbf{Z}_i^{(NOC)}\right)^T - \hat{\mathbf{Z}}_i^{(NOC)}\left(\hat{\mathbf{Z}}_i^{(NOC)}\right)^T$$
$$i = 1, \ldots, N_w - m + 1. \tag{7.4}$$

The average reconstruction distance $e^{(NOC)}$ (scaled by embedding dimension m and number of samples $N_w - (m-1)k$) for the samples in $\mathbf{Z}^{(NOC)}$ is given by (since $k = 1$)

$$e^{(NOC)} = \frac{1}{m\left(N_w - (m-1)k\right)}\sum_{i=1}^{(m-1)k} e_i. \tag{7.5}$$

7.3.2 *Test Stage with Test Data*

The phase space characterization models (defined by $\mu^{(\mathrm{NOC})}$, $\sigma^{(\mathrm{NOC})}$, k, m, $\mathcal{F}_{\mathbf{Z}\to\mathbf{T}^*}$, $\mathcal{F}_{\mathbf{T}^*\to\tilde{\mathbf{Z}}}$ and $e^{(\mathrm{NOC})}$) can be applied to test time series data $\mathbf{x}^{(\mathrm{test})}$ to determine whether changes in the dynamic process generating the time series data have occurred. While the training stage was applied on N_{w} samples of $\mathbf{x}^{(\mathrm{NOC})}$, the test stage can be applied on individual samples of $\mathbf{x}^{(\mathrm{test})}$, given that each test sample considered has at least $k(m-1)$ past samples, required for embedding.

Consider a test sample $\mathbf{x}_\iota^{(\mathrm{test})}$ at time index ι. A scaled test sample $\mathbf{Z}_\iota^{(\mathrm{test})}$ (and its $m-1$ past samples) is calculated from the training stage scaling parameters:

$$z_i^{(\mathrm{test})} = \frac{x_i^{(\mathrm{test})} - \mu^{(\mathrm{test})}}{\sigma^{(\mathrm{test})}}; \quad i = \iota - m + 1, \ldots, \iota. \tag{7.6}$$

The scaled test time series data $\mathbf{z}_i^{(\mathrm{test})}$ $(i = \iota - m + 1, \ \ldots, \ \iota)$ are embedded (according to embedding parameters k and m) to give the lag-trajectory vector $\mathbf{Z}_\iota^{(\mathrm{test})}$:

$$\mathbf{Z}_\iota^{(\mathrm{test})} = \left(z_{\iota-m+1}^{(\mathrm{test})}, z_{\iota-m+2}^{(\mathrm{test})}, z_{\iota-m+3}^{(\mathrm{test})}, \ldots, z_\iota^{(\mathrm{test})} \right). \tag{7.7}$$

The d-dimensional feature vector for $\mathbf{Z}_\iota^{(\mathrm{test})}$ is calculated by applying the mapping models $\mathcal{F}_{\mathbf{Z}\to\mathbf{T}^*}$. From this feature vector, the reconstruction $\tilde{\mathbf{Z}}_\iota^{(\mathrm{test})}$ can be determined by applying the demapping models $\mathcal{F}_{\mathbf{T}^*\to\tilde{\mathbf{Z}}}$. The reconstruction distance $e_\iota^{(\mathrm{test})}$ from the lag-trajectory vector $\mathbf{Z}_\iota^{(\mathrm{test})}$ to its reconstruction $\hat{\mathbf{Z}}_\iota^{(\mathrm{test})}$ is then given by

$$e_\iota^{(\mathrm{test})} = \mathbf{Z}_\iota^{(\mathrm{test})} \left(\mathbf{Z}_\iota^{(\mathrm{test})} \right)^T - \hat{\mathbf{Z}}_\iota^{(\mathrm{test})} \left(\hat{\mathbf{Z}}_\iota^{(\mathrm{test})} \right)^T. \tag{7.8}$$

The average, scaled diagnostic statistic for dynamic monitoring ($\tilde{e}_\iota^{(\mathrm{test})}$) is given by

$$\tilde{e}_\iota^{(\mathrm{test})} = \frac{e_\iota^{(\mathrm{test})}/m}{e^{(\mathrm{NOC})}}. \tag{7.9}$$

7.3.3 *Validation Stage to Determine Threshold*

The diagnostic statistic described above can be calculated for each sample of a validation time series data set, $\mathbf{x}^{(\mathrm{valid})}$, giving a distribution of validation statistics $\tilde{e}^{(\mathrm{valid})}$. The validation time series must represent expected normal operating conditions, as with the training NOC data. The diagnostic statistic threshold can then be set as the $((1-\alpha) \times 100)$th percentile of this distribution, where α represents a design selection for expected false alarm rates.

7.4 Feature Extraction and Reconstruction: Methods

The details of singular spectrum analysis, random forest feature extraction and inverse nonlinear principal components analysis for dynamic monitoring are given here.

7.4.1 Singular Spectrum Analysis

Singular spectrum analysis is a popular linear phase space quantification technique and has been applied in change point detection (Moskvina and Zhigljavsky 2003; Salgado and Alonso 2006; Palomo et al. 2003; Auret and Aldrich 2010). Here, the application of SSA for dynamic monitoring is considered (i.e. NOC data are not updated, as in change point detection).

During the training stage, NOC time series data (with N_w samples) are scaled and embedded. Embedding parameters of $k = 1$ and $m = \lfloor N_w \div 2 \rfloor$ are used. These embedding parameters (rather than those determined by average mutual information and false nearest neighbour approaches) are used, as the feature extractive step of SSA automatically determines the relevant weights of lag variables for optimal phase space characterization. Singular value decomposition of the lag-trajectory matrix requires the lag-covariance matrix $\mathbf{C}^{(\mathrm{NOC})}$:

$$\mathbf{C}^{(\mathrm{NOC})} = \left(\mathbf{Z}^{(\mathrm{NOC})}\right)^T \left(\mathbf{Z}^{(\mathrm{NOC})}\right). \tag{7.10}$$

The eigenvalues $(\lambda_1, \lambda_2, \ldots, \lambda_m)$ and eigenvectors $(U_1, U_1, \ldots, U_m)$ of $\mathbf{C}^{(\mathrm{NOC})}$ are then determined by singular value decomposition. The first d eigenvalues $(\lambda_1, \lambda_2, \ldots, \lambda_d)$ and d eigenvectors $(\mathbf{U}^* = \{\mathbf{U}_1, \mathbf{U}_2, \ldots, \mathbf{U}_d\})$ are retained, where d is the number of eigenvalues that account for 90 % of the total of m eigenvalues. An example of the eigenvector weights for embedded lag variables is shown in Fig. 7.5.

The first d features $(\mathbf{T}^* = \{\mathbf{T}_1, \mathbf{T}_2, \ldots, \mathbf{T}_d\})$ are calculated by projection the embedded data $\mathbf{Z}^{(\mathrm{NOC})}$ onto each eigenvector:

$$\mathbf{T}_j = \mathbf{Z}^{(\mathrm{NOC})}\mathbf{U}_j. \tag{7.11}$$

The lag-trajectory matrix can be constructed $\mathbf{Z}^{(\mathrm{NOC})}$ as $\hat{\mathbf{Z}}^{(\mathrm{NOC})}$ from the d-subspace of features $\mathbf{T}^*$ and associated d eigenvectors $\mathbf{U}^*$:

$$\widehat{\mathbf{Z}}^{(\mathrm{NOC})} = \mathbf{T}^*\left(\mathbf{U}^*\right)^T. \tag{7.12}$$

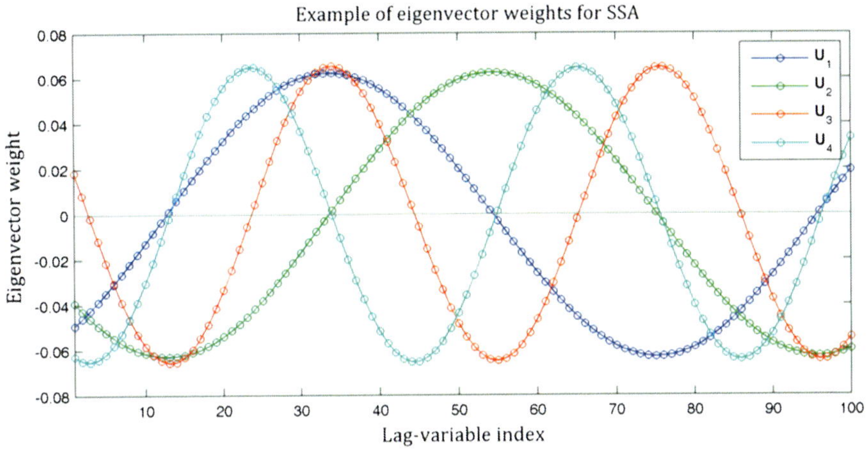

Fig. 7.5 Example of eigenvector weights for lag variables for four retained eigenvectors (plot is truncated to first 100 of 500 lag variables)

A filtered approximation ($\hat{\mathbf{z}}^{(\mathrm{NOC})}$) of the original scaled time series $\mathbf{z}^{(\mathrm{NOC})}$ can be calculated by averaging the diagonals (with $i + j =$ constant; i indicating the row index and j the column index of $\hat{\mathbf{Z}}^{(\mathrm{NOC})}$) of $\hat{\mathbf{Z}}^{(\mathrm{NOC})}$.

The reconstruction distance $e_i^{(\mathrm{NOC})}$ from an actual entry of the lag-trajectory matrix $\mathbf{Z}_i^{(\mathrm{NOC})}$ to its reconstruction $\hat{\mathbf{Z}}_i^{(\mathrm{NOC})}$ is given by

$$
\begin{aligned}
e_i^{(\mathrm{NOC})} &= \mathbf{Z}_i^{(\mathrm{NOC})}\left(\mathbf{Z}_i^{(\mathrm{NOC})}\right)^{\mathrm{T}} - \widehat{\mathbf{Z}}_i^{(\mathrm{NOC})}\left(\widehat{\mathbf{Z}}_i^{(\mathrm{NOC})}\right)^{\mathrm{T}} \\
&= \mathbf{Z}_i^{(\mathrm{NOC})}\left(\mathbf{Z}_i^{(\mathrm{NOC})}\right)^{\mathrm{T}} - \left(\mathbf{T}^*(\mathbf{U}^*)^{\mathrm{T}}\right)\left(\mathbf{T}^*(\mathbf{U}^*)^{\mathrm{T}}\right)^{\mathrm{T}} \\
&= \mathbf{Z}_i^{(\mathrm{NOC})}\left(\mathbf{Z}_i^{(\mathrm{NOC})}\right)^{\mathrm{T}} - \left(\mathbf{Z}_i^{(\mathrm{NOC})}\mathbf{U}^*(\mathbf{U}^*)^{\mathrm{T}}\right)\left(\mathbf{Z}_i^{(\mathrm{NOC})}\mathbf{U}^*(\mathbf{U}^*)^{\mathrm{T}}\right)^{\mathrm{T}} \\
&= \mathbf{Z}_i^{(\mathrm{NOC})}\left(\mathbf{Z}_i^{(\mathrm{NOC})}\right)^{\mathrm{T}} - \mathbf{Z}_i^{(\mathrm{NOC})}\mathbf{U}^*(\mathbf{U}^*)^{\mathrm{T}}\left(\mathbf{Z}_i^{(\mathrm{NOC})}\right)^{\mathrm{T}}(\mathbf{U}^*)^{\mathrm{T}}\mathbf{U}^* \\
&= \mathbf{Z}_i^{(\mathrm{NOC})}\left(\mathbf{Z}_i^{(\mathrm{NOC})}\right)^{\mathrm{T}} - \mathbf{Z}_i^{(\mathrm{NOC})}\left(\mathbf{U}^*(\mathbf{U}^*)^{\mathrm{T}}\right)\left(\mathbf{Z}_i^{(\mathrm{NOC})}\right)^{\mathrm{T}} \\
i &= 1, \ldots, N_{\mathrm{w}} - m + 1.
\end{aligned}
\tag{7.13}
$$

Above, $(\mathbf{U}^*)^{\mathrm{T}}\mathbf{U}^*$ is equal to the identity matrix, since the eigenvectors $\mathbf{U}^*$ are orthonormal. It should be noted that the reconstruction distance calculation given in Eq. 7.13 does not require the calculation of the d features $\mathbf{T}^*$. (Both the mapping and demapping calculations, $\mathcal{F}_{\mathbf{Z}\to\mathbf{T}}$ and $\mathcal{F}_{\mathbf{T}^*\to\tilde{\mathbf{Z}}}$, are captured by $\mathbf{U}^*$.) Once the reconstruction has been done, the average reconstruction distance $e^{(\mathrm{NOC})}$ can be calculated.

Table 7.1 Default settings of unsupervised random forests for feature extraction

Setting	Default value/method
Number of samples for contrast data	Same as number of samples for $\mathbf{Z}^{(\text{NOC})}$
Generation of contrast data	Sampling from product of marginal distributions of $\mathbf{Z}^{(\text{NOC})}$
Number of randomly selected variables	Floored square root of number of variables in $\mathbf{Z}^{(\text{NOC})}$
Minimum leaf node size	1
Number of trees	1,000

The SSA phase space characterization model (defined by $\mu^{(\text{NOC})}$, $\sigma^{(\text{NOC})}$, k, m, $\mathbf{U}^*$ and $e^{(\text{NOC})}$) can be applied to test time series data $\mathbf{x}^{(\text{test})}$ through scaling, embedding and reconstruction. The reconstruction distance $e_t^{(\text{test})}$ from the lag-trajectory vector $\mathbf{Z}_t^{(\text{test})}$ to its reconstruction $\hat{\mathbf{Z}}_t^{(\text{test})}$ is given by

$$e_t^{(\text{test})} = \mathbf{Z}_t^{(\text{test})}\left(\mathbf{Z}_t^{(\text{test})}\right)^{\mathrm{T}} - \mathbf{Z}_t^{(\text{test})}\left(\mathbf{U}^*\left(\mathbf{U}^*\right)^{\mathrm{T}}\right)\left(\mathbf{Z}_t^{(\text{test})}\right)^{\mathrm{T}}. \quad (7.14)$$

The average, scaled diagnostic statistic for SSA dynamic monitoring ($\tilde{e}_t^{(\text{test})}$) is calculated from Eq. 7.9. The threshold for the diagnostic statistic is calculated from validation data, as discussed above.

7.4.2 *Random Forest Feature Extraction*

The feature extraction and reconstruction approach of dynamic monitoring with SSA can be extended by applying nonlinear feature extraction methods. A dynamic monitoring framework with random forest feature extraction (and random forest regression for mapping and demapping) is considered here. Auret and Aldrich (2010) considered change point detection with random forest feature extraction. As with SSA, RF dynamic monitoring consists of a training stage with NOC data, an application stage on test data and a validation stage to determine a diagnostic statistic threshold.

Scaling and embedding of training NOC time series data (with N_{w} samples) delivers a lag-trajectory matrix $\mathbf{Z}^{(\text{NOC})}$, where the embedding parameters $k = 1$ and $m = \lfloor N_{\mathrm{w}} \div 2 \rfloor$ are used, similar to SSA. Random forest features are extracted from the NOC embedding $\mathbf{Z}^{(\text{NOC})}$ by applying classical multidimensional scaling to the average proximity data from a set of five unsupervised random forests $\mathcal{F}_{\mathbf{Z}\to(0,1)}$, each with 1,000 trees and default settings. The default settings for these unsupervised random forests are given in Table 7.1. A number (d) of significant features are retained to characterize the phase space structure.

To determine the number of random forest features to retain, parallel analysis (Ku et al. 1995) is applied. Parallel analysis (as applied to CMDS results) consists of comparing the descending eigenvalues associated with CMDS of the dissimilarity

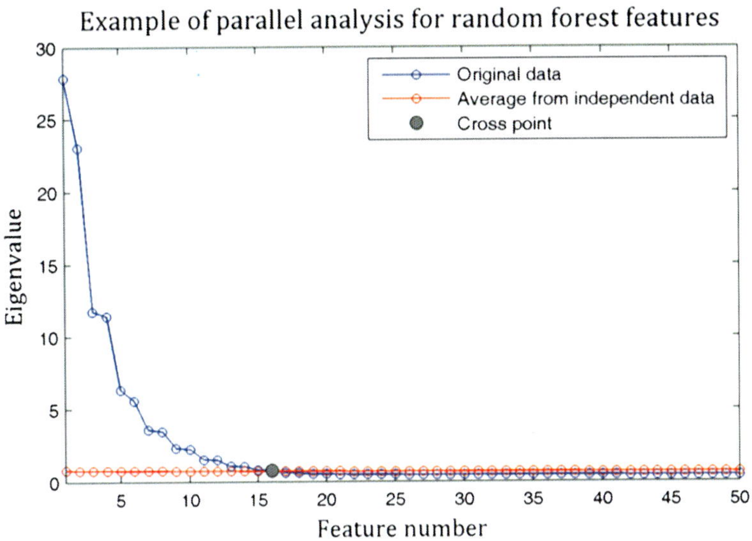

Fig. 7.6 Example of parallel analysis to determine the number of random forest features to retain (plot is truncated to first 50 of 500 features)

Table 7.2 Default settings of regression random forests for mapping and demapping

Setting	Default value/method
Number of randomly selected variables	Number of variables in $\mathbf{Z}^{(NOC)}$ divided by 3
Minimum leaf node size	5
Number of trees	100

data from $\mathcal{F}_{\mathbf{Z}\to(0,1)}$, with the descending eigenvalues associated with CMDS of the dissimilarity data of ten forests ($\mathcal{F}_{\mathbf{Z}r\to(0,1)}$) trained on data representing independent variables, of the same size as $\mathbf{Z}^{(NOC)}$. The average of ten eigenvalue profiles is used for comparison. The point where the curves cross gives an estimate of the number of features d to retain. This is intuitive, as features associated with eigenvalues smaller than eigenvalues from independent data can be considered uninformative. For example, the result of parallel analysis is shown in Fig. 7.6.

Since features cannot be calculated directly from the set of trained unsupervised forests $\mathcal{F}_{\mathbf{Z}\to(0,1)}$ for new data, mapping regression forests $\mathcal{F}_{\mathbf{Z}\to\mathbf{T}^*}$ are learnt, with one forest for each of the d retained features $\mathbf{T}^*$. The mapping forests serve as feature calculation models for new lag-trajectory data. A set of demapping regression forests $\mathcal{F}_{\mathbf{T}^*\to\tilde{\mathbf{Z}}}$ are also learnt, one forest for each of the m variables of $\mathbf{Z}^{(NOC)}$. The demapping forests serve as reconstruction models, taking calculated random forest features as input. The default settings for these mapping and demapping random forests are given in Table 7.2.

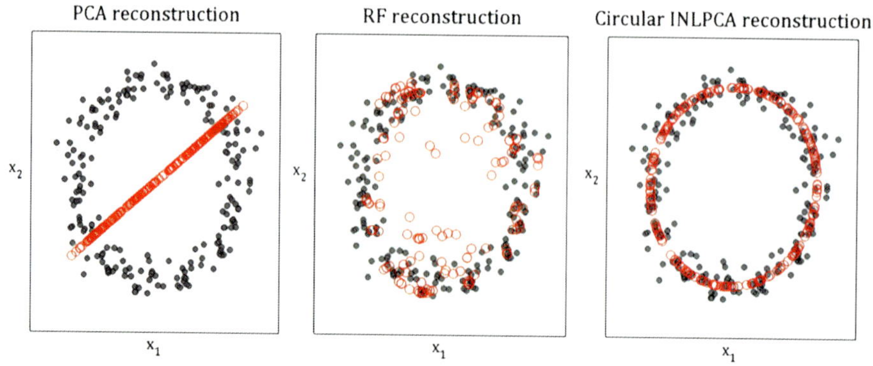

Fig. 7.7 Comparison of one-dimensional manifolds obtained from noisy circular data with linear PCA, random forest feature extraction and circular inverse NLPCA

The reconstruction distance $e_i^{(NOC)}$ from an actual entry of the lag-trajectory matrix $\mathbf{Z}_i^{(NOC)}$ to its reconstruction $\hat{\mathbf{Z}}_i^{(NOC)}$ is calculated from 7.8. The average reconstruction distance $e^{(NOC)}$ (scaled by embedding dimension m) for the samples in $\mathbf{Z}^{(NOC)}$ is obtained from 7.9.

The RF phase space characterization model (defined by $\mu^{(NOC)}$, $\sigma^{(NOC)}$, k, m, $\mathcal{F}_{\mathbf{Z}\to\mathbf{T}^*}$, $\mathcal{F}_{\mathbf{T}^*\to\tilde{\mathbf{Z}}}$ and $e^{(NOC)}$) can be applied to test time series data $\mathbf{x}^{(test)}$ through scaling, embedding and reconstruction. The reconstruction distance $e_i^{(test)}$ from the lag-trajectory vector $\mathbf{Z}_i^{(test)}$ to its reconstruction $\hat{\mathbf{Z}}_i^{(test)}$ is then given by Eq. 7.8, while the average, scaled diagnostic statistic for RF dynamic monitoring ($\tilde{e}_i^{(test)}$) is calculated from Eq. 7.9. The threshold for the diagnostic statistic is calculated from validation data, as discussed above.

The randomForest package (Liaw and Wiener 2002) in the R statistical environment (R Development Core Team 2010) is used for all random forest training and application steps.

7.4.3 Inverse Nonlinear Principal Component Analysis

Random forest feature extraction is a powerful nonlinear extension of SSA. Another powerful nonlinear feature extraction technique is inverse nonlinear principal component analysis, INLPCA (Scholz et al. 2005, 2008). In particular, circular INLPCA is considered here. Circular nodes in neural networks allow the explicit characterization of periodic behaviour in an input space (Scholz 2007). Such circular nodes allow a one-dimensional manifold (for one extracted feature) to be a closed curve. Figure 7.7 compares the one-dimensional manifolds obtained by extraction with linear PCA, random forest feature extraction and circular INLPCA. The PCA manifold does not follow the closed curve of the data; the linear structure clearly being unsuited to this task. The RF manifold does better in capturing the circularity of the

Table 7.3 Default settings of inverse nonlinear principal component analysis (*INLPCA*) networks for feature extraction, mapping and demapping

Setting	Default value/method
Number of hidden layers	1
Number of nodes in hidden layer	6
Activation function of hidden layer nodes	Hyperbolic tangent
Maximum number of iterations	$\min\{3{,}000;\ 5 \times [N_w - (m-1)k]\}$
Pre-scaling	To give maximum standard deviation of 0.1
Weight decay for regularization	Weight decay coefficient of 0.001

data, albeit with a noisy manifold with reconstructions outside the data support. The circular INLPCA manifold fits the noisy circle the best, with a simple closed curve.

Bottleneck autoassociative networks estimate the weights of the hidden layer(s) mapping from $\mathbf{Z}^{(NOC)}$ to $\mathbf{T}^*$, as well as the weights of the hidden layer(s) mapping from $\mathbf{T}^*$ to $\hat{\mathbf{Z}}^{(NOC)}$ in order to minimize the errors $\mathbf{Z}^{(NOC)} - \hat{\mathbf{Z}}^{(NOC)}$. Compared to this, inverse networks (Scholz et al. 2005) estimate the network inputs (feature values $\mathbf{T}^*$) and the weights of the hidden layer(s) mapping from $\mathbf{T}^*$ to $\hat{\mathbf{Z}}^{(NOC)}$, in order to minimize the errors $\mathbf{Z}^{(NOC)} - \hat{\mathbf{Z}}^{(NOC)}$. Unlike bottleneck networks, there is no direct mapping from the input space $\mathbf{Z}^{(NOC)}$ to the feature space $\mathbf{T}^*$. To determine the feature values for an input vector, an optimal feature vector is searched for (through gradient optimization) to minimize the reconstruction error.

The application of circular INLPCA to dynamic monitoring is now considered. The periodicity of certain dynamic systems makes circular INLPCA an attractive option for monitoring through the feature extraction and reconstruction framework. The extraction of a one-dimensional feature (to represent a closed-curve one-dimensional manifold in the phase space) is considered here to represent a dense attractor.

Scaling and embedding of training NOC time series data (with N_w samples) delivers a lag-trajectory matrix $\mathbf{Z}^{(NOC)}$. Here, the embedding parameters k and m are determined through average mutual information and false nearest neighbour errors, respectively. A circular INLPCA is trained to extract a single feature from the NOC embedding $\mathbf{Z}^{(NOC)}$. The training of the network is subject to certain default parameters, as presented in Table 7.3. The trained network represents both the mapping and demapping models ($\mathcal{F}_{\mathbf{Z}\to\mathbf{T}^*}$ and $\mathcal{F}_{\mathbf{T}^*\to\tilde{\mathbf{Z}}}$).

The reconstruction distance $e_i^{(NOC)}$ from an actual entry of the lag-trajectory matrix $\mathbf{Z}_i^{(NOC)}$ to its reconstruction $\hat{\mathbf{Z}}_i^{(NOC)}$ is calculated from 7.8. The average reconstruction distance $e^{(NOC)}$ (scaled by embedding dimension m and number of samples $N_w - (m-1)k$) for the samples in $\mathbf{Z}^{(NOC)}$ is given by

$$e^{(NOC)} = \frac{1}{m\left(N_w - (m-1)\,k\right)} \sum_{i=1}^{N_w-(m-1)k} e_i. \tag{7.15}$$

As with SSA and RF methods, the circular INLPCA phase space characterization model (defined by $\mu^{(NOC)}$, $\sigma^{(NOC)}$, k, m, $\mathcal{F}_{\mathbf{Z}\to\mathbf{T}^*}$, $\mathcal{F}_{\mathbf{T}^*\to\tilde{\mathbf{Z}}}$ and $e^{(NOC)}$) can be applied to test time series data $\mathbf{x}^{(\text{test})}$ through scaling, embedding and reconstruction. The reconstruction distance $e_t^{(\text{test})}$ from the lag-trajectory vector $\mathbf{Z}_t^{(\text{test})}$ to its reconstruction $\hat{\mathbf{Z}}_t^{(\text{test})}$ is then given by 7.8, while the average, scaled diagnostic statistic for RF dynamic monitoring ($\tilde{e}_t^{(\text{test})}$) is calculated from 7.9. The threshold for the diagnostic statistic is calculated from validation data, as discussed above. The nonlinear PCA toolbox for MATLAB (Scholz 2011) is used to construct and apply all circular INLPCA models.

7.5 Feature Space Characterization Approaches

7.5.1 Phase Space Distribution Estimation

Where circular INLPCA attempts to characterize an attractor by a closed-curve manifold, one could also view phase space characterization as a data support estimation problem. The one-class support vector machine (1-SVM) model is a powerful support estimation approach, which combines the versatility of kernel functions with the rigorous statistical learning framework of support vectors. 1-SVM estimates the support (analogous to the probability distribution) of an unlabelled data set. In a dynamic monitoring framework, a 1-SVM model can be trained to characterize the support of the phase space representing normal operating conditions. New data in the phase space can then be compared to the 1-SVM NOC support: Where the new data lie outside the NOC support, it may indicate that the data generating process has undergone a change from NOC.

As with the previous approaches, NOC time series data $\mathbf{x}^{(\text{NOC})}$ (with N_{w} samples) are scaled and embedded to create a lag-trajectory matrix $\mathbf{Z}^{(\text{NOC})}$. The embedding parameters k and m are determined through average mutual information and false nearest neighbour errors, respectively. In the application of 1-SVM considered here, a Gaussian kernel function was used with the ν-SVM version of 1-SVM. Recall that the ν-parameter enforces a lower bound on the fraction of support vectors and an upper bound on the fraction of outliers. The user-defined confidence limit (or designed false alarm rate expectation) α is thus a natural choice for the ν-parameter.

The Gaussian kernel width σ affects the complexity of the support. When σ is too small, all training points are support vectors, and the support is a set of small spheres around each training point. As σ increases, the support boundary becomes smoother (less intricate). When σ is too large, the support becomes too smooth, essentially an ellipse across the range of all data. An estimate of an optimal kernel width can be made by searching a grid of possible σ values. The range of possible σ values can be restricted according to the distribution of interpoint distances of $\mathbf{Z}^{(\text{NOC})}$ (Belousov et al. 2002). The smallest σ considered is the first percentile of these interpoint distances, while the largest σ considered the 50th percentile of the interpoint distances. An evenly spaced grid with ten σ values is used in conjunction with fivefold cross-validation.

The optimal kernel width for the 1-SVM model is selected based on a trade-off between the mean fraction of false alarms and mean fraction of support vectors. As a heuristic approach, only σ values delivering mean false alarm rates and fraction of support vectors below 0.1 are considered. The largest valid σ value for these conditions is selected as the optimal kernel width. Selecting the largest valid σ gives an affirmative nod in the direction of model complexity control. If these conditions cannot be satisfied, σ is set to the mean interpoint distance of the training data set. Minimizing the mean fraction of false alarms ensures accuracy of the support, while minimizing the mean fraction of support vectors controls the complexity (and associated propensity to overfitting) of the support. With the selection of the optimal kernel width made, the 1-SVM model can be trained with this kernel on all the training data.

This 1-SVM model represents a mapping from the phase space to the diagnostic statistic space, $\mathcal{F}_{\mathbf{Z}\to e}$. For a lag-trajectory vector $\mathbf{Z}_i^{(\mathrm{NOC})}$ (or, in the test stage, $\mathbf{Z}_i^{(\mathrm{test})}$), membership to the 1-SVM support is calculated from the 1-SVM decision function:

$$f\left(\mathbf{Z}_i^{(\mathrm{NOC})}\right) = \mathrm{sgn}\left(\mathbf{w}^{\mathrm{T}}\Phi\left(\mathbf{Z}_i^{(\mathrm{NOC})}\right) - \rho\right), \tag{7.16}$$

where $\mathbf{w}$ is the learnt weight vector of the support hyperplane and ρ is the learnt bias defining the location of the support hyperplane. The 1-SVM decision function is $+1$ where the lag-trajectory vector falls within the NOC support and -1 outside the NOC support. However, a simple $-1/+1$ diagnostic would be too restrictive for dynamic monitoring purposes. A continuous variable would allow the diagnostic statistic threshold to be another potential optimization parameter. To convert the 1-SVM decision function into such a continuous variable (somewhat analogous to reconstruction distance in the feature extraction methods), the diagnostic statistic $e_i^{(\mathrm{NOC})}$ for a lag-trajectory vector $\mathbf{Z}_i^{(\mathrm{NOC})}$ ($e_i^{(\mathrm{test})}$ and $\mathbf{Z}_i^{(\mathrm{test})}$, respectively, for the test stage) is defined as

$$e_i^{(\mathrm{NOC})} = -\left(\mathbf{w}^{\mathrm{T}}\Phi\left(\mathbf{Z}_i^{(\mathrm{NOC})}\right) - \rho\right). \tag{7.17}$$

To determine a diagnostic statistic threshold, a similar approach as for the feature extraction and reconstruction methods is applied. The diagnostic statistic described above is calculated for each sample of a validation time series data set, $\mathbf{x}^{(\mathrm{valid})}$, giving a distribution of validation statistics $e^{(\mathrm{valid})}$. The diagnostic statistic threshold can then be set as the $((1-\alpha)\times 100)$-th percentile of this distribution, where α represents a design selection for expected false alarm rates.

7.5.2 *Recurrence Quantification Analysis*

Recurrence quantification analysis (RQA) characterizes an attractor in the phase space by identifying the repeated occurrences (recurrences) of points in the same neighbourhood. Deterministic dynamic systems show particular recurrent state behaviours, and the nature of these recurrences can be tracked over time as identifiers of the system states. Recurrence information can be quantified as a scalar diagnostic statistic in terms of the recurrence rate: The number of times that points in an attractor is arbitrarily close in the phase space, given a set length of time series data. Increasing or decreasing recurrence rates are indicative of changes in the dynamics of the generating process.

The existence of a recurrence event R_{ij} between two samples i and j in a time series $\mathbf{x}$ is calculated as follows:

$$R_{ij} = I\left(\varepsilon - \left\|\mathbf{x}_i - \mathbf{x}_j\right\|\right). \tag{7.18}$$

In Eq. 7.18, ε is a threshold distance, and $I(\cdot)$is the indicator function: +1 when its argument is positive and 0 otherwise. When the two samples i and j are less than ε distance apart (in terms of the Euclidian norm), a recurrence event exists, i.e. $R_{ij} = 1$ (otherwise, $R_{ij} = 0$). A recurrence plot can be created by indicating recurrence events for all sample combinations. Examples of recurrence plots for the Lorenz system are shown in Fig. 7.8 (see Lindfield and Penny 2000, for details of the Lorenz system of ordinary differential equations).

To allow comparison between recurrence plots, the recurrence rate $\dot{R}$ quantifies the fraction of recurrence events that occur for a specific extent (N samples) of a time series:

$$\dot{R} = \frac{1}{N^2}\sum_{i=1}^{N}\sum_{j=1}^{N} R_{ij}. \tag{7.19}$$

The recurrence rate is a continuous variable that can be exploited for dynamic monitoring purposes.

In the RQA dynamic monitoring approach considered here, the recurrence rate of a scaled, embedded window of a time series is treated as the diagnostic statistic. During the training stage, a NOC time series $\mathbf{x}^{(\mathrm{NOC})}$ with N_{w} is scaled to $\mathbf{z}^{(\mathrm{NOC})}$, from the scaling parameters $\mu^{(\mathrm{NOC})}$ and $\sigma^{(\mathrm{NOC})}$. The scaled time series data are embedded in a lag-trajectory matrix $\mathbf{Z}^{(\mathrm{NOC})}$, with embedding parameters k and m determined with average mutual information and false nearest neighbour approaches. The neighbourhood size threshold ε is calculated as the mean of the interpoint distances of $\mathbf{Z}^{(\mathrm{NOC})}$. The diagnostic statistic is the recurrence rate of the lag-trajectory matrix, given the threshold ε.

The monitoring parameters obtained from training ($\mu^{(\mathrm{NOC})}$, $\sigma^{(\mathrm{NOC})}$, k, m and ε) can be applied to moving windows (with N_{w} samples) of test time series data $\mathbf{x}^{(\mathrm{test})}$ through scaling, embedding and recurrence calculations.

Since both an increase and a decrease in the recurrence rate may signal a change in the dynamic structure of a time series, upper and lower limits to the RQA diagnostic must be defined. These diagnostic statistic thresholds can be determined from validation normal operating conditions data. The RQA monitoring parameters

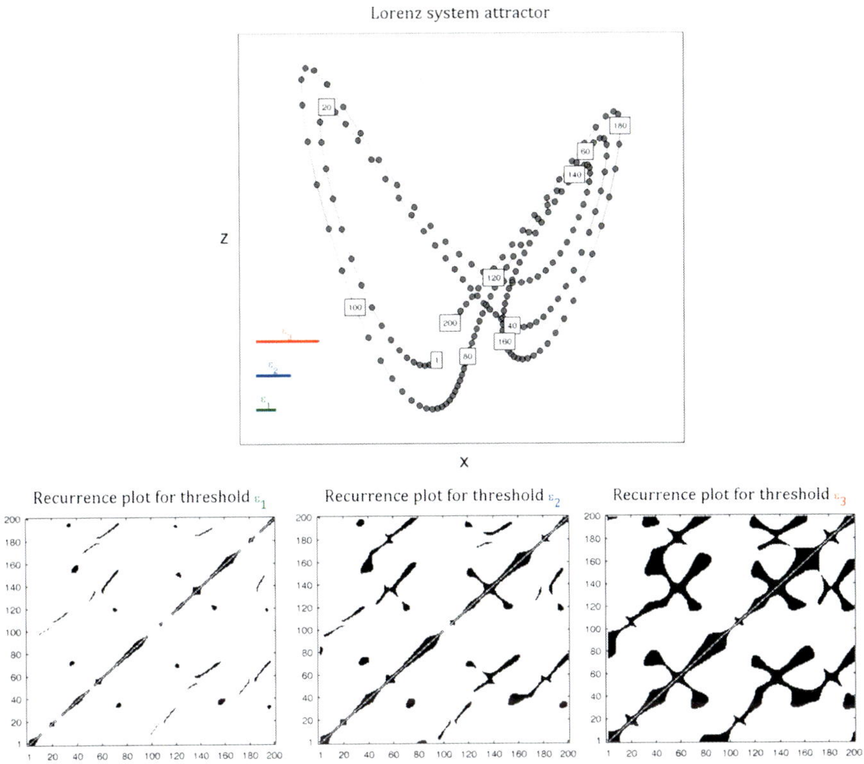

Fig. 7.8 Lorenz system attractor and its associated recurrence plots, for three different neighbourhood threshold distances. The *text* in the attractor plot indicates specific time indices. In the recurrence plots, index combinations that correspond to recurrence events are coloured *black*, while non-events are coloured *white*

obtained during the NOC training steps are applied to N_w-sized windows of the validation portion of the time series, $\mathbf{x}^{(\text{valid})}$. The distribution of the validation diagnostic statistic allows for the definition of diagnostic statistic limits: The lower limit can be defined at the $((\alpha/2) \times 100)$-th percentile, and the upper limit at $((1 - \alpha/2) \times 100)$-th percentile. Here, α is a user-defined expected false alarm rate.

7.6 Dynamic Monitoring Case Studies

As an illustration of data-driven dynamic monitoring approaches, a number of data sets with dynamic characteristics are considered. The predator–prey, Belousov–Zhabotinsky reaction and autocatalytic data sets are generated from ordinary differential equations, summarized here.

7.6.1 Lotka–Volterra Predator–Prey Model

The Lotka–Volterra model was one of the first to describe the interactions between predator and prey species and has acquired broader relevance since then in, for example, biophysical coastal ecosystem models (Dowd 2005), atmospheric chemistry models (Wang et al. 2002), juvenile salmon migration models by Anderson et al. (2005) and a model on plankton predation rates by Lewis and Bala (2006). The predator–prey model proposed by Lindfield and Penny (2000) was considered in this study. It is based on the following two differential equations:

$$\frac{\mathrm{d}x}{\mathrm{d}t} = k_1 x - Cxy \tag{7.20}$$

$$\frac{\mathrm{d}y}{\mathrm{d}t} = -k_2 y + Dxy. \tag{7.21}$$

The number of prey is represented by x, while the number of predators is represented by y. The rate of prey population growth is given by k_1, the predator mortality rate is k_2 and C is the frequency of contact between the two species (reaction rate), while D is the efficiency of predators in converting food into offspring (conversion efficiency).

The above system was simulated with initial conditions of 5,000 prey and 100 predators, with parameters of $k_1 = 2$, $k_2 = 10$, $C = 0.0010$ and $D = 0.0020$ for an initial period of 400 time units (with each time step equal to 0.02 time units). A change in the system was introduced after 400 time units by adjusting the k_1, k_2, C and D parameters. The parameter changes involved a linear ramp function from 400 time units to 600 time units, with k_1 increasing linearly from 2 to 3, k_2 increasing linearly from 10 to 11, C from 0.0010 to 0.0011 and D from 0.0020 to 0.0021. A second change in the system was introduced after 600 time units: The parameters k_1, k_2, C and D were kept constant at their new values (3, 11, 0.0011 and 0.0021, respectively). The time-varying parameter profiles are shown in Fig. 7.9.

The predator–prey data set for dynamic monitoring was created from 10,000 equidistant samples of x between 200 and 400 time units (representing normal operating conditions), 10,000 equidistant samples of x between 400 and 600 time units (representing the first change condition) and 10,000 equidistant samples of x between 600 and 800 time units (representing the second change condition). Data from the first 200 time units are thus discarded, serving as a period to allow steady state of the dynamic attractor: $t \leftarrow t–200$. The state space of the predator–prey model is shown in Fig. 7.10, and the predator–prey data to be monitored are shown in Fig. 7.11.

The state space of the predator–prey model (Fig. 7.10) shows that the attractor expands as the model parameters are changed. The attractor has a wide global range or extent, but a very narrow local range ("thickness").

To investigate the effect of noise, various levels of uniform noise were added to the predator–prey variable of interest, viz. 0, 5, 10 and 20 % of the range of the

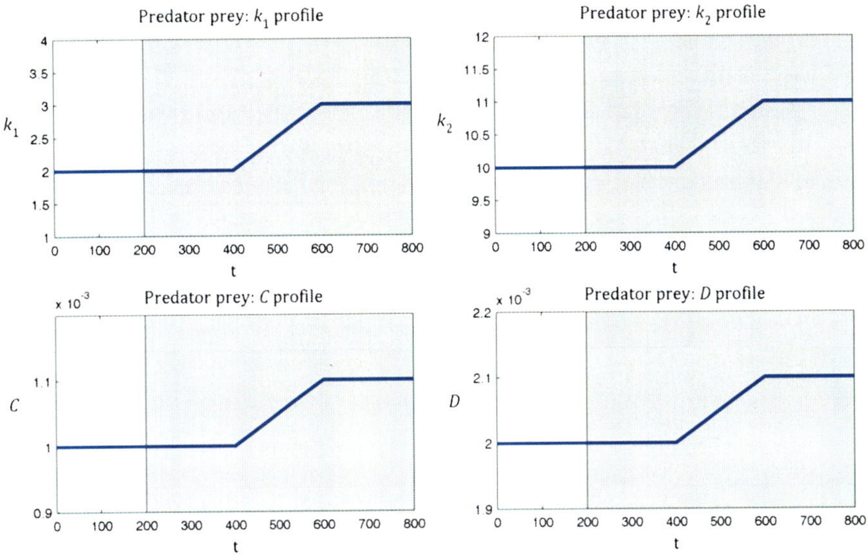

Fig. 7.9 Predator–prey parameter profiles. Changes to the simulated system were introduced at 400 and 600 time units. The *highlighted* regions indicate the time span relevant to the final data set

Fig. 7.10 State space for predator–prey model

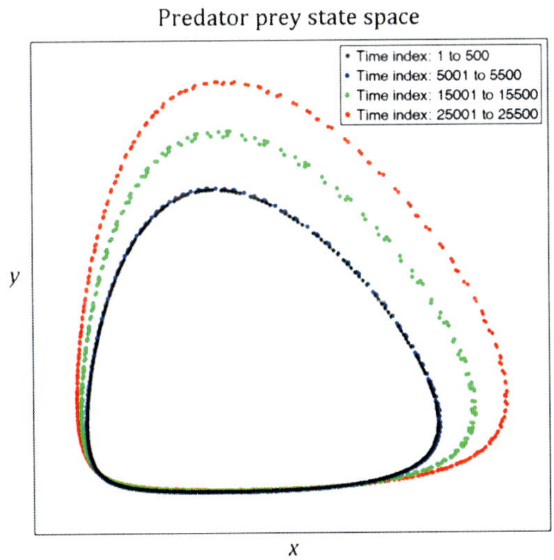

variable over the first 10,000 samples. The optimal lags for each of these four data sets were determined from average mutual information profiles, while the optimal embedding dimensions were determined with false nearest neighbour ratios. The lag and embedding profiles are shown in Fig. 7.12. Figure 7.12 also shows the first two PCA features of the reconstructed attractors, according to the optimal lag and embedding parameters shown.

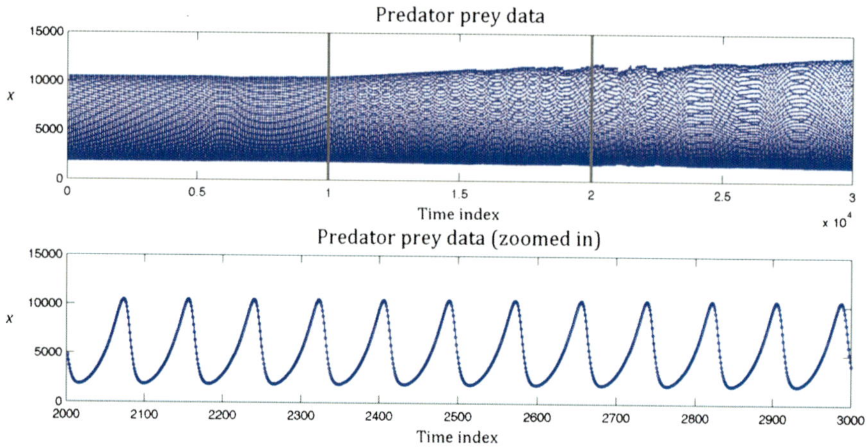

Fig. 7.11 Plots of predator–prey data. Changes to the simulated system are indicated after 10,000 and 20,000 time indices

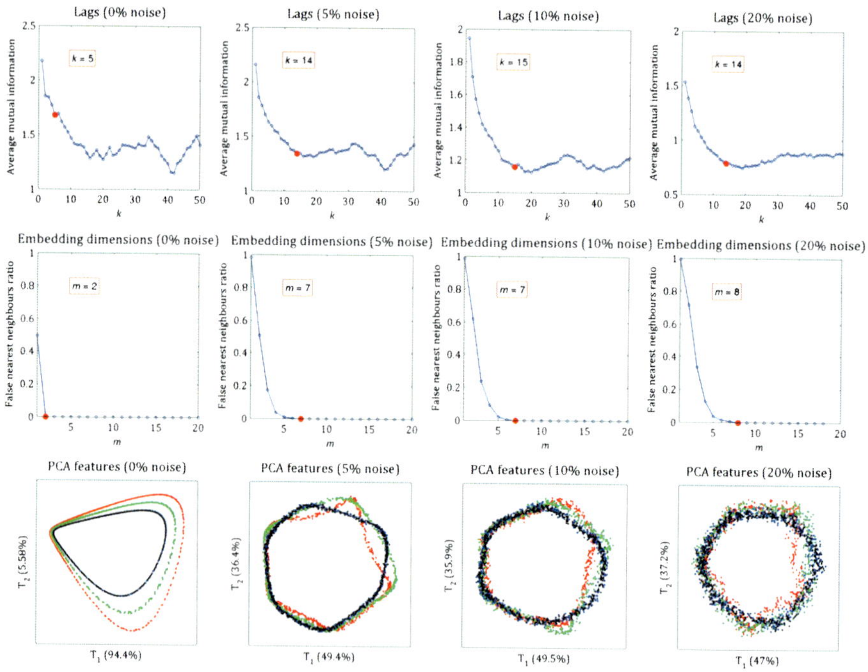

Fig. 7.12 Lag profiles, embedding dimension profiles and projected phase space attractors for the predator–prey data set, with four different levels of noise. PCA models were built based on the first 2,000 samples of each data set. *Black markers* show projected samples for time indices 1–500, *blue markers* for time indices 5,001–5,500, *green markers* for time indices 15,001–15,500 and *red markers* for time indices 25,001–25,500. The percentage variance accounted for by each PCA component is shown

Table 7.4 Symbols for the Belousov–Zhabotinsky reaction system

Symbol	Chemical species	Concentration
A	BrO_3^-	Constant
C	Total Ce^{3+} and Ce^{4+}	Constant
M	$CH_2(COOH)_2$	Constant
H	H^+	Constant
V	$BrCH(COOH)_2$	Variable
X	$HBrO_2$	Variable
Y	Br^-	Variable
Z	Ce^{4+}	Variable

From Fig. 7.12, it can be seen that the optimal embedding dimension (and the lag, to a lesser extent) increases as noise is added. This shows that a larger embedding dimension is required to separate false neighbours associated with noise. These false neighbours are due to the noisy nature of the data and not due to insufficient embedding. The effect of the larger embedding dimensions on monitoring performance will be discussed along with the monitoring results.

The phase space projection for data with no noise added (Fig. 7.12) shows considerable agreement with the state space of the generating model (Fig. 7.10), supporting the choice of lag and embedding dimension. Apart from some distortion of the attractor shape, the relative change of the attractor as the model parameters change is similar to the state space. Both the state and phase spaces show "thin" attractor shapes.

The projections in Fig. 7.12 further show that the local range of the attractor (its "thickness") increases with added noise. As more noise is added, there is more overlap between the attractors of the different time period. This suggests that detection of changes in the attractor should be more difficult. It must be noted that for visualization purposes, only the first two components are shown; less overlap of the attractors is likely for the actual (higher-dimensional) embedded data.

7.6.2 Belousov–Zhabotinsky Reaction

The Belousov–Zhabotinsky (BZ) reaction is an unstable chemical reaction that maintains self-oscillations and propagating waves, which may display chaos under certain conditions. The reaction consists of the transition-metal-ion-catalysed oxidation and bromination of an organic dicarboxylic acid by bromate ions BrO_3^- in an acidic aqueous medium. A simplified three-variable reaction model for the reactions taking place in a continuous stirred tank reactor (CSTR) is presented here (Zhang et al. 1993). For this simulated reaction system, malonic acid $CH_2(COOH)_2$ is used as the dicarboxylic acid substrate, while the redox pair Ce^{3+}/Ce^{4+} is the metal-ion catalyst.

Given the definitions in Table 7.4, the seven considered chemical reactions (in the $CH_2(COOH)_2$ substrate), reaction rates and rate constants are:

(1) $X + Y + H \rightarrow 2V$

$$r_1 = k_1 [X][Y][H]; \quad k_1 = 4.0 \times 10^6, \text{units}: \frac{1}{\text{M}^2\text{s}} \tag{7.22}$$

(2) $Y + A + 2H \rightarrow V + X$

$$r_2 = k_2 [Y][A][H]^2; \quad k_2 = 2.0, \text{units}: \frac{1}{\text{M}^3\text{s}} \tag{7.23}$$

(3) $2X \rightarrow V$

$$r_3 = k_3[X]^2; \quad k_3 = 3,000, \text{units}: \frac{1}{\text{Ms}} \tag{7.24}$$

(4) $0.5X + A + H \rightarrow X + Z$

$$r_4 = k_4[X]^{0.5}[A]^{0.5}[H]^{1.5}([C] - [Z]);$$
$$k_4 = 55.2, \text{units}: \frac{1}{\text{M}^{2.5}\text{s}} \tag{7.25}$$

(5) $X + Z \rightarrow 0.5X$

$$r_5 = k_5 [X][Z]; \quad k_5 = 7,000, \text{units}: \frac{1}{\text{Ms}} \tag{7.26}$$

(6) $V + Z \rightarrow Y$

$$r_6 = \alpha k_6 [V][Z]; \quad k_6 = 0.09, \text{units}: \frac{1}{\text{Ms}} \tag{7.27}$$

(7) $M + Z \rightarrow$ products

$$r_7 = \beta k_7 [M][Z]; \; k_7 = 0.23, \text{units}: \frac{1}{\text{Ms}}. \tag{7.28}$$

Above, α and β are adjustable parameters. The four-variable model represented by Eqs. 7.22, 7.23, 7.24, 7.25, 7.26, 7.27 and 7.28 can be stated in terms of a scaled three-variable model, if Y is stated as the fast variable y through quasi-steady-state approximation.[1] The scaled variables and their scaling factors are given in the following equations:

$$t = \frac{\tau}{T_0}; \quad T_0 = \frac{1}{10k_2 [A][H][C]} \tag{7.29}$$

$$v = \frac{[V]}{[V_0]}; \quad [V_0] = \frac{4 [A][H][C]}{[M]^2} \tag{7.30}$$

[1] In a system of differential equations, a fast variable is a variable that shows dynamic variation at a much shorter time scale than the other variables in the system. Its derivative is set to zero to obtain an algebraic equation for substitution into the other differential equations of the system.

$$x = \frac{[X]}{[X_0]};\ [X_0] = \frac{k_2 [A] [H]^2}{k_5} \tag{7.31}$$

$$\tilde{y} = \frac{\frac{\alpha k_6 [V_0][Z_0] vz}{k_1 [H][X_0] x + k_2 [A][H]^2 + k_f}}{[Y_0]};\ [Y_0] = \frac{4k_2 [A] [H]^2}{k_5} \tag{7.32}$$

$$z = \frac{[Z]}{[Z_0]};\ [Z_0] = \frac{[C] [A]}{40 [M]}. \tag{7.33}$$

Reaction time is given by τ, and scaled reaction time by t. The parameter k_f is known as the flow rate and represents the inverse of the CSTR residence time.

The BZ reaction system is represented by the following three differential equations:

$$\frac{dv}{dt} = [T_0] \left\{ \begin{array}{c} \frac{2k_1 [H] [X_0] [Y_0]}{[V_0]} x \tilde{y} + \frac{k_2 [A] [H]^2 [Y_0]}{[V_0]} \tilde{y} \ldots \\ + \frac{k_3 [X_0]^2}{[V_0]} x^2 - \alpha k_6 [Z_0] zv - k_f v \end{array} \right\} \tag{7.34}$$

$$\frac{dx}{dt} = [T_0] \left\{ \begin{array}{c} -k_1 [H] [Y_0] x \tilde{y} + \frac{k_2 [A] [H]^2 [Y_0]}{[X_0]} \tilde{y} - 2k_3 [X_0] x^2 \ldots \\ +0.5k_4 [A]^{0.5} [H]^{1.5} \times [X_0]^{-0.5} ([C] - [Z_0] z) x^{0.5} \ldots \\ -0.5k_5 [Z_0] xz - k_f x \end{array} \right\} \tag{7.35}$$

$$\frac{dz}{dt} = [T_0] \left\{ \begin{array}{c} k_4 [A]^{0.5} [H]^{1.5} [X_0]^{0.5} \left(\frac{[C]}{[Z_0]} - z \right) x^{0.5} \ldots \\ -k_5 [X_0] xz - \alpha k_6 [V_0] zv - \beta k_7 [M] z - k_f z \end{array} \right\}. \tag{7.36}$$

The above system was simulated with initial conditions of $v = 0.4582$, $x = 0.0099$ and $z = 2.2001$. Time-independent parameters were set as follows: $[A] = 0.1$, $[C] = 0.000833$, $[H] = 0.26$, $[M] = 0.25$, $\alpha = 6{,}000/9$ and $\beta = 8/23$. The flow rate parameter k_f was set to 0.00045 for an initial period of 5 time units (with each time step equal to 0.00025 time units). A change in the system was introduced after 5 time units by adjusting k_f. The parameter change involved a linear ramp function from 5 time units to 7.5 time units, with k_f increasing linearly from 0.00045 to 0.00050. A second change in the system was introduced after 7.5 time units: k_f was kept constant at its new value (0.00050). The time-varying parameter profile of k_f is shown in Fig. 7.13.

The BZ reaction data set for dynamic monitoring was created from 10,000 equidistant samples of z between 2.5 and 5 time units (representing normal operating conditions), 10,000 equidistant samples of z between 5 and 7.5 time units (representing the first change condition) and 10,000 equidistant samples of x between 7.5 and 10 time units (representing the second change condition). Data

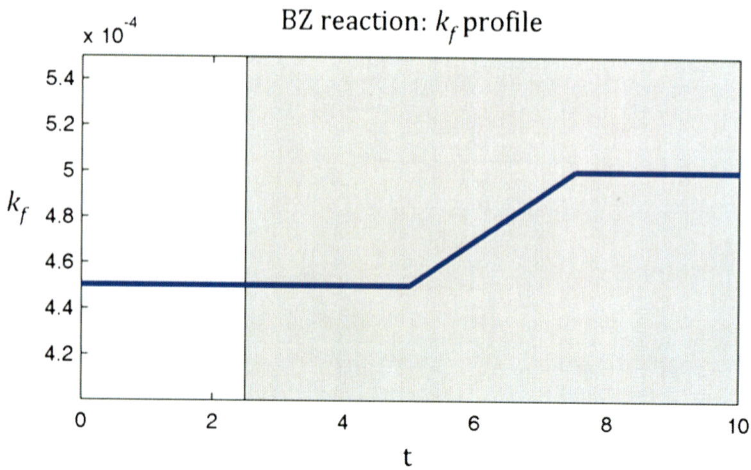

Fig. 7.13 BZ reaction parameter profile. Changes to the simulated system were introduced at 2.5 and 7.5 time units. The *highlighted* regions indicate the time span relevant to the final data set

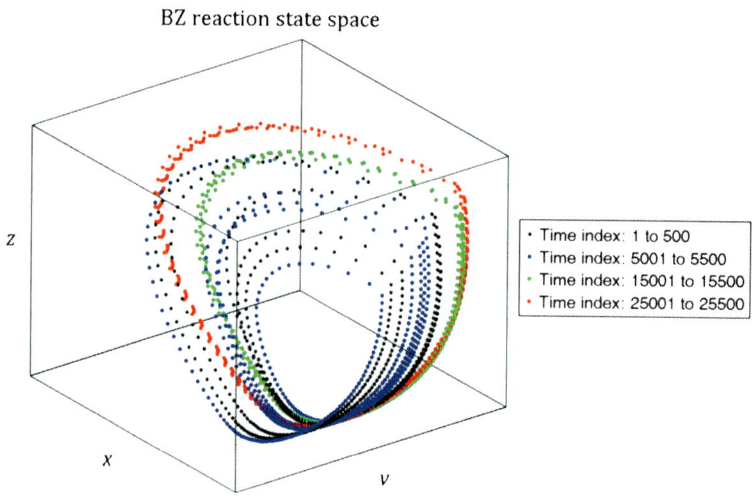

Fig. 7.14 State space for BZ reaction model

from the first 2.5 time units are thus discarded, serving as a period to allow steady state of the dynamic attractor: $t \leftarrow t - 2.5$. The BZ reaction data are shown in Fig. 7.15.

The state space of the BZ reaction model (Fig. 7.14) shows a more complex attractor than in the case of the predator–prey model. The attractor has not collapsed to a single region within 500 time indices, as can be seen from the separate

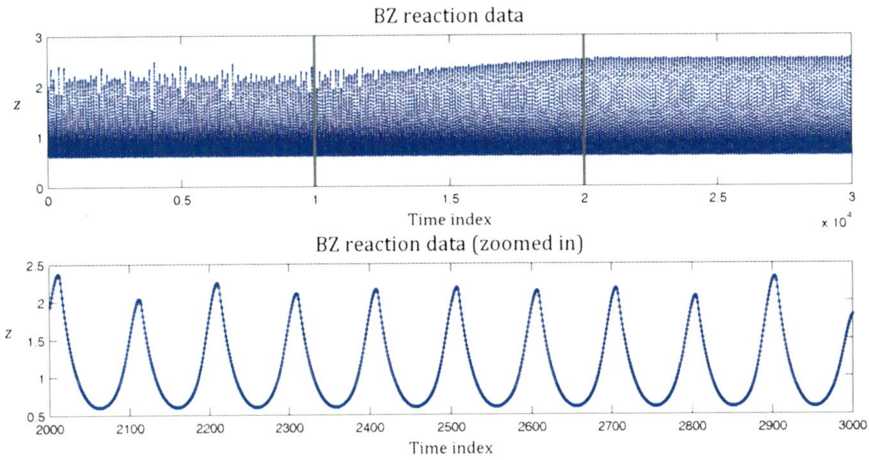

Fig. 7.15 Plots of BZ reaction data. Changes to the simulated system are indicated after 10,000 and 20,000 time indices

trajectories of series 1–500 (black markers) and series 5,001–5,500 (blue markers). In some regions, the trajectories for changed parameter conditions show a definite shift from the initial conditions trajectories, while some overlap is evident in other regions of the state space.

To investigate the effect of noise, various levels of uniform noise were added to the BZ reaction variable of interest, viz. 0, 5, 10 and 20 % of the range of the variable over the first 10,000 samples. The optimal lags for each of these four data sets were determined from average mutual information profiles, while the optimal embedding dimensions were determined with false nearest neighbour ratios. The lag and embedding profiles are shown in Fig. 7.15. Figure 7.15 also shows the first two PCA features of the reconstructed attractors, according to the optimal lag and embedding parameters shown.

As with the predator–prey data, the optimal embedding dimension (and the lag, to a lesser extent) increases as noise is added (Fig. 7.16). From the average mutual information profile for noiseless data, it appears that the first average mutual information minimum may be spurious, an erroneous local minimum. This may lead to a bad estimate of the optimal lag. With the addition of noise, the optimal lag selection appears to be more robust to local minima.

The phase space projection for data with no noise added (Fig. 7.16) shows less agreement with the state space of the generating model (Fig. 7.14) than is the case with the predator–prey data. However, the phase space projection is two dimensional, while the state space is three dimensional. Certain aspects of the phase space projection do agree with the state space: The black and blue trajectories do not completely overlap, and there is a shift in the trajectory for a fault condition (for time indices from 25,001 to 25,500, represented by the red markers). This shift in trajectory is not clear for data from time indices 15,001–15,500.

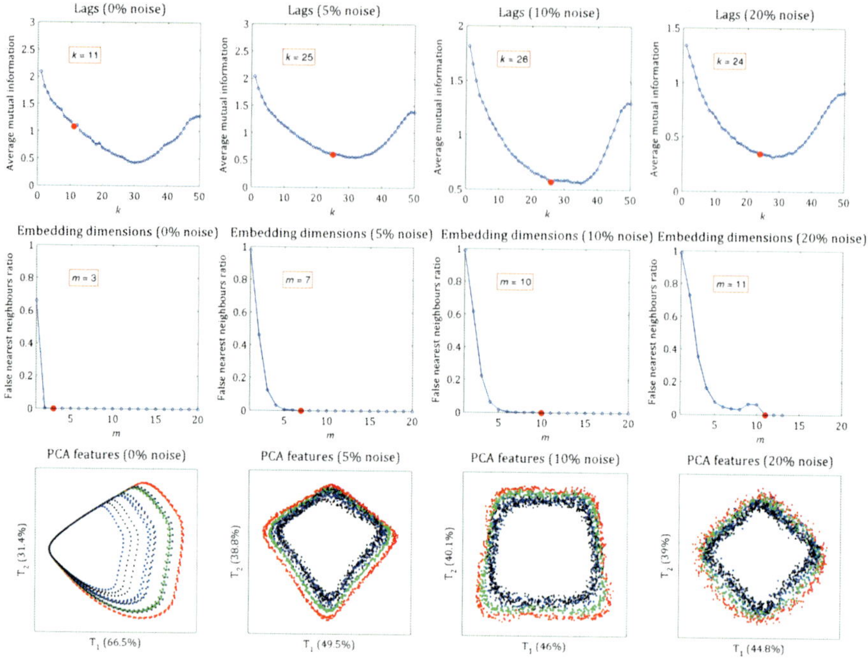

Fig. 7.16 Lag profiles, embedding dimension profiles and projected phase space attractors for the BZ reaction data set, with four different levels of noise. PCA models were built based on the first 2,000 samples of each data set. *Black markers* show projected samples for time indices 1–500, *blue markers* for time indices 5,001–5,500, *green markers* for time indices 15,001–15,500 and *red markers* for time indices 25,001–25,500. The percentage variance accounted for by each PCA component is shown

From Fig. 7.16, it is interesting to note that there is less overlap between attractors of normal and fault conditions as more noise is added. The added noise may prevent the selection of lags that are too low and embedding dimensions, as discussed above. Adding noise may be an approach to more robust embedding parameter selection. However, the effect of added noise in apparently overlapping trajectories for the predator–prey data serves as a caveat for this noise-addition approach.

7.6.3 *Autocatalytic Process*

An autocatalytic process is considered which consists of two parallel, isothermal autocatalytic reactions taking place in a continuous stirred tank reactor (CSTR) (Lee and Chang 1996). The system is capable of producing self-sustained oscillations based on cubic autocatalysis with catalyst decay at certain parameters. Chemical species B is involved in two autocatalytic reactions with chemical species A and D, separately, while B is also converted to chemical species C. The chemical reactions and reaction rates of this system are given in the following equations (with k_1, k_2 and k_3 the rate constants for the three reactions):

$$\begin{aligned}&(1)\quad A + 2B \rightarrow 3B\\ &\qquad r_1 = k_1 [A][B]^2\end{aligned} \tag{7.37}$$

$$\begin{aligned}&(2)\quad B \rightarrow C\\ &\qquad r_2 = k_2 [B]\end{aligned} \tag{7.38}$$

$$\begin{aligned}&(3)\quad D + 2B \rightarrow 3B\\ &\qquad r_3 = k_3 [D][B]^2.\end{aligned} \tag{7.39}$$

The concentration variables above are converted to scaled variables, and the system is further defined by scaled time and ratios of feed concentrations:

$$t = \frac{\tau Q}{V} \tag{7.40}$$

$$x = \frac{[A]}{[A_0]};\quad y = \frac{[D]}{[D_0]};\quad z = \frac{[B]}{[B_0]} \tag{7.41}$$

$$a = \frac{k_1 [B_0]^2 V}{Q};\quad b = \frac{k_3 [B_0]^2 V}{Q};\quad c = \frac{k_2 V}{Q} \tag{7.42}$$

$$\gamma_1 = \frac{[A_0]}{[B_0]};\quad \gamma_2 = \frac{[D_0]}{[B_0]}. \tag{7.43}$$

The autocatalytic process is then represented by the following three differential equations:

$$\frac{dx}{dt} = 1 - x - axz^2 \tag{7.44}$$

$$\frac{dy}{dt} = 1 - y - byz^2 \tag{7.45}$$

$$\frac{dz}{dt} = 1 - (1 + c)z + \gamma_1 axz^2 + \gamma_2 byz^2. \tag{7.46}$$

For $a = 18{,}000$, $b = 400$, $c = 80$, $\gamma_1 = 1.5$, $\gamma_2 = 4.2$ and initial conditions of $x = 0$, $y = 0$ and $z = 0$, the system exhibits chaotic behaviour (Lee and Chang 1996).

The autocatalytic process was simulated with initial conditions as mentioned above. The feed ratio parameters γ_1 and γ_2 were set to 1.50 and 4.20 for an initial period of 100 time units (with each time step equal to 0.005 time units). A change

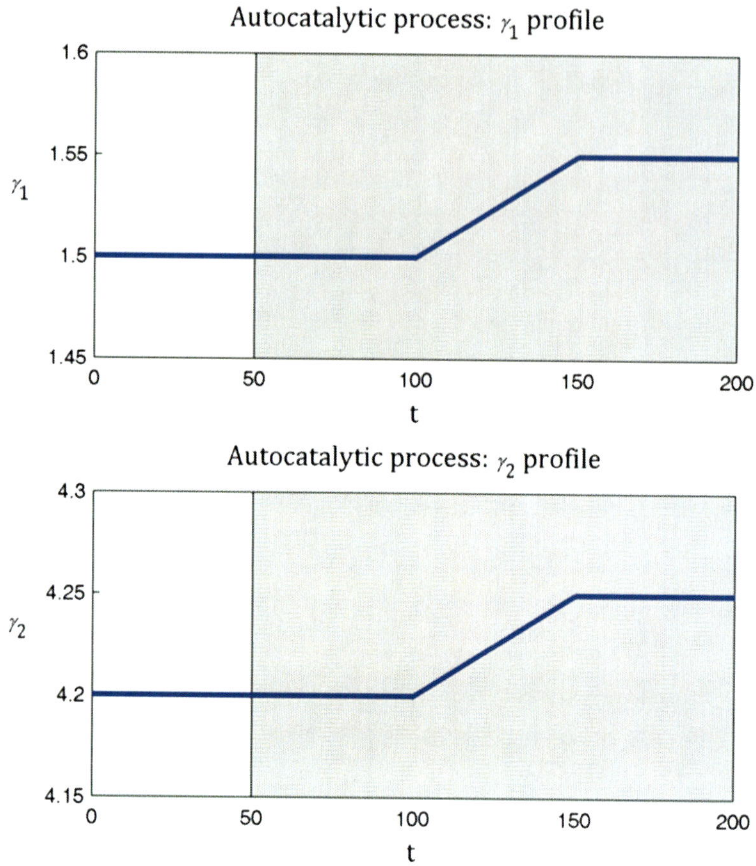

Fig. 7.17 Autocatalytic process parameter profiles. Changes to the simulated system were introduced at 100 and 150 time units. The *highlighted* regions indicate the time span relevant to the final data set

in the system was introduced after 100 time units by adjusting γ_1 and γ_2. The parameter changes involved a linear ramp function from 100 time units to 150 time units, with γ_1 increasing linearly from 1.50 to 1.55 and γ_2 increasing linearly from 4.20 to 4.25. A second change in the system was introduced after 150 time units: γ_1 and γ_2 were kept constant at their new values (1.55 and 4.25, respectively). The time-varying parameter profiles of γ_1 and γ_2 are shown in Fig. 7.17.

The autocatalytic process data set for dynamic monitoring was created from 10,000 equidistant samples of x between 50 and 100 time units (representing normal operating conditions), 10,000 equidistant samples of x between 100 and 150 time units (representing the first change condition) and 10,000 equidistant samples of x between 150 and 200 time units (representing the second change condition). Data from the first 50 time units are thus discarded, serving as a period to allow steady

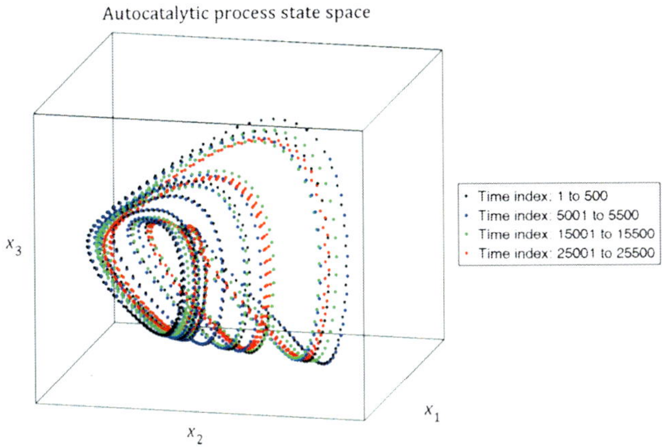

Fig. 7.18 State space for autocatalytic process model

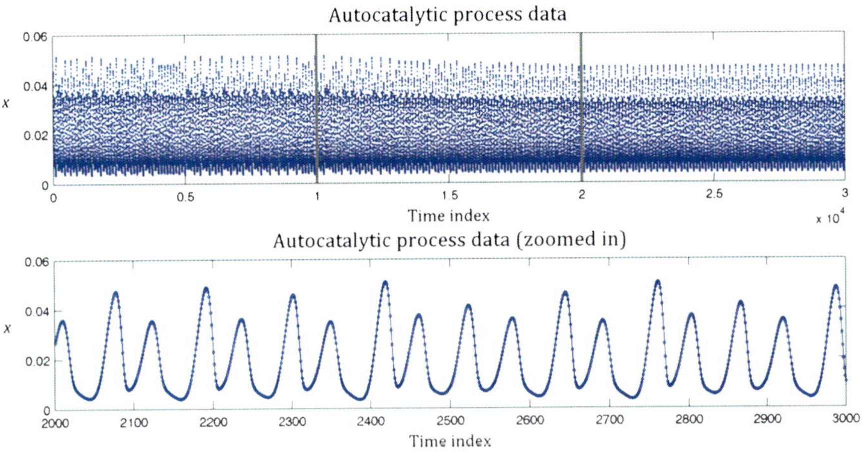

Fig. 7.19 Plots of autocatalytic process data. Changes to the simulated system are indicated after 10,000 and 20,000 time indices

state of the dynamic attractor: $t \leftarrow t{-}50$. The state space of the autocatalytic process model is shown in Fig. 7.18, and the autocatalytic process data to be monitored are shown in Fig. 7.19.

The state space of the autocatalytic process model (Fig. 7.18) shows a complex attractor. The changed parameter data (green markers for time indices 15,001–15,500, red markers for time indices 25,001–25,500) have trajectories that are intertwined with the attractor of the initial conditions (shown by black and blue markers). Not only is the shape of the attractor complex, but much overlap is visible between the different parameter conditions.

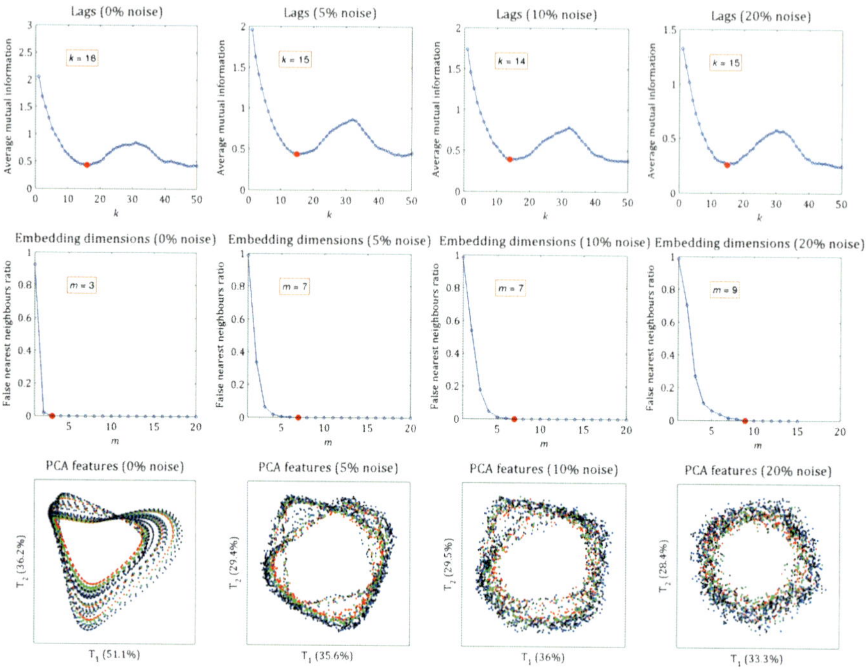

Fig. 7.20 Lag profiles, embedding dimension profiles and projected phase space attractors for the autocatalytic process data set, with four different levels of noise. PCA models were built based on the first 2,000 samples of each data set. *Black markers* show projected samples for time indices 1–500, *blue markers* for time indices 5,001–5,500, *green markers* for time indices 15,001–15,500 and *red markers* for time indices 25,001–25,500. The percentage variance accounted for by each PCA component is shown

To investigate the effect of noise, various levels of uniform noise were added to the autocatalytic process variable of interest, viz. 0, 5, 10 and 20 % of the range of the variable over the first 10,000 samples. The optimal lags for each of these four data sets were determined from average mutual information profiles, while the optimal embedding dimensions were determined with false nearest neighbour ratios. The lag and embedding profiles are shown in Fig. 7.20. Figure 7.20 also shows the first two PCA features of the reconstructed attractors, according to the optimal lag and embedding parameters shown.

Contrary to the predator–prey and BZ reaction data, the optimal lag for the autocatalytic process data stays fairly constant as noise is added (Fig. 7.20), while the optimal embedding dimension increases (as with the previous two data sets). The problem of spurious local minima in the lag profiles (as for the BZ reaction data) is not present for the autocatalytic data.

The phase space projection for data with no noise added (Fig. 7.16) shows some agreement with the state space of the generating model (Fig. 7.18), considering that a three-dimensional state space is shown in two-dimensional phase space and that the attractor in the original state space is quite complex. Certain aspects of the phase space projection do agree with the state space: the intertwined nature of the different parameter conditions trajectories as well as certain geometrical features.

From Fig. 7.16, the overlap of the projected trajectories becomes even worse as noise is added. No comment can be made on the overlap of the trajectories in the actual, higher-dimensional embedded data sets.

7.7 Performance Metrics for Fault Detection

Chapter 6 gave an overview of alarm rates, alarm run lengths and receiver operator characteristic curves as performance metrics for fault detection. These performance metrics will also be utilized in this chapter. In addition, the extension of receiver operator characteristic curves to incorporate both upper and lower limits is now considered.

An ROC curve is a one-dimensional curve in the ROC space, where its single degree of freedom is associated with the single (upper limit) diagnostic threshold. As such, it is applicable to the SSA, RF, NLPCA and 1-SVM approaches discussed before. The RQA diagnostic statistic is monitored in terms of both an upper and lower limit. The true alarm rates and false alarm rates for a RQA monitoring system are thus functions of two parameters, the upper and lower thresholds (θ_1 and θ_2, respectively):

$$\mathrm{TAR} = f_{\mathrm{TAR}}\left(\theta_1, \theta_2\right) \tag{7.47}$$

$$\mathrm{FAR} = f_{\mathrm{FAR}}\left(\theta_1, \theta_2\right). \tag{7.48}$$

The parametric equations above are used to construct the ROC curve. The system of two equations (7.47 and 7.48) with four unknowns (TAR, FAR and θ_1 and θ_2) results in a two-dimensional parametric surface. Therefore, one cannot directly compare monitoring systems with only a single-sided limit (SSA, RF, NLPCA and 1-SVM) to a monitoring system with double-sided limits (RQA). To enable comparison of ROC curves, a simplification of the RQA double-sided limits is enforced: The upper and lower limits are restricted to be symmetrical around the mean of the diagnostic statistic distribution. The relation of the upper and lower limit adds another parametric equation:

$$\theta_1 = f_\theta\left(\theta_2\right). \tag{7.49}$$

This system of three equations (7.47, 7.48 and 7.49) and four unknowns (TAR, FAR and θ_1 and θ_2) now gives a one-dimensional parametric curve. This simplifying assumption allows comparison of all the monitoring systems. However, the limitation of such a simplification must be kept in mind when interpreting ROC curve and AUC results.

7.8 Dynamic Monitoring Results

The results of application of the five dynamic monitoring approaches on the three data sets are given in the following subsection. Results are presented as alarm rates and receiver operating characteristic (ROC) curves (Fawcett 2006), as well as examples of diagnostic statistic monitoring sequences and attractor reconstructions.

For all monitoring approaches and data sets, a window size of 1,000 samples was used. The first 1,000 samples corresponded to NOC training data, while the first 2,000 samples were employed as the NOC validation data. A confidence limit of 99 %, i.e. $\alpha = 0.01$, was applied to calculate thresholds.

7.8.1 *Optimal Embedding Lag and Dimension Parameters*

For the NLPCA, 1-SVM and RQA approaches, the time series data were embedded according to the optimal lag and embedding dimension parameters shown in Figs. 7.12, 7.16 and 7.20 (summarized in Table 7.5). For the SSA and RF approaches, a lag of 1 and embedding dimension of 501 were employed, as these methods implicitly calculate an optimal embedding through lag-variable weighting.

As the embedding parameters of SSA and RF differ from NLPCA, 1-SVM and RQA, care must be taken in comparing these two groups in terms of monitoring performance. The difference in monitoring performance may be due to the adequacy or otherwise of embedding and not only the potential discrimination power of the core algorithms in the different monitoring techniques. SSA and RF might have an advantage in terms of having more information available (with a small lag and large embedding dimension), although the availability of information is dependent on the ability of SSA and RF to extract informative features.

Another caveat worth mentioning is the effect of noise on the embedding parameters. The NLPCA, 1-SVM and RQA monitoring performances for different noise levels are influenced not only by the effect of noise on the core algorithms of these monitoring schemes but also by the amount of information available in the embedded data. From Table 7.5, the optimal embedding dimension

Table 7.5 Summary of optimal lag and embedding dimension parameters for the three simulated data sets, as used in the NLPCA, 1-SVM and RQA dynamic monitoring schemes

Simulated system	Embedding parameter	0 % noise	5 % noise	10 % noise	20 % noise
Predator–prey	Lag k	5	14	15	14
	Dimension m	2	7	7	8
BZ reaction	Lag k	11	25	26	24
	Dimension m	3	7	10	11
Autocatalytic process	Lag k	16	15	14	15
	Dimension m	3	7	7	9

Table 7.6 Summary statistics for dynamic monitoring of predator–prey data sets

Added noise	Method	FAR	ARL (false)	MAR	ARL (true)	AUC
0 %	*Univariate*	**0.01**	–	**0.76**	–	–
	SSA	**0.01**	51	0.04	273	0.98
	RF	0.04	0	**0.02**	206	**0.99**
	NLPCA	0.01	**83**	0.91	61	0.63
	1-SVM	0.03	1	0.24	**33**	0.84
	RQA	0.35	75	0.04	43	0.97*
5 %	*Univariate*	**0.01**	–	**0.85**	–	–
	SSA	**0.01**	612	0.04	273	0.98
	RF	0.05	0	0.02	37	**0.99**
	NLPCA	0.01	147	0.26	177	0.93
	1-SVM	0.02	**1001**	0.06	138	0.97
	RQA	0.53	312	**0.01**	**1**	0.97*
10 %	*Univariate*	**0.01**	–	**0.88**	–	–
	SSA	**0.00**	93	0.05	439	0.98
	RF	0.02	269	**0.03**	242	**0.99**
	NLPCA	0.01	7	0.22	605	0.92
	1-SVM	0.03	**969**	0.09	20	0.96
	RQA	0.76	0	**0.03**	**1**	0.94*
20 %	*Univariate*	**0.01**	–	**0.92**	–	–
	SSA	0.13	934	**0.02**	**1**	**0.99**
	RF	**0.01**	520	0.03	7	**0.99**
	NLPCA	**0.01**	21	0.21	11	0.94
	1-SVM	**0.01**	**970**	0.14	210	0.95
	RQA	0.34	33	0.06	247	0.97*

Best performances per noise level are highlighted with *bold text*; *asterisks* (*) serve as a reminder that AUC for RQA is calculated with a heuristic parameterization

and lag parameters increase when a small amount of noise (5 %) is added. This increase in lag variables for the phase space may increase the information content of the phase space, above and beyond the masking effect of the added noise.

7.8.2 *Results: Predator–Prey Data Sets*

Table 7.6 presents summary statistics comparing the performance of five dynamic monitoring approaches for the predator–prey data sets. These results are presented visually in Fig. 7.21, while Fig. 7.22 presents the ROC curves. To place these results in context, the false alarm rates and missing alarm rates of a simple univariate monitoring scheme, with upper and lower limits calculated as with the RQA diagnostic statistic, are included.

Except for the NLPCA results for a noise level of 0 %, all dynamic monitoring techniques perform better (in fact, considerably so) than the simple univariate

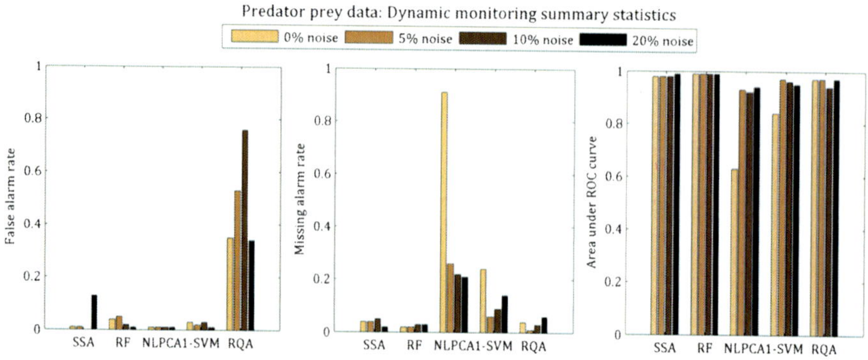

Fig. 7.21 Dynamic monitoring summary statistics for predator–prey data

approach, as seen by comparing missing alarm rates. This shows that a simple univariate monitoring approach cannot capture the dynamic behaviour (and shift in this behaviour) of the predator–prey system.

A number of interesting observations are made from these results. The monitoring approach with the worst performance was NLPCA. The NLPCA alarm system is too conservative as a classifier, with high missing alarm rates (and associated low false alarm rates). This might suggest that the detection threshold is too high. This is not the only shortcoming of the NLPCA approach on the predator–prey data sets: The low AUC values for the NLPCA alarm system (compared to the other methods) also suggest that the NLPCA diagnostic statistic captures less useful discriminatory information than the other methods.

Figure 7.23 shows the NLPCA diagnostic sequence for the predator–prey data set with 0 % noise added. The conservative NLPCA threshold arises from the long tail of the diagnostic statistic under NOC conditions (as apparent from the sparse distribution of diagnostic statistic values between 1 and 8).

Figure 7.24 shows the phase space with NLPCA reconstructions for three different time windows of the predator–prey data set with 0 % noise added. The NLPCA reconstructed attractor is not an accurate representation of the NOC data, with the reconstructed attractor being much narrower than the actual NOC attractor. The reconstruction is also skewed towards one side of the actual attractor. This inaccurate reconstruction leads to the long-tailed distribution of the diagnostic statistic: The side of the actual attractor farthest removed from the reconstruction shows larger reconstruction errors than the side of the actual attractor closer to the reconstruction. The biased reconstruction causes the skewed statistic distribution, which leads to the conservative threshold and subsequent poor performance of the NLPCA monitoring method on this data.

Whereas the NLPCA is too conservative, the RQA approach is too liberal: High false alarm rates (with low missing alarm rates) suggest that the thresholds are too tight. The good performance of the RQA approach in terms AUC values does

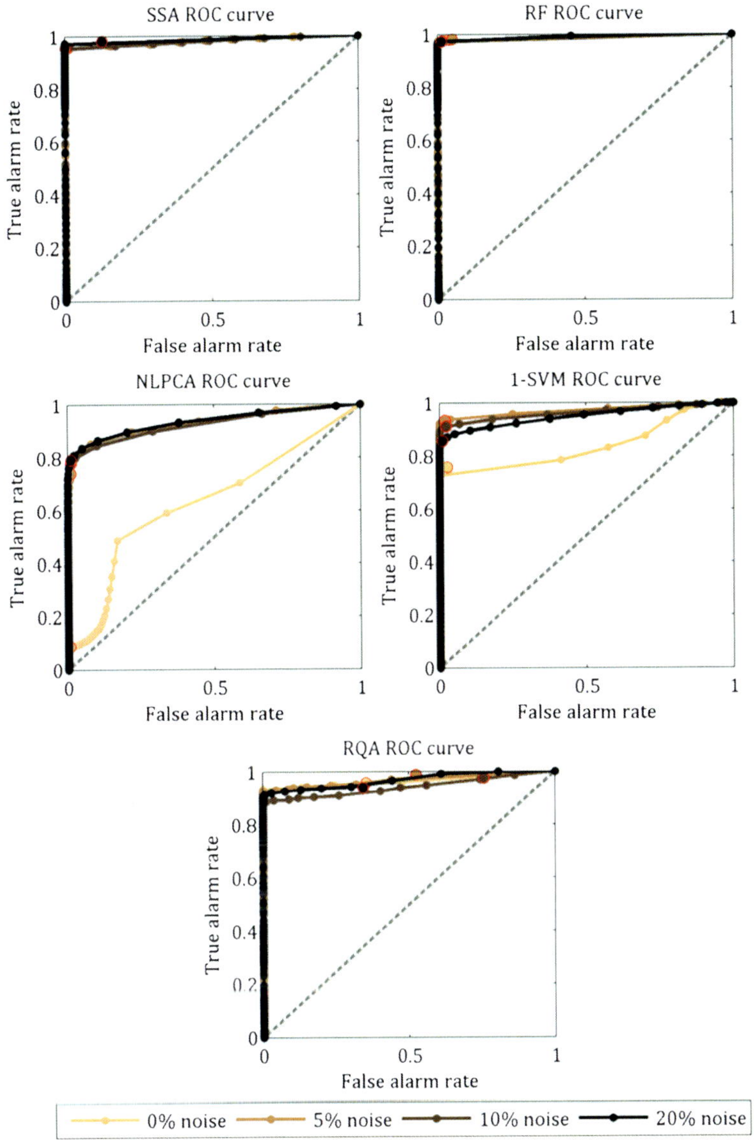

Fig. 7.22 Dynamic monitoring ROC curves for predator–prey data (*circles* indicate alarm rates for thresholds selected with percentile approach)

suggest that the RQA diagnostic statistic is informative. An improved detection performance could thus be had for RQA by a better selection of diagnostic thresholds.

Figure 7.25 shows the RQA diagnostic sequence for the predator–prey data set with 0 % noise added. The false alarms due to the tight thresholds can be seen

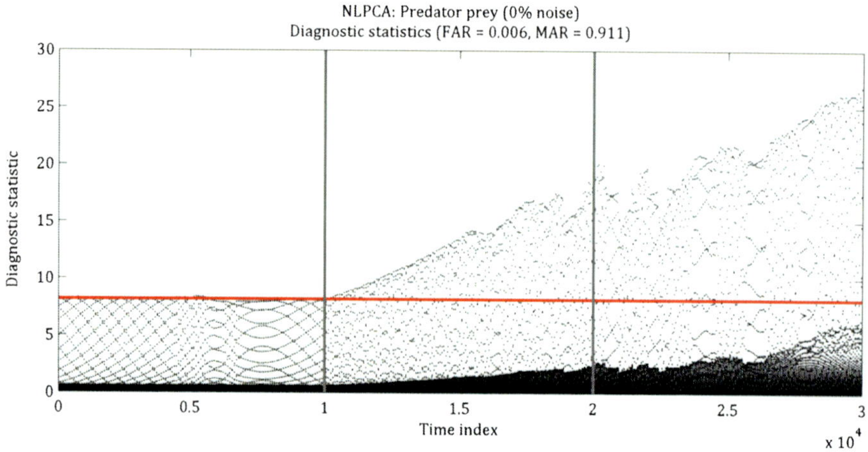

Fig. 7.23 NLPCA diagnostic statistic sequence for the predator–prey data set with 0 % noise added (threshold shown in *red*; location of change points shown with *vertical lines*)

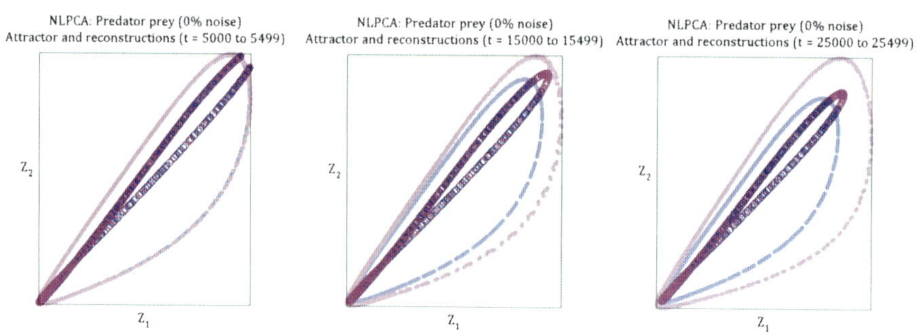

Fig. 7.24 Attractors and NLPCA reconstructed attractors in phase space for the predator–prey data set with 0 % noise added (*light blue* and *light pink dots* represent original NOC and test data, while *blue* and *pink circles* represent reconstructed NOC and test data)

after 5,000 time steps. The change in the RQA statistic (recurrence rate) after 5,000 time steps might suggest a low-frequency process in the predator–prey system, manifesting some change in the system variables (and associated phase space) after 20,000 time steps of the original simulation (the initial 15,000 time steps included). Such possible low-frequency dynamics also constitute a change and should thus be detected.

Regardless of the presence of low-frequency dynamics, the RQA diagnostic statistic captures valuable information of the system considered, with the clear change in its monitoring statistic after simulated changes evident. The selection of better diagnostic thresholds may be had though more extensive validation data sets.

To illustrate the change in dynamics of the predator–prey system, the recurrence plots for three different time windows of the predator–prey data set with 0 % noise

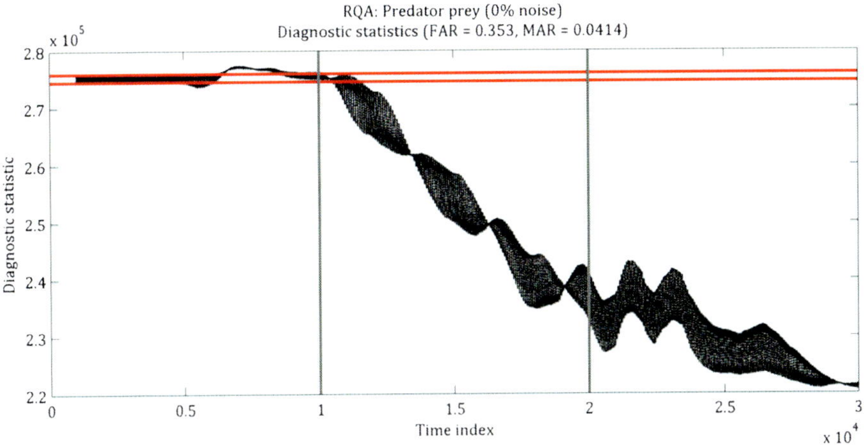

Fig. 7.25 RQA diagnostic statistic sequence for the predator–prey data set with 0 % noise added (upper and lower thresholds shown in *red*; location of change points shown with *vertical lines*)

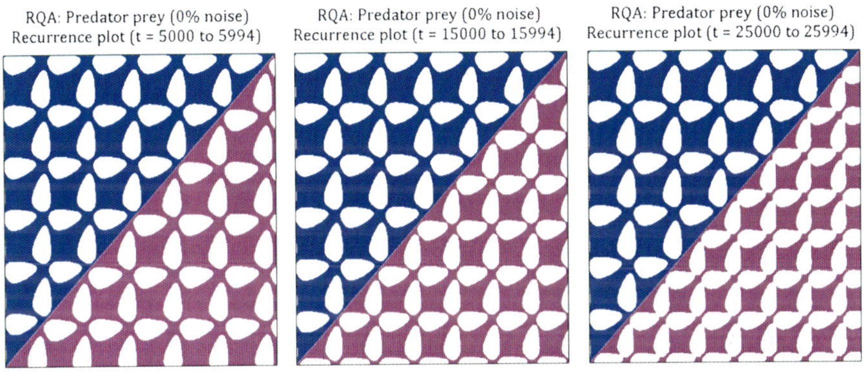

Fig. 7.26 Phase space recurrence plots for the predator–prey data set with 0 % noise added (*blue upper left triangle* representing NOC data and *pink lower right triangle* representing test data)

added are shown in Fig. 7.26. A subtle change in the recurrence plots of time steps 5,000–5,994, time steps 15,000–15,994 and time steps 25,000–25,994 is the increased frequency (and reduced size) of the petal structures. Recalling the phase space plots of Fig. 7.12, the attractor shape of the predator–prey data increases in circumference after 10,000 and 20,000 time steps, while its period stays constant. This implied increased "rotational speed" of the attractor is seen in the shrinking petal structures.

In terms of the effect of noise, adding a small amount of noise (5 %) improved the detection performance of both the NLPCA and 1-SVM approaches, as quantified by lower missing alarm rates and higher AUC. The ROC curves of NLPCA and 1-SVM also show dramatic improvement with the addition of noise to the

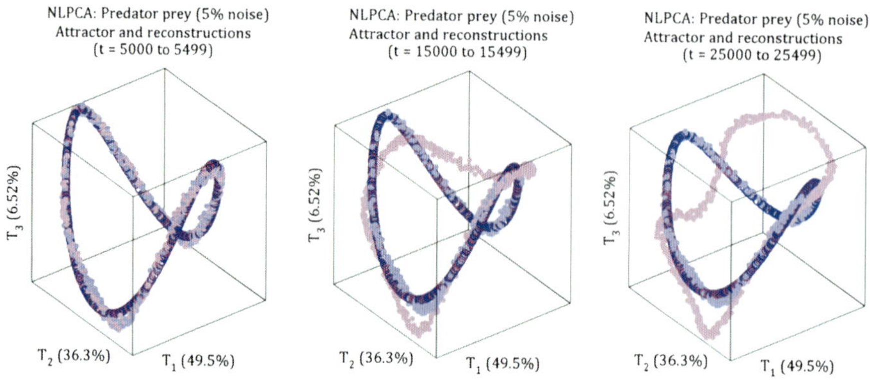

Fig. 7.27 PCA projection of attractors and NLPCA reconstructed attractors in phase space for the predator–prey data set with 5 % noise added (*light blue* and *light pink dots* represent original NOC and test data, while *blue* and *pink circles* represent reconstructed NOC and test data; *percentages* in brackets indicate variance accounted for by each component)

predator–prey data. The false alarm rate of the RQA approach is somewhat sensitive to added noise (5 and 10 %), while other RQA summary statistics show little change. As mentioned before, the effect of noise on the embedding parameters may contribute to the improved performance for NLPCA, 1-SVM and RQA. Both SSA and RF appear to be more robust to noise, showing little change in summary statistics for different levels of noise.

Apart from the embedding effect, adding small amounts of noise to input data has been shown to improve feed forward neural networks, as the added noise serves as a type of regularization of the complexity of the trained network (Seghouane et al. 2004). The beneficial effect of input noising can be extended to other statistical learners as well. The input noise regularization effect prevents overfitting by enforcing a smoothing of the response function. However, adding too much noise can destroy the functional information in the data, leading to poor fitting.

To illustrate the effect of adding 5 % noise to the predator–prey set on the performance of NLPCA, Fig. 7.27 shows the first three PCA components of phase space attractors and their NLPCA reconstructions. Compared to Fig. 7.24, the NLPCA reconstruction of the NOC attractor is much more accurate for the case with 5 % noise (based on the visualized PCA subspace). Adding a small amount of noise dramatically improved the accuracy of the NLPCA reconstruction and, from this, its monitoring performance.

Inspecting the 1-SVM results for the predator–prey data set with 0 % noise added, the NOC support in Fig. 7.28 is a simple, slightly irregular ellipse, where all points interior to the NOC attractor are considered as belonging to the NOC support. For the simulation of changes in the predator–prey system considered here, the inclusion of the interior region of the attractor does not prove disadvantages, as

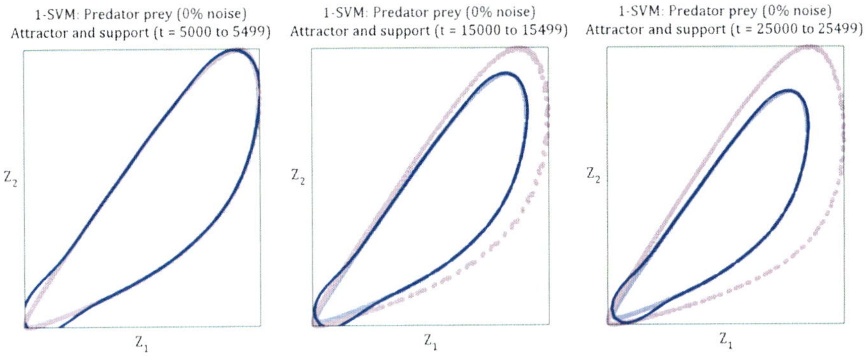

Fig. 7.28 Attractors and 1-SVM support boundary in phase space for the predator–prey data set with 0 % noise added (*light blue* and *light pink dots* represent original NOC and test data, while the *blue line* represents the 1-SVM support boundary)

the introduced changes only increase the attractor trajectory. However, a change that would decrease the attractor trajectory would be undetectable by the NOC support as shown in Fig. 7.28. To obtain an annulus 1-SVM support (excluding the interior of the attractor shape), a smaller kernel width would be required.

Overall, the SSA and RF approaches perform the best for all noise levels, with good performance on all summary statistic measures. The RF approach does marginally better than SSA, but no conclusion can be made as to the significance of this marginal improvement. As an example of the results, Figs. 7.29 and 7.30 show the SSA diagnostic statistic sequence for the predator–prey data set with 10 % noise as well as examples of the attractors and reconstructed attractors at different locations in the time series.[2] Figures 7.31 and 7.32 show the same graphs for the RF monitoring scheme.

What is notable about the SSA and RF diagnostic statistic sequences are their similarity. Both sequences show clear changes for the simulated changes in the data sets. The SSA approach appears to be less accurate than the RF approach in reconstructing the NOC attractor, but this does not degrade the SSA performance. From the attractor plots, it is noticeable that the reconstructions for both SSA and RF after the simulated changes have much smaller extents than the NOC attractor and the actual attractors for the relevant time windows. These inaccurate reconstructions lead to successful detections, indicating that the original feature extraction models (and by implication, system dynamics) are no longer valid.

[2]For consistency, the SSA and RF reconstructed attractors are obtained from the filtered approximations of a specified window of the time series, with this time series approximation embedded according to the optimal embedding parameters used for the NLPCA, 1-SVM and RQA monitoring approaches.

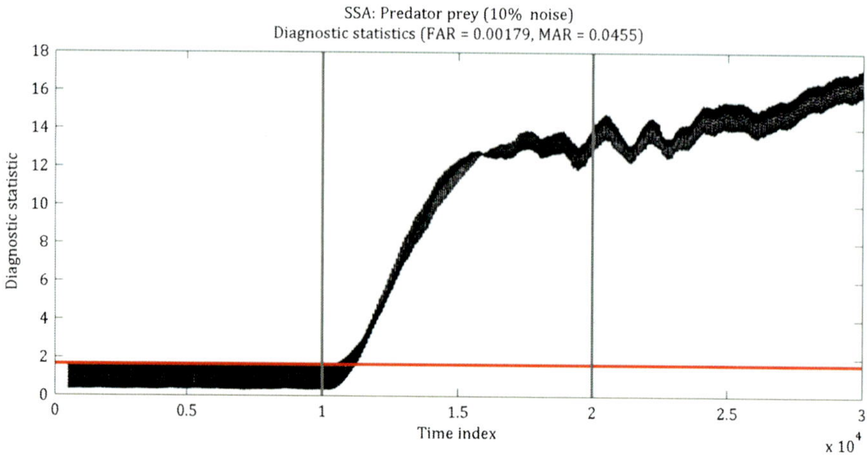

Fig. 7.29 SSA diagnostic statistic sequence for the predator–prey data set with 10 % noise added (threshold shown in *red*; location of change points shown with *vertical lines*)

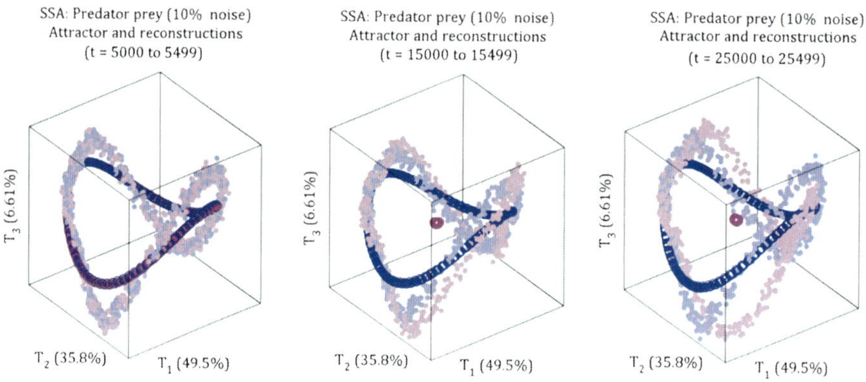

Fig. 7.30 PCA projection of attractors and SSA reconstructed attractors in phase space for the predator–prey data set with 10 % noise added (*light blue* and *light pink dots* represent original NOC and test data, while *blue* and *pink circles* represent reconstructed NOC and test data; *percentages* in brackets indicate variance accounted for by each component)

7.8.3 Results: BZ Reaction Data Sets

Table 7.7 presents summary statistics comparing the performance of five dynamic monitoring approaches for the BZ reaction data sets. These results are presented visually in Fig. 7.33, while Fig. 7.34 presents the ROC curves. To place these results in context, the false alarm rates and missing alarm rates of a simple univariate monitoring scheme, with upper and lower limits calculated as with the RQA diagnostic statistic, are included.

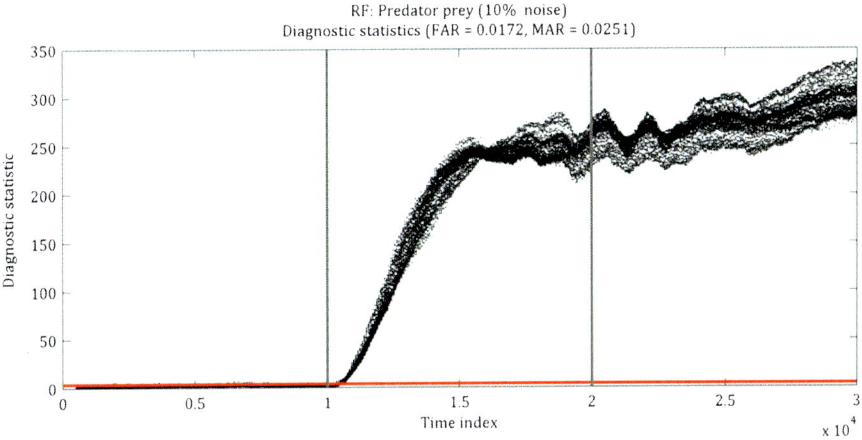

Fig. 7.31 RF diagnostic statistic sequence for the predator–prey data set with 10 % noise added (threshold shown in *red*; location of change points shown with *vertical lines*)

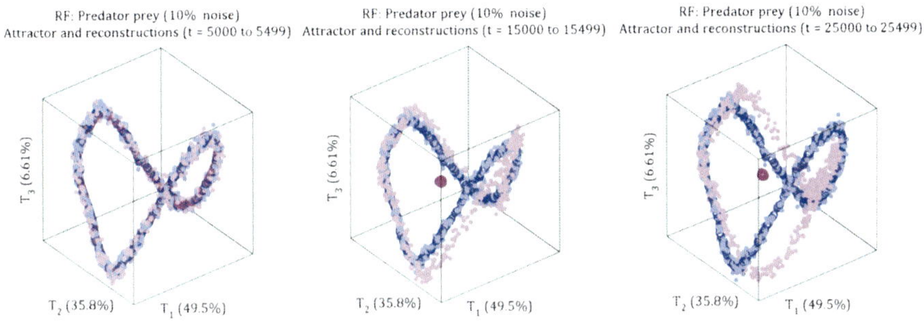

Fig. 7.32 PCA projection of attractors and RF reconstructed attractors in phase space for the predator–prey data set with 10 % noise added (*light blue* and *light pink dots* represent original NOC and test data, *while blue* and *pink circles* represent reconstructed NOC and test data; *percentages* in brackets indicate variance accounted for by each component)

For the BZ reaction system, all dynamic monitoring techniques have lower missing alarm rates than the simple univariate approach. As with the predator–prey system, the changes in the BZ reaction system cannot be captured by a simple univariate monitoring approach. Judged solely on missing alarm rates, the RQA approach performs the best for all noise levels. Taking false alarm rates into account (incorporated in AUC), SSA and RF showed comparable performance to RQA.

The NLPCA (as with the predator–prey results) and 1-SVM approaches are conservative, with low false alarm rates and high missing alarm rates. Also similar to the predator–prey results, adding some noise (5 %) dramatically improved the missing alarm rates and AUCs of both NLPCA and 1-SVM. The effect of improved embedding, as well as the regularizing effect of input noise on statistical learners, may again be at play.

Table 7.7 Summary statistics for dynamic monitoring of BZ reaction data sets

Added noise	Method	FAR	ARL (false)	MAR	ARL (true)	AUC
0 %	*Univariate*	**0.01**	–	**0.93**	–	–
	SSA	0.08	122	0.16	2192	**0.96**
	RF	0.06	0	0.11	630	**0.96**
	NLPCA	**0.01**	**317**	0.92	713	0.62
	1-SVM	0.02	145	0.65	**33**	0.73
	RQA	0.16	228	**0.06**	627	**0.96***
5 %	*Univariate*	**0.01**	–	**0.93**	–	–
	SSA	0.08	122	0.16	2190	0.96
	RF	0.05	**570**	0.13	258	0.96
	NLPCA	**0.03**	270	0.69	6255	0.84
	1-SVM	0.06	248	0.32	**3**	0.90
	RQA	0.07	274	**0.05**	723	**0.98***
10 %	*Univariate*	**0.01**	–	**0.93**	–	–
	SSA	0.07	121	0.17	3275	0.95
	RF	0.06	**563**	0.12	253	0.95
	NLPCA	**0.04**	111	0.67	29	0.83
	1-SVM	0.05	59	0.29	**3**	0.91
	RQA	0.10	179	**0.04**	249	**0.98***
20 %	*Univariate*	**0.01**	–	**0.95**	–	–
	SSA	0.07	24	0.09	172	**0.98**
	RF	0.06	**564**	0.11	635	0.96
	NLPCA	**0.04**	191	0.56	83	0.87
	1-SVM	**0.04**	12	0.31	**11**	0.90
	RQA	0.22	176	**0.05**	32	0.97*

Best performances per noise level are highlighted with *bold text*; *asterisks* (*) serve as a reminder that AUC for RQA is calculated with a heuristic parameterization

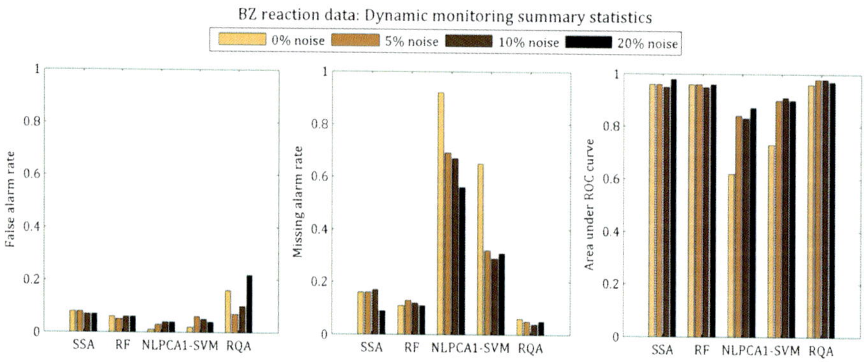

Fig. 7.33 Dynamic monitoring summary statistics for BZ reaction data

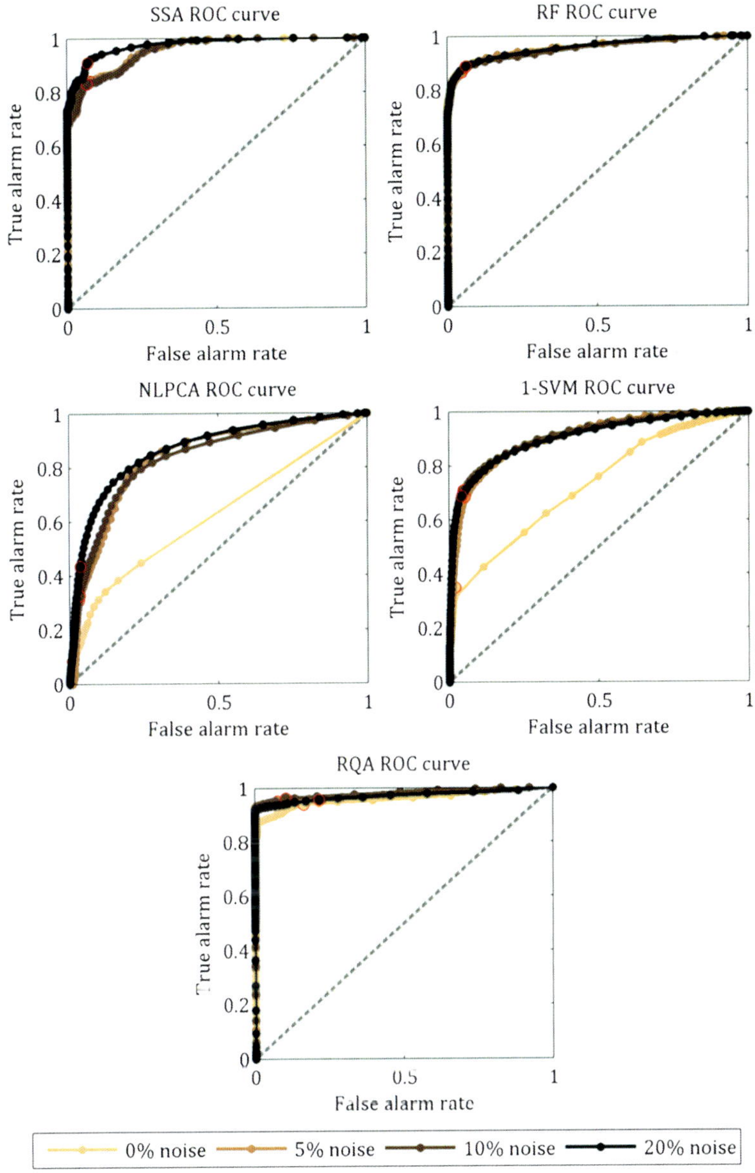

Fig. 7.34 Dynamic monitoring ROC curves for BZ reaction data (*circles* indicate alarm rates for thresholds selected with percentile approach)

Figure 7.35 shows the phase space and NLPCA reconstructed attractors for the BZ reaction system (0 % noise added) at different time windows. The NOC data follows several paths, and not a single manifold, as with the predator–prey data. The one-dimensional NLPCA manifold cannot sufficiently capture the

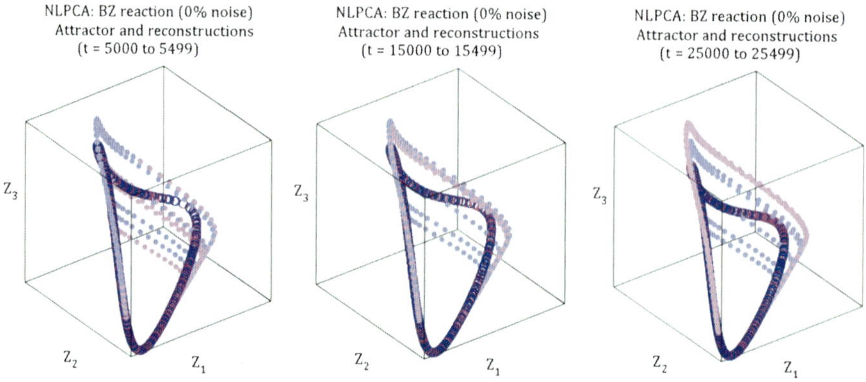

Fig. 7.35 Attractors and NLPCA reconstructed attractors in phase space for the BZ reaction data set with 0 % noise added (*light blue* and *light pink dots* represent original NOC and test data, *while blue* and *pink circles* represent reconstructed NOC and test data)

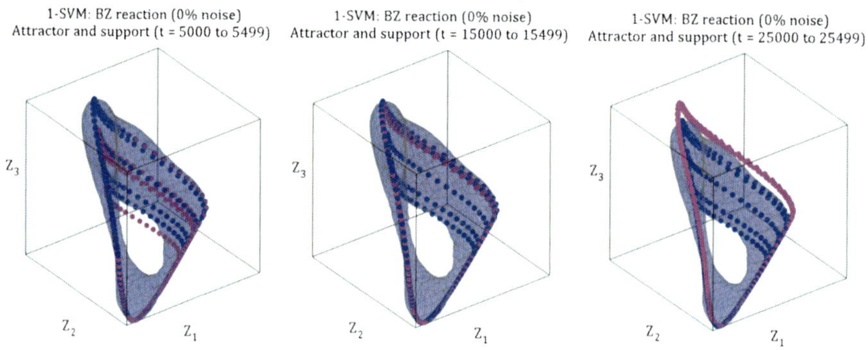

Fig. 7.36 Attractors and 1-SVM support boundary in phase space for the BZ reaction data set with 0 % noise added (*blue* and *pink dots* represent original NOC and test data, *while* the *blue* meshed surface represents the 1-SVM support boundary)

higher-dimensional manifold of the NOC attractor. A two-dimensional NLPCA manifold could have better presented the NOC data. It is thus evident that the assumption of a one-dimensional manifold restricts the data-fitting complexity of the NLPCA monitoring approach.

Figure 7.36 shows the 1-SVM support boundaries for the BZ reaction data set (0 % noise added), at different time steps. Where the NLPCA manifold was unable to capture the several pathways of the NOC attractor, the 1-SVM support enfolds these pathways. The 1-SVM support shows an annulus shape, correctly excluding the central region of the phase space (compare this with the erroneous inclusion of the central region for the predator–prey data set; Fig. 7.28). In some regions of the attractor, the 1-SVM support boundary is too conservative, enclosing a large region where no NOC data are present. This may account for the conservative nature of the 1-SVM monitoring approach, mentioned earlier. Although a remedy for the

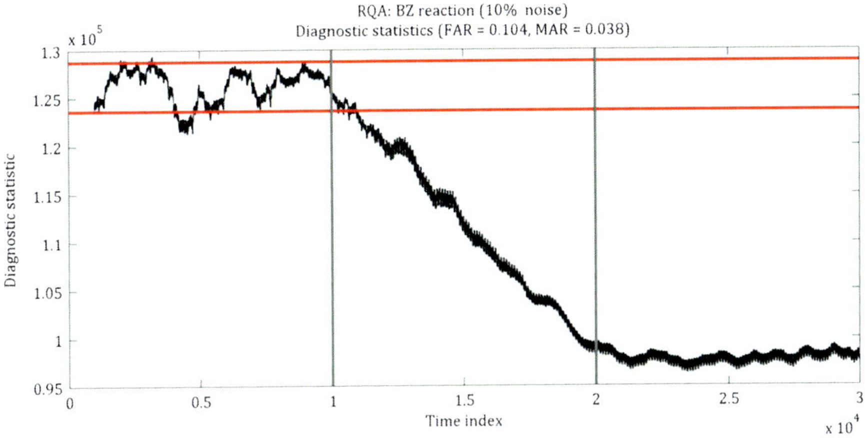

Fig. 7.37 RQA diagnostic statistic sequence for the BZ reaction data set with 10 % noise added (*upper* and *lower* thresholds shown in *red*; location of change points shown with *vertical lines*)

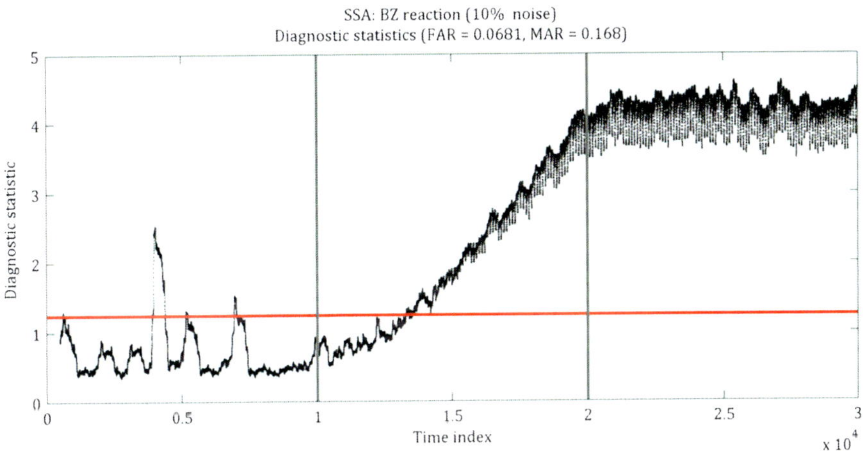

Fig. 7.38 SSA diagnostic statistic sequence for the BZ reaction data set with 10 % noise added (threshold shown in *red*; location of change points shown with *vertical lines*)

conservative 1-SVM support may be to employ a smaller kernel width, this may lead to a fragmented support in data-sparse regions of the NOC attractor. The sharp corners of the NOC attractor appear difficult for accurate support estimation. (This sharp-corner effect was also evident in Fig. 7.28.) Possible solutions may be variable kernel widths or a different choice of kernel type.

Figures 7.37, 7.38 and 7.39 show the diagnostic statistic sequences for RQA, SSA and RF, respectively, for the BZ reaction data with 10 % noise added. All three sequences show a clear change after the first simulated change and a change

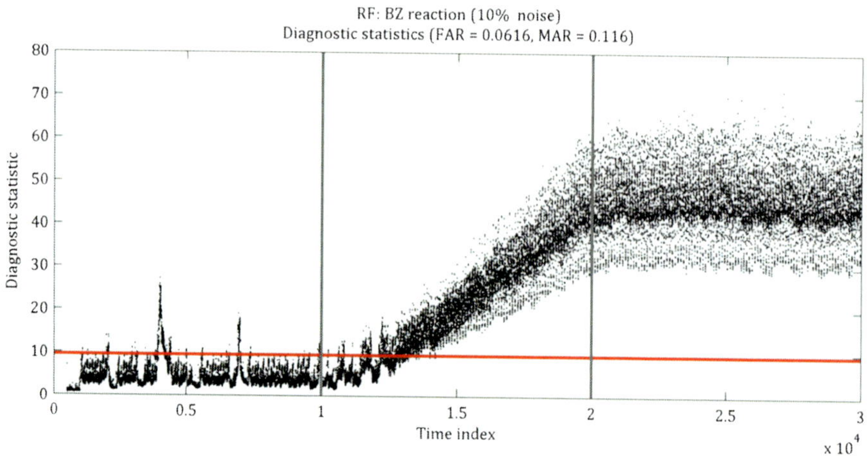

Fig. 7.39 RF diagnostic statistic sequence for the BZ reaction data set with 10 % noise added (threshold shown in *red*; location of change points shown with *vertical lines*)

in profile for the second simulated change. A set of false alarms occur for all three methods after around 4,000 time steps. As with the predator–prey data, this could indicate some lower-frequency dynamics which is not represented in the 1,000 NOC data points or the 2,000 validation data points. Since all three monitoring approaches show an agreement of exceeded limits at 4,000 time steps, it would be insightful to investigate such an alarm in real-world monitoring applications.

PCA projections of the attractors and their reconstructions with SSA and RF for certain time windows of the BZ reaction system (10 % noise added) are shown in Figs. 7.40 and 7.41. While the SSA reconstructions show the expanding true attractor as merely shifting, the RF reconstructions show a shrinking attractor. The increasing discrepancy between the true attractor and its reconstruction (regardless of the nature of the discrepancy) indicates that the original system dynamics, as captured by the feature extraction methods, are no longer valid, thus indicating a change in the system.

7.8.4 Results: Autocatalytic Process Data Sets

Table 7.8 presents summary statistics comparing the performance of five dynamic monitoring approaches for the autocatalytic process data sets. These results are presented visually in Fig. 7.42, while Fig. 7.43 presents the ROC curves. To place these results in context, the false alarm rates and missing alarm rates of a simple univariate monitoring scheme, with upper and lower limits calculated as with the RQA diagnostic statistic, are included.

In contrast to the predator–prey and BZ reaction systems (where SSA, RF, NLPCA and 1-SVM showed decent to good results), only the RQA approach is

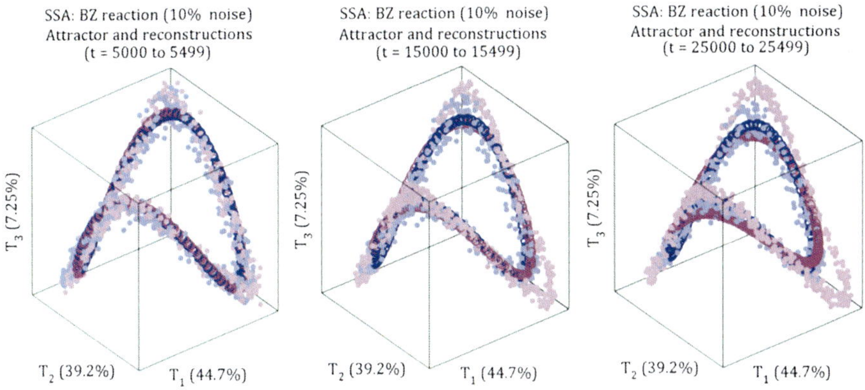

Fig. 7.40 PCA projection of attractors and SSA reconstructed attractors in phase space for the BZ reaction data set with 10 % noise added (*light blue* and *light pink dots* represent original NOC and test data, while *blue* and *pink circles* represent reconstructed NOC and test data; *percentages* in brackets indicate variance accounted for by each component)

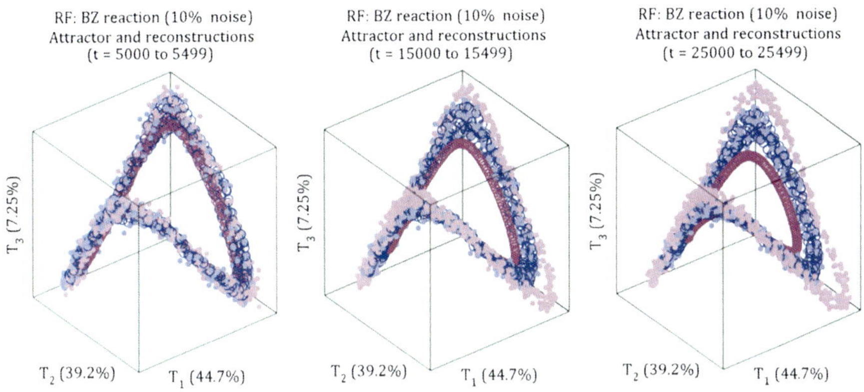

Fig. 7.41 PCA projection of attractors and RF reconstructed attractors in phase space for the BZ reaction data set with 10 % noise added (*light blue* and *light pink dots* represent original NOC and test data, while *blue* and *pink circles* represent reconstructed NOC and test data; *percentages* in brackets indicate variance accounted for by each component)

consistently able to successfully detect the change in dynamics of the autocatalytic process data. All other techniques (including the simple univariate approach) failed dismally in detecting the simulated changes.

The poor performance of SSA, RF, NLPCA and 1-SVM is also evident from the ROC curves (Fig. 7.43), where SSA, RF and 1-SVM curves lie below the random performance lines (diagonals). As mentioned earlier, reversing the alarm signals of the SSA, RF and 1-SVM approaches would improve their monitoring performance. Such reversed alarm systems would have AUC of one minus the AUC of the original alarm system, which would lead to better-than-random-guessing results.

Table 7.8 Summary statistics for dynamic monitoring of autocatalytic process data sets

Added noise	Method	FAR	ARL (false)	MAR	ARL (true)	AUC
0 %	*Univariate*	**0.01**	–	**1.00**	–	–
	SSA	0.02	**1351**	1.00	293	0.22
	RF	**0.01**	1123	1.00	**63**	0.39
	NLPCA	**0.01**	26	0.99	211	0.50
	1-SVM	0.02	45	0.81	225	0.49
	RQA	0.08	147	**0.11**	721	**0.96***
5 %	*Univariate*	**0.01**	–	**1.00**	–	–
	SSA	0.02	**1351**	1.00	292	0.22
	RF	**0.01**	1148	1.00	63	0.40
	NLPCA	**0.01**	100	1.00	**1**	0.51
	1-SVM	**0.01**	988	0.99	109	0.33
	RQA	0.10	152	**0.15**	383	**0.93***
10 %	*Univariate*	**0.01**	–	**1.00**	–	–
	SSA	**0.01**	**1351**	1.00	294	0.24
	RF	0.02	1148	1.00	63	0.37
	NLPCA	**0.01**	113	1.00	**6**	0.50
	1-SVM	**0.01**	918	0.99	15	0.37
	RQA	0.16	153	**0.15**	695	**0.92***
20 %	*Univariate*	**0.01**	–	**1.00**	–	–
	SSA	**0.01**	**1350**	1.00	291	0.30
	RF	**0.01**	1119	1.00	58	0.32
	NLPCA	**0.01**	90	1.00	**17**	0.47
	1-SVM	**0.01**	892	1.00	28	0.35
	RQA	0.13	142	**0.11**	410	**0.95***

Best performances per noise level are highlighted with *bold text*; *asterisks* (*) serve as a reminder that AUC for RQA is calculated with a heuristic parameterization

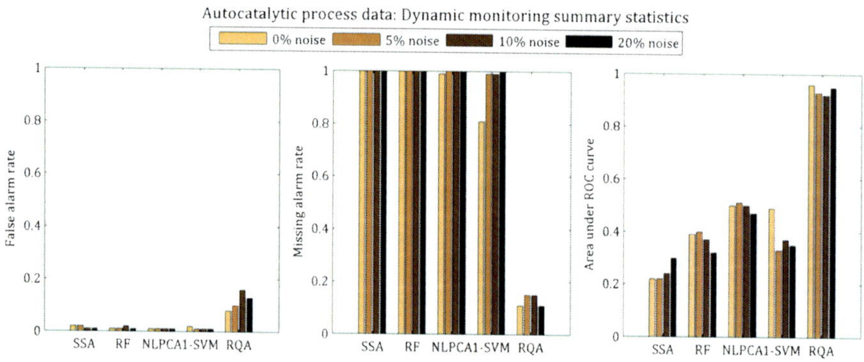

Fig. 7.42 Dynamic monitoring summary statistics for autocatalytic process data

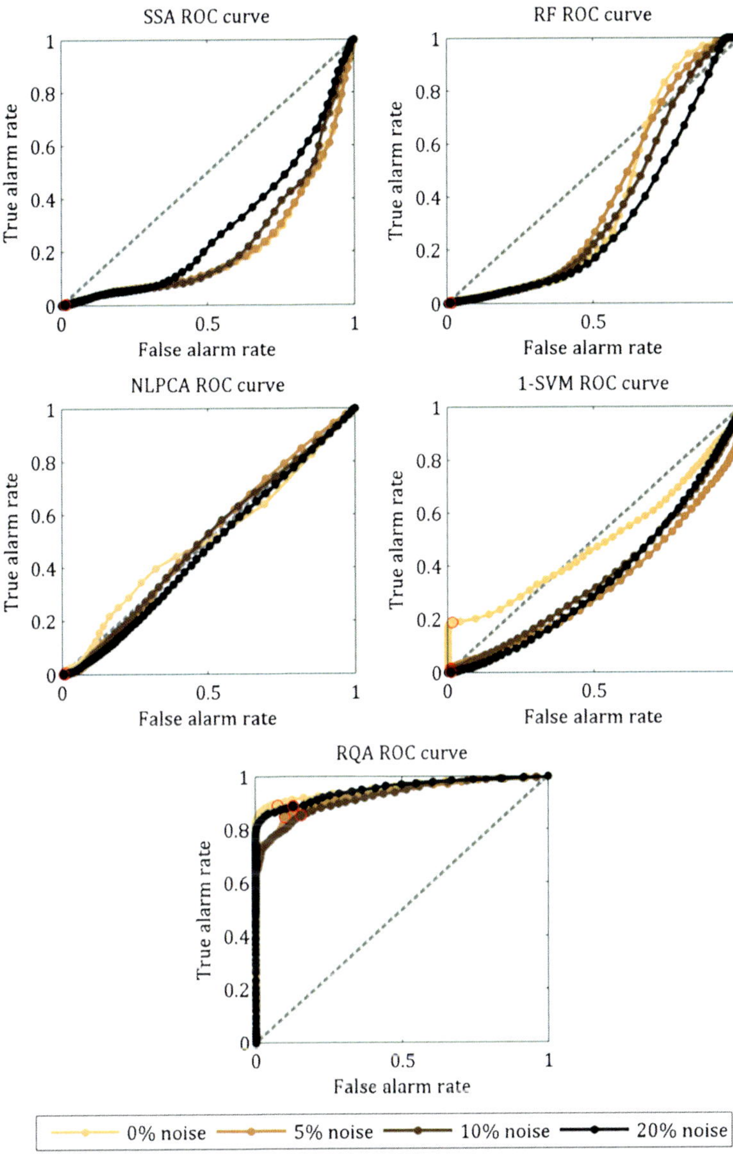

Fig. 7.43 Dynamic monitoring ROC curves for autocatalytic process data (*circles* indicate alarm rates for thresholds selected with percentile approach)

However, simply reversing alarm signals would be inconsistent ad hockery. The original premise of the diagnostic statistics of SSA and RF is that the reconstruction distance (i.e. summed squared errors between actual attractor and expected attractor) increases as the models capturing the dynamics of a system become invalid.

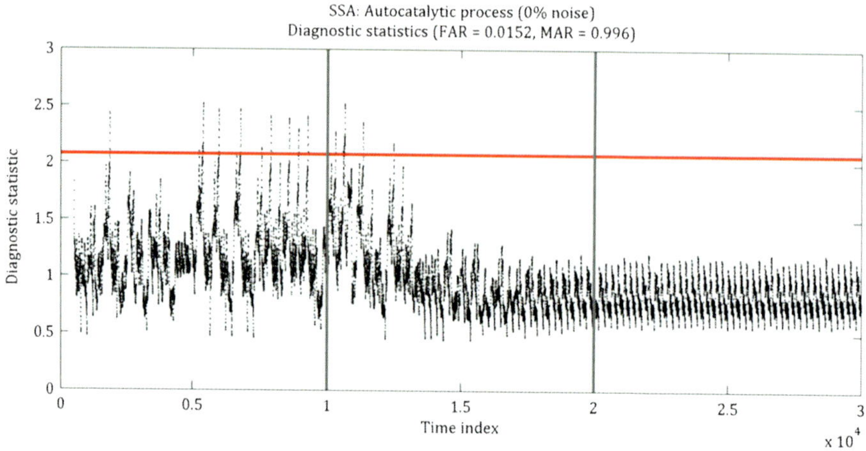

Fig. 7.44 SSA diagnostic statistic sequence for the autocatalytic process data set with 0 % noise added (threshold shown in *red*; location of change points shown with *vertical lines*)

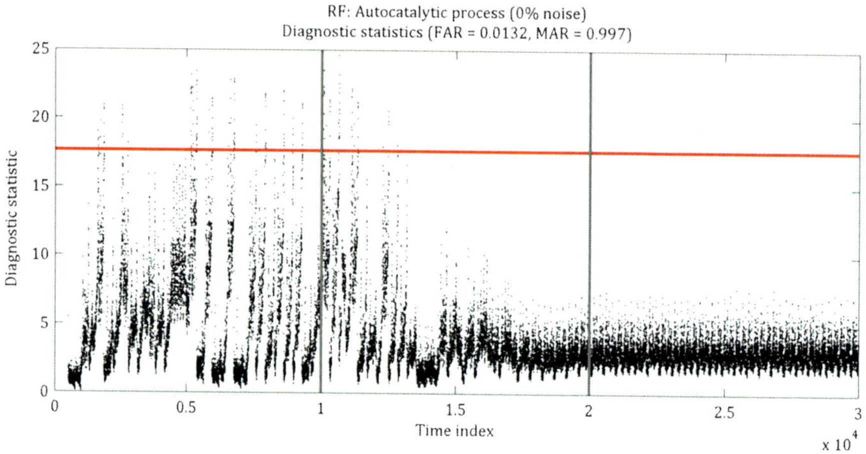

Fig. 7.45 RF diagnostic statistic sequence for the autocatalytic process data set with 0 % noise added (threshold shown in *red*; location of change points shown with *vertical lines*)

Figures 7.44 and 7.45 show the SSA and RF diagnostic sequences for the autocatalytic process data with 0 % noise added. Although the reconstruction errors do not increase after the simulated changes, the distributions of the reconstruction errors do change. The SSA and RF monitoring algorithms are thus able to capture some relevant information on the dynamics of the autocatalytic process, but in a manner which is not exploitable by an alarm system that relies on an upper limit of the reconstruction errors. Monitoring the distribution of the diagnostic statistic for change is an inefficient, Russian nesting doll approach.

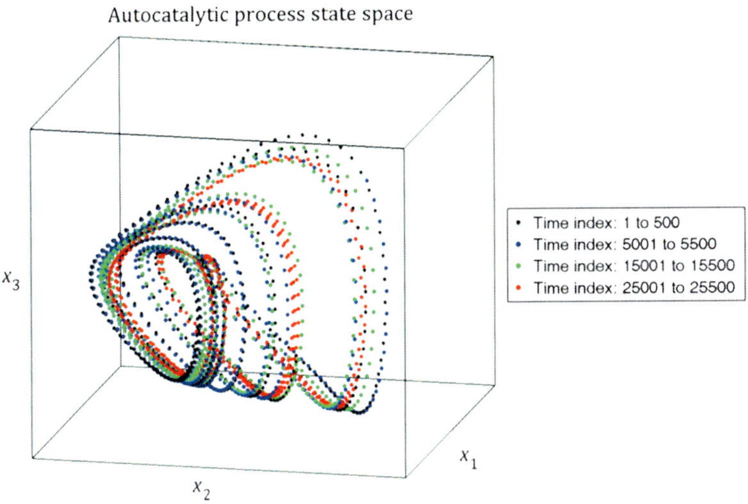

Fig. 7.46 State space for autocatalytic process model

The decrease of the reconstruction distances for the SSA and RF approaches after the simulated changes can be related to the attractor structure, as depicted in the state space of the autocatalytic process (Fig. 7.18, repeated here for ease of reference as Fig. 7.46).

After the simulated changes, the complex attractor of the autocatalytic process shifts to a narrower expanse within in the NOC phase space. If a feature extraction algorithm led to the estimation of a wide ribbon-like manifold (instead of separate strings for the NOC attractor paths), the shifted attractors of simulated change conditions would still lie on this ribbon manifold, resulting in low reconstruction errors. A change could be detected in the distribution of the feature vectors projected on the manifold, but not in its reconstruction. A natural extension of feature extraction dynamic monitoring approaches is then to monitor feature space distributions as well.

The ribbon-like manifolds of the SSA and RF approaches can be seen in the reconstructed attractors of the autocatalytic process data with 0 % noise added (see Figs. 7.47 and 7.48). Especially the RF reconstructed attractors (Fig. 7.48) suggest a smooth, wide ribbon manifold.

Excluding the monitoring of the feature space, more effort can be expended in achieving an informative resolution for the manifold: in the case of the autocatalytic attractor, strings rather than ribbons. This resolution is related to the dimensionality of the manifold or extracted feature space. Evaluating the suitability of the manifold structure and dimensionality can generally only be done in comparison with fault data (available *post priori*) and when the structure of a low-dimensional manifold can be visualized. Where such an approach is not feasible (due to automation pressures and inadequate quantification of suitability of a manifold), the two-pronged

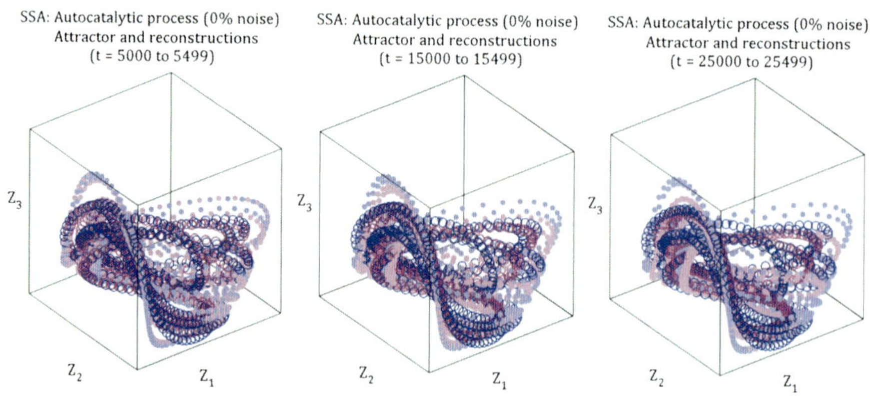

Fig. 7.47 Attractors and SSA reconstructed attractors in phase space for the autocatalytic process data set with 0 % noise added (*light blue* and *light pink dots* represent original NOC and test data, while *blue* and *pink circles* represent reconstructed NOC and test data)

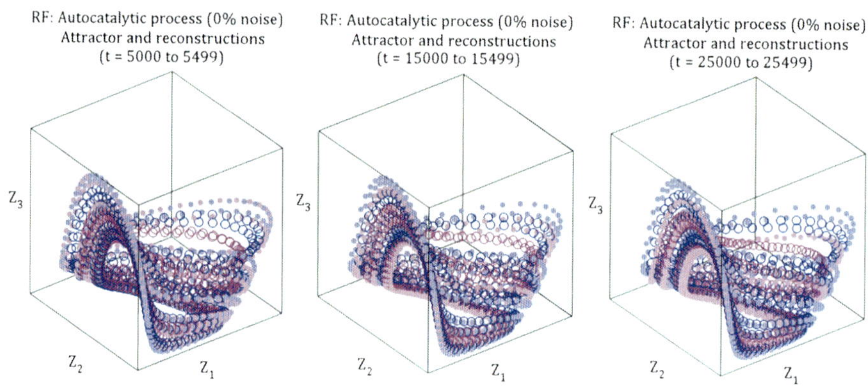

Fig. 7.48 Attractors and RF reconstructed attractors in phase space for the autocatalytic process data set with 0 % noise added (*light blue* and *light pink dots* represent original NOC and test data, while *blue* and *pink circles* represent reconstructed NOC and test data)

approach of monitoring both the feature space and residual (reconstruction) space is a valid monitoring option.

Where the SSA and RF approaches might have floundered due to overly high-dimensional manifolds (analogous to ribbons versus strings), the NLPCA approach with only one component (resulting in a one-dimensional manifold) was too simple to capture the complex autocatalytic attractor. Figure 7.49 shows the phase space and reconstructed NLPCA attractors for the autocatalytic data with 0 % noise added. The one-dimensional NLPCA manifold is clearly insufficient in capturing the multimodal paths of the NOC data.

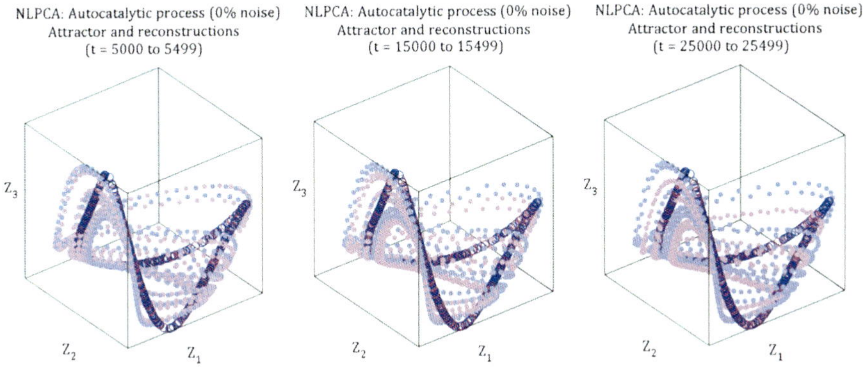

Fig. 7.49 Attractors and NLPCA reconstructed attractors in phase space for the autocatalytic data set with 0 % noise added (*light blue* and *light pink dots* represent original NOC and test data, while *blue* and *pink circles* represent reconstructed NOC and test data; *percentages* in brackets indicate variance accounted for by each component)

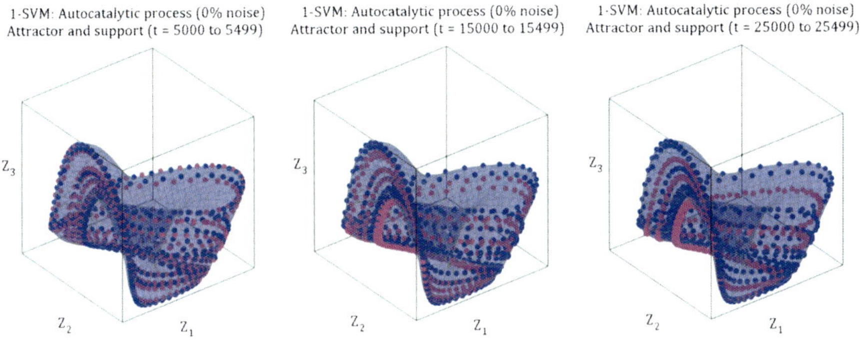

Fig. 7.50 Attractors and 1-SVM support boundary in phase space for the autocatalytic process data set with 0 % noise added (*blue* and *pink dots* represent original NOC and test data, while the *blue* meshed surface represents the 1-SVM support boundary)

Figure 7.50 shows the 1-SVM support boundaries for the autocatalytic process data set (0 % noise added), at different time steps, from a different viewing angle with Fig. 7.51. As with the BZ reaction data, the 1-SVM method is able to create a continuous, smooth, annulus-like support boundary around the multimodal NOC attractor paths. By enfolding all paths in one support region, a similar "string versus ribbon" restriction as with the SSA and RF methods is created.

As seen in Fig. 7.50, the autocatalytic data attractors after the simulated changes, which are closely entwined with the NOC attractor, rarely leave the confines of the estimated 1-SVM support. Certain extents of the attractors do leave the support area and correspond to diagnostic statistics exceeding the 1-SVM statistic limit. These alarms are visible in the 1-SVM diagnostic sequence for the autocatalytic process data with 0 % noise added, as shown in Fig. 7.52.

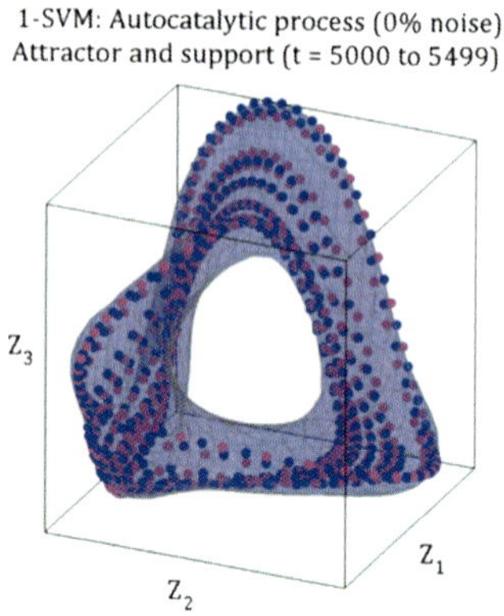

Fig. 7.51 Another view of the attractors and 1-SVM support boundary in phase space for the autocatalytic process data set with 0 % noise added (*blue* and *pink dots* represent original NOC and test data, while the *blue* meshed surface represents the 1-SVM support boundary)

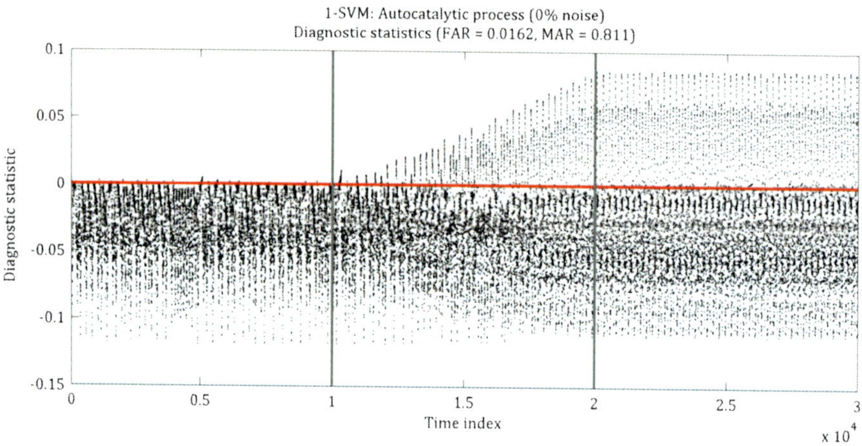

Fig. 7.52 1-SVM diagnostic statistic sequence for the autocatalytic process data set with 0 % noise added (threshold shown in *red*; location of change points shown with *vertical lines*)

A modification that would allow more complex support structures, such as supports that surround individual NOC attractor paths, is to use a smaller kernel width. However, without visual inspection or some other form of confirmation of support suitability, automating the selection of an optimum kernel width for a myriad of possible attractor configurations remains a challenge. In essence, the nature of the estimated support determines the nature of changes in process dynamics that can be detected.

As is evident from Fig. 7.53, the RQA diagnostic responded strongly to the change in the parameters of the autocatalytic process, unlike the other diagnostics discussed previously. The reason for this is that the RQA diagnostic is derived from the recurrence plots of the data, shown in Fig. 7.54. More specifically, the

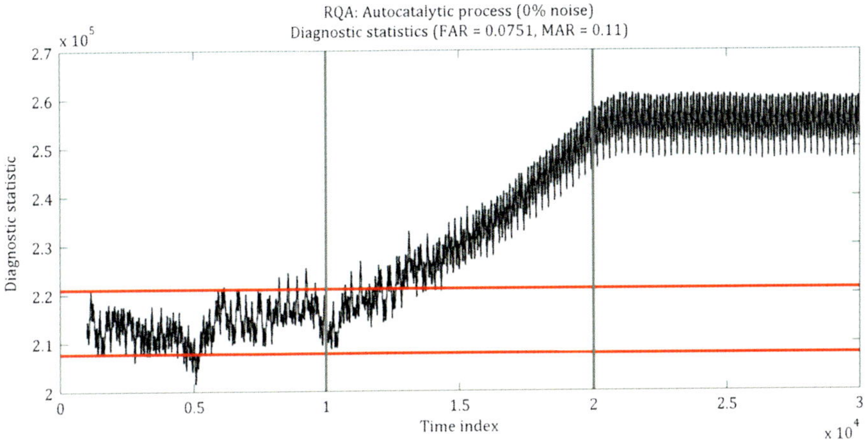

Fig. 7.53 RQA diagnostic statistic sequence for the autocatalytic process data set with 0 % noise added (upper and lower thresholds shown in *red*; location of change points shown with *vertical lines*)

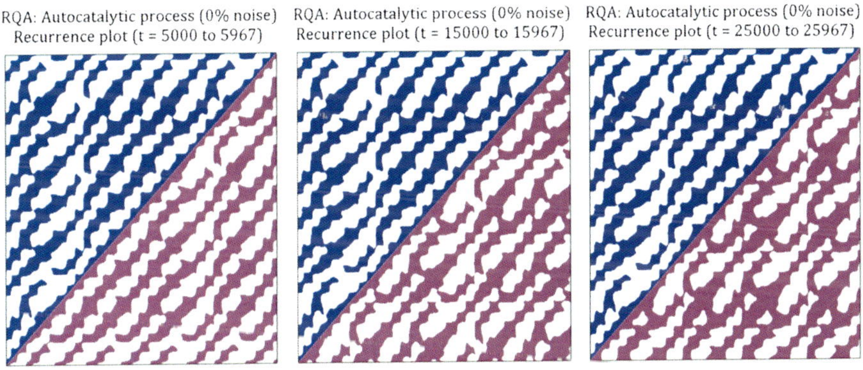

Fig. 7.54 Phase space recurrence plots for the autocatalytic process data set with 0 % noise added (*blue upper left triangle* representing NOC data *and pink lower right triangle* representing test data)

textural changes in these plots arise not only from changes in the macrostructures of the attractor geometries but also from potentially subtle changes in the density distributions of the data in the phase space.

Some of the changes in the lower left (pink) triangles of the recurrence plots in Fig. 7.54 are visible to the naked eye, but in general, the RQA diagnostics are capable of detecting changes that may escape such perception. Even so, other approaches at textural modelling, such as the use of textons to develop additional RQA diagnostic variables, may also be useful.

Table 7.9 Number of features retained for SSA and RF dynamic monitoring schemes applied to the predator–prey, BZ reaction and autocatalytic process data sets

Simulated system	Number of features retained d	0 % noise	5 % noise	10 % noise	20 % noise
Predator–prey	SSA	3	3	3	4
	RF	30	22	20	16
BZ reaction	SSA	4	4	4	6
	RF	43	32	29	20
Autocatalytic process	SSA	13	13	14	19
	RF	66	63	55	43

7.8.5 Number of Retained Features

The number of retained features for the SSA and RF approaches (based on 90 % retained variance and parallel analysis, respectively) is shown in Table 7.9. Note that one feature was extracted in all cases for NLPCA, in order to represent a closed-curve one-dimensional manifold in the phase space.

It is noticeable that the RF approach results in many more features to be retained than the SSA approach. This may indicate that the parallel analysis may be inadequate in determining the optimal number of features to retain.

As the noise level increases, the number of retained RF features generally decreases, which may indicate another example of input noise regulation. This is not the case with SSA, where the number of features stays relatively constant with noise addition. The linear character of SSA may impose regularization, in itself.

7.9 Concluding Remarks

In this chapter, the performance of nonlinear dynamic monitoring approaches was compared via case studies on simulated systems showing complex behaviour. Monitoring of the systems was essentially based on detecting changes in the geometrical structures of the attractors of the systems after embedding the data in phase space.

The different strategies considered are all in principle capable of detecting changes in complicated data structures, by fitting the NOC data with decision envelopes that serve as control limits. When new data breach the envelope, a change is signalled. In the case studies associated with the autocatalytic process, detection of change was difficult, since the change in the geometry of the attractor was very subtle. The trajectory of the new data remained mostly within the decision envelopes of the monitoring schemes, although the density distribution of the new data within this envelope changed.

It is for this reason that the monitoring scheme based on recurrence quantitative analysis was better able to detect these changes. Changes in the density of the data in phase space are reflected in the recurrence plots of the data, and therefore, these changes are also captured in the diagnostic variables derived from the recurrence plots. Variation in the density distributions of the data can be seen as microscale changes in the geometrical structures of the data that cannot be captured efficiently by fitting of decision surfaces to NOC data.

On the other hand, recurrence plots would in principle not be able to detect changes associated with simple translation of the attractors in phase space. The development of more advanced process or manifold monitoring schemes could therefore be based on incorporating the best of both approaches, for example, by monitoring the recurrence plot and random forest diagnostic variables.

References

Anderson, J., Gurarie, E., & Zabel, R. (2005). Mean free-path length theory of predator–prey interactions: Application to juvenile salmon migration. *Ecological Modelling, 186*(2), 196–211.

Auret, L., & Aldrich, C. (2010). Change point detection in time series data with random forests. *Control Engineering Practice, 18*(8), 990–1002.

Belousov, A. I., Verzakov, S. A., & von Frese, J. (2002). Applicational aspects of support vector machines. *Journal of Chemometrics, 16*(8–10), 482–489.

Dowd, M. (2005). A bio-physical coastal ecosystem model for assessing environmental effects of marine bivalve aquaculture. *Ecological Modelling, 183*(2–3), 323–346.

Fawcett, T. (2006). An introduction to ROC analysis. *Pattern Recognition Letters, 27*(8), 861–874.

Ku, W., Storer, R. H., & Georgakis, C. (1995). Disturbance detection and isolation by dynamic principal component analysis. *Chemometrics and Intelligent Laboratory Systems, 30*(1), 179–196.

Lee, J. S., & Chang, K. S. (1996). Applications of chaos and fractals in process systems engineering. *Journal of Process Control, 6*(2), 71–87.

Lewis, D. M., & Bala, S. I. (2006). Plankton predation rates in turbulence: A study of the limitations imposed on a predator with a non-spherical field of sensory perception. *Journal of Theoretical Biology, 242*(1), 44–61.

Liaw, A., & Wiener, M. (2002). Classification and regression by randomForest. *R News, 2*(3), 18–22.

Lindfield, G. R., & Penny, J. E. T. (2000). *Numerical methods using Matlab*. Upper Saddle River: Prentice Hall.

Moskvina, V., & Zhigljavsky, A. (2003). An algorithm based on singular spectrum analysis for change point detection. *Communications in Statistics: Simulation and Computation, 32*(2), 319–352.

Palomo, M. J., Sanchis, R., Verdú, G., & Ginestar, D. (2003). Analysis of pressure signals using a Singular System Analysis (SSA) methodology. *Progress in Nuclear Energy, 43*(1–4), 329–336.

R Development Core Team. (2010). *R: A language and environment for statistical computing*. Vienna: R Foundation for Statistical Computing. Available at: http://www.R-project.org

Salgado, D. R., & Alonso, F. J. (2006). Tool wear detection in turning operations using singular spectrum analysis. *Journal of Materials Processing Technology, 171*(3), 451–458.

Scholz, M. (2007). Analysing periodic phenomena by circular PCA. In S. Hochreiter & R. Wagner (Eds.), *Bioinformatics research and development* (pp. 38–47). Berlin/Heidelberg: Springer. Available at: http://www.springerlink.com/index/10.1007/978-3-540-71233-6_4. Accessed 23 June 2011.

Scholz, M. (2011). Nonlinear PCA toolbox for Matlab – Matthias Scholz. *Nonlinear PCA*. Available at: http://www.nlpca.de/matlab.html. Accessed 22 June 2011.

Scholz, M., Kaplan, F., Guy, C. L., Kopka, J., & Selbig, J. (2005). Non-linear PCA: a missing data approach. *Bioinformatics, 21*(20), 3887–3895.

Scholz, M., Fraunholz, M., & Selbig, J. (2008). Nonlinear principal component analysis: Neural network models and applications. In A. N. Gorban, B. Kégl, D. C. Wunsch, & A. Y. Zinovyev (Eds.), *Principal manifolds for data visualization and dimension reduction* (pp. 44–67). Berlin/Heidelberg: Springer. Available at: http://www.springerlink.com/index/10.1007/978-3-540-73750-6_2. Accessed 22 June 2011.

Seghouane, A.-K., Moudden, Y., & Fleury, G. (2004). Regularizing the effect of input noise injection in feedforward neural networks training. *Neural Computing and Applications, 13*(3), 248–254.

Wang, K.-Y., Shallcross, D. E., Hadjinicolaou, P., & Giannakopoulos, C. (2002). An efficient chemical systems modelling approach. *Environmental Modelling & Software, 17*(8), 731–745.

Zhang, D., Gyorgyi, L., & Peltier, W. R. (1993). Deterministic chaos in the Belousov–Zhabotinsky reaction: Experiments and simulations. *Chaos: An Interdisciplinary Journal of Nonlinear Science, 3*(4), 723.

Nomenclature

Symbol	Description
$\widehat{\mathbf{Z}}_i^{(NOC)}$	jth column of reconstructed zero mean unit variance scaled lagged trajectory matrix associated with normal operating conditions (NOC)
$\tilde{e}_t^{(test)}$	Average scaled diagnostic statistic
$\mathbf{C}^{(NOC)}$	Covariance matrix associated with normal operating conditions
$\mathbf{T}^*$	Optimal score matrix, $\mathbf{T}^* \in \Re^{n \times d}$
$\mathbf{T}_j$	jth column of score matrix
$\mathbf{U}^*$	Optimal set of eigenvectors, $\mathbf{U}^* \in \Re^{m \times d}$
$\mathbf{U}_j$	jth eigenvector of the covariance matrix associated with normal operating conditions
$\mathbf{Z}^{(NOC)}$	Zero mean unit variance scaled lagged trajectory matrix associated with normal operating conditions (NOC)
$\mathbf{Z}_j^{(NOC)}$	jth column of zero mean unit variance scaled lagged trajectory matrix associated with normal operating conditions (NOC)
R_{ij}	Recurrence event between two samples i and j
$e^{(NOC)}$	Residual of reconstructed sample vector associated with normal operating conditions (NOC)
$e_i^{(NOC)}$	Residual of ith reconstructed variable associated with normal operating conditions (NOC)
$x_i^{(NOC)}$	jth sample vector associated with normal operating conditions (NOC)
$z_j^{(NOC)}$	jth zero mean unit variance scaled variable associated with normal operating conditions (NOC)
$\mu^{(NOC)}$	Mean scaling parameter for variables associated with normal operating conditions (NOC)
$\sigma^{(NOC)}$	Unit variance scaling parameter for variables associated with normal operating conditions (NOC)

(continued)

(continued)

Symbol	Description
$\mathcal{F}_{\mathbf{T}^* \to \tilde{\mathbf{Z}}}$	Feature space of mapping of $\mathbf{T}^*$ to $\mathbf{Z}$
$\mathcal{F}_{\mathbf{Z} \to \mathbf{T}^*}$	Feature space of mapping of $\mathbf{Z}$ to $\mathbf{T}^*$
$I(\cdot)$	Indicator function
N_{w}	Size of sliding window in time series
$\mathbf{x}^{(\mathrm{NOC})}$	Sample vector associated with normal operating conditions (NOC)
$\mathbf{z}^{(\mathrm{NOC})}$	Lagged zero mean unit variance scaled sample vector associated with normal operating conditions (NOC)
$\mathbf{U}$	Set (matrix) of eigenvectors, $\mathbf{T}^* \in \Re^{m \times m}$
R	Recurrence rate
ε	Threshold distance

Chapter 8
Process Monitoring Using Multiscale Methods

8.1 Introduction

As discussed previously, the need for these advanced control methods is becoming critical in the successful management and operation of modern plants, whose control and monitoring is inherently difficult because of their complex large-scale configurations as well as the tendency towards plant-wide integration. Multivariate statistical process control is based on the statistical projection methods of principal component analysis (PCA) and partial least squares (PLS). These methods are based on projecting highly correlated data into a rotated coordinate space in which the latent variables are uncorrelated. In particular, PCA projects data into two orthogonal subspaces: (1) the principal component subspace captured by a few latent variables that describe most of the variability in the original data and (2) a residual subspace that describes unexplained variation that is usually attributable to measurement noise and model errors.

More formally, given a data matrix **X** with N samples of dimension d, the PCA model can be expressed as

$$\mathbf{X} = \sum_{i=1}^{k} \mathbf{t}_i \mathbf{p}_i^{\mathrm{T}} + \mathbf{E} \quad (8.1)$$

where $\mathbf{p}_i$ are principal directions or eigenvectors of the covariance matrix of **X**, k $(<d)$ is the number of retained principal directions and **E** the residual subpace. Control charts can be obtained for both the principal component and residual subspaces: Hotelling's T^2-statistic is used to characterize the common cause variation in the low-dimensional plane defined by a few latent variables, while the squared prediction error (SPE) or Q-statistic describes the distance of an observation to the plane of the principal component subspace. Abnormal events affecting all variables and yet consistent with the PCA model are detected using the T^2-statistic. On the other hand, the SPE or Q-statistic detects disturbances or process changes

C. Aldrich and L. Auret, *Unsupervised Process Monitoring and Fault Diagnosis with Machine Learning Methods*, Advances in Computer Vision and Pattern Recognition, DOI 10.1007/978-1-4471-5185-2_8, © Springer-Verlag London 2013

inconsistent with the PCA model. In each case, control limits are determined using historical data collected under normal operating conditions, that is, the so-called in-control data (Jackson 1991; MacGregor and Kourti 1995).

As was considered in more detail in Chap. 2, the basic PCA/PLS-based approach to statistical process monitoring has been extended to many variants to handle specific cases. For example, multiblock PCA/PLS was introduced for the modelling and analysis of very large collinear data sets with variables that can block into conceptually meaningful blocks (Westerhuis et al. 1998). Multiway projection methods provide a monitoring scheme for the analysis of time-varying batch processes (Nomikos and MacGregor 1995a, b).

A major limitation in the application of these multivariate projection methods to dynamic physical systems, for example, chemical processes, is the assumption of minimal autocorrelation in each measured variable. In the presence of autocorrelation in observed signals (arising from, e.g. process dynamics or high sampling frequency), assumptions underlying classical SPC methods are violated (Runger and Willemain 1995). Failure to adequately compensate for the autocorrelation invariably leads to reduced reliability of corresponding control scheme because of a high rate of false alarms. An approach that has been successfully used in univariate SPC is the use of residuals from an appropriate time series model (Montgomery and Mastrangelo 1991; Harris and Ross 1991; Montgomery 1996). Although a similar approach can also be used for multivariate time series (Wilson 1973; Tiao and Box 1981; Tjostheim and Paulsen 1982), it is generally a complex task, particularly for large-dimensional data. Dynamic PCA has been proposed as a direct approach for including dynamic information in an otherwise linear static model extracted by PCA (Kresta et al. 1991; Ku et al. 1995). More precisely, the dimensionality of each of the variables in a given data matrix $\mathbf{X} \in \mathbb{R}^{N \times d}$ is enlarged to include dynamic dependence information provided by the l past values such that the following holds:

$$\begin{aligned}\mathbf{0} &= \mathbf{X}(l)\mathbf{b} \\ &= \left[\mathbf{X}\left(t-l\right)\cdots\mathbf{X}\left(t-1\right)\ \ \mathbf{X}(t)\right]\mathbf{b} \\ &= \begin{bmatrix} \mathbf{x}^{\mathrm{T}}\left(1-l\right) & \cdots & \mathbf{x}^{\mathrm{T}}(1) \\ \vdots & \ddots & \vdots \\ \mathbf{x}^{\mathrm{T}}\left(n-l\right) & \cdots & \mathbf{x}^{\mathrm{T}}(n) \end{bmatrix}\mathbf{b}\end{aligned} \tag{8.2}$$

where $\mathbf{x}(t)$ is a d-dimensional vector of observations at time t. The solution to Eq. 8.2 is the null space of the augmented data matrix $\mathbf{X}(l)$, that is, the subspace spanned by right singular vectors whose singular values are zero. The null space describes linear constraining relationships among variables (Strang 2009). Dynamic PCA, which exploits this idea, involves selecting an appropriate value for l, and, subsequently, performing PCA on the corresponding time-lagged matrix (Ku et al. 1995). Linear static and dynamic constraining relations among the variables appear in the "approximate null space" or residual subspace, while the principal component subspace implicitly describes a multivariate autoregressive model. With this approach, SPE statistics evaluated from the residual subspace are guaranteed to be independent. Dynamic PCA has also been applied to the problem of sensor identification (Lee et al. 2004).

An alternative to using linear time series models to capture the dynamics is multiscale modelling using wavelet analysis (Bakshi 1998, 1999; Aradhye et al. 2003; Yoon and MacGregor 2004). Although the history of wavelet analysis is quite old, theoretical advances (Daubechies 1992), as well as development of efficient algorithms (Mallat 1989, 1999), have seen an increase in the use of wavelets in diverse applications in recent years. The appeal to wavelets can be attributed to their ability to define a representation that highlights specific signal characteristics such as local structures and singularities. Given a wavelet basis function φ, the wavelet transform of a signal $x(t)$ is obtained by projecting the signal onto translated and dilated copies of φ, that is,

$$W_{\varphi} = \frac{1}{\sqrt{a}} \int_{-\infty}^{\infty} x(t)\varphi\left(\frac{t-b}{a}\right) \tag{8.3}$$

where a and b are dilation and translation parameters, respectively. The basis function or *mother wavelet* φ, which must satisfy general admissibility properties, is localized in both time and frequency (Mallat 1989; Daubechies 1992). In applications to discretely sampled signals, an orthogonal wavelet decomposition using Eq. 8.3 can be obtained by dyadic sampling of the dilation and translation parameters a and b. Since wavelets are approximate eigenfunctions of many operators and, also, because of the time–frequency localization property, wavelets have been used for various tasks including compact representation of multiscale features, reducing autocorrelation in stochastic processes as well as denoising (Donoho et al. 1995).

A multiscale approach to statistical process control (MSSPC) was introduced by Bakshi (1998) that combines advantages of PCA and multiscale representation using wavelets. In particular, PCA is used to extract linear relations among variables, whereas wavelets separate deterministic features from stochastic influences. The methodology adapts to local structures at particular scales that are a result of special or abnormal events affecting a process. By adjusting the nature of the transform filter according to the scale of the abnormal event as well as the detection limits for a target control limit, MSSPC subsumes classical SPC charting methods such as Shewhart, cumulative sum (CUSUM) and exponentially weighted moving average (EWMA) (Aradhye et al. 2003; Ganesan et al. 2004).

The multiscale approach to statistical process control using wavelet methods involves choosing the mother wavelet from a large library of admissible functions and, subsequently mapping each monitored process variable into the wavelet domain for a selected depth of decomposition L, resulting in the detail signals $\{\mathbf{G}_m\mathbf{X}\}_{m=1}^{L}$. The last approximate signal $\mathbf{H}_L\mathbf{X}$ is a coarse approximation of the original signal and constitutes an additional scale whose basis function is nominally

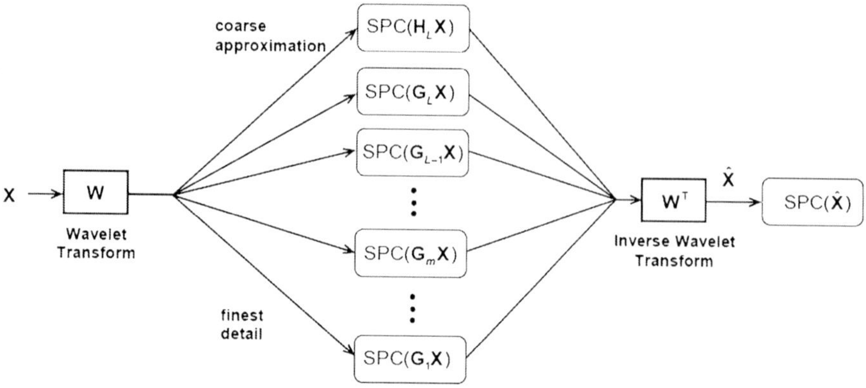

Fig. 8.1 Wavelet-based multiscale process monitoring methodology

referred to as the *father wavelet*. Here, $\mathbf{G}_m$ and $\mathbf{H}_L$ denote the wavelet operators for the different decomposition levels, and $\mathbf{X}$ is a matrix of process measurements. The coefficients from the decomposition of each variable are grouped into $L + 1$ matrices on the basis of scale as illustrated in Fig. 8.1. These coefficient matrices are subsequently monitored separately using statistical projection methods such as PCA or classical control charts, for example, the Shewhart chart, for univariate SPC. Using control limits determined from in-control data at each scale, scales at which coefficients are larger than these detection limits are identified. An inverse wavelet transform is then applied only for the scales at which a violation has been detected to recover the reconstructed signal $\mathbf{X}$. The monitoring method is next applied to the matrix of reconstructed signals to verify process deviation. The detection limits for the reconstructed signal are obtained by performing inverse wavelet transform on the normal data only at the same scales violating control limits in the coefficient space for the monitored stream of data, thus ensuring that MSSPC is adaptive to the nature of the signal (Bakshi 1998).

In this chapter, multivariate statistical process monitoring based on singular spectrum analysis (SSA) is considered. Similar to dynamic PCA (Ku et al. 1995), SSA is based on the singular value decomposition of a covariance matrix of time-lagged or trajectory matrix (Broomhead and King 1986). Unlike the dynamic PCA approach, however, SSA is not restricted to extracting linear static and dynamic relations. Using data-adaptive filters, a time series can be separated into statistically independent constituents (at zero lag), namely, linear or nonlinear trends, amplitude- and phase-modulated oscillatory patterns and an aperiodic or noise component (Vautard and Ghil 1989; Vautard et al. 1992; Elsner and Tsonis 1996; Golyandina et al. 2001). This is of important significance in SPC applications, owing to the limited error reduction capabilities of the widely used PCA, whereas SSA can enhance signal/noise separation. In the case of PCA, the model always contains an intrinsic noise component of magnitude proportional to the number of retained components. SSA has proved particularly useful in the study of fractal systems in nonlinear time series analysis applications in climatology (Ghil et al. 2002, 2011) and econometrics (Hassani and Zhigljavsky 2009), as well as change-point detection (Moskvina and Zhigljavsky 2003). As in many physical systems, the observed data in these systems are invariably short, noisy and nonstationary.

In the following sections, the motivation and concepts underlying the SSA methodology are explained. A statistical process control based on SSA with multiblock PCA characteristics is proposed. The performance of the SSA-based SPC method is empirically compared with some established approaches using the average run length (ARL) criterion across mean shifts of different magnitudes on simulated systems for both univariate and multivariate data. A reliability-based analysis is then used to assess the performance of the proposed technique on a simulated multivariate correlated system for both mean shifts as well as parameter changes. Finally, major findings are highlighted as well as recommendations for further work.

8.2 Singular Spectrum Analysis

Singular spectrum analysis is a time series analysis method whose main objective is non-parametric decomposition of a time series x_t of length N according to

$$x_t = \sum_{i=1}^{m} x_t^{(m)}, \quad t = 1, \ldots, N, \tag{8.4}$$

where each member of the set $\left\{x_t^{(1)}, \ldots, x_t^{(m)}\right\}$ is an independent and identifiable time series, that is, a trend, periodic or quasi-periodic amplitude-modulated series or an aperiodic component. Establishing the independence or separability of extracted series $x_t^{(k)}$ is fundamental in SSA theory. SSA decomposition according to 8.4 is considered successful if the obtained additive components are separable from each other, at least approximately. Hence, significant research effort has gone into understanding separability in SSA for the purposes of evaluating the quality and predicting the results of a decomposition. In applications, however, the focus has mainly been on identifying and extracting trends and oscillatory patterns in observed data. These have been mostly in climatology and meteorology (Vautard and Ghil 1989; Vautard et al. 1992; Elsner and Tsonis 1996; Ghil et al. 2002).

The SSA methodology involves the two main stages of decomposition and reconstruction, each of which can be described in two steps as outlined in the following:

8.2.1 Decomposition

Embedding

The first step in SSA is to embed a given time series $x_t, t = 1, \ldots, N$ in a vector space of dimensionality L, resulting in K lagged vectors $\mathbf{x}_n \in \mathbb{R}^L$, that is,

$$\mathbf{x}_n = [x_n \ x_{n+1} \ldots \ x_{n+L+1}]^{\mathrm{T}}, \quad n = 1, \ldots, K, \tag{8.5}$$

where $K = N - M + 1$ is the number lagged vectors from a time series of length N. These embedded vectors are augmented into a multidimensional time series commonly referred to as the trajectory matrix in the study of nonlinear dynamic systems since it represents a trajectory of the evolution of the dynamical system represented by the observed time series (Broomhead and King 1986; Ghil et al. 2002):

$$\mathbf{X} = [\mathbf{x}_1 \ \mathbf{x}_2 \ldots \mathbf{x}_K]^{\mathrm{T}} = \begin{bmatrix} x_1 & x_2 & \cdots & x_L \\ x_2 & x_3 & \cdots & x_{L+1} \\ \vdots & \vdots & \ddots & \vdots \\ x_K & x_{K+1} & \cdots & x_N \end{bmatrix}. \tag{8.6}$$

Similar to the dynamic PCA method (Eq. 8.1), the augmented matrix $\mathbf{X}$ is in the form of a Hankel matrix (Strang 2009).

Singular Value Decomposition

Singular value decomposition (SVD) of the covariance matrix of the trajectory matrix yields left and right singular vectors $(\mathbf{u}_k, \mathbf{v}_k)$, and the singular values σ_k according to

$$\mathbf{C}_{\mathbf{X}} = \mathbf{U}\boldsymbol{\Sigma}\mathbf{V}^{\mathrm{T}} \tag{8.7}$$

where $\mathbf{U}, \mathbf{V}$ are matrices with left and right singular vectors as columns, $\boldsymbol{\Sigma}$ is a diagonal matrix, with $\sigma_1 > \sigma_2 \geq \cdots \geq \sigma_L > 0$ on the diagonal, and $\mathbf{C}_{\mathbf{X}}$ is the covariance matrix defined as

$$\mathbf{C}_{\mathbf{X}} = \frac{1}{N - L + 1}\mathbf{X}^{\mathrm{T}}\mathbf{X}. \tag{8.8}$$

The ordered set of singular values is called the singular spectrum, from which SSA derives its name. The tuple $(\mathbf{u}_k, \sigma_k, \mathbf{v}_k)$ is often referred to as the eigentriple.

The SVD of the trajectory matrix can be expressed in terms of elementary matrices (i.e. matrices with unity rank) defined by each eigentriple:

$$\mathbf{X} = \sum_{i=1}^{p} \mathbf{X}_i = \sum_{i=1}^{p} \sigma_i \mathbf{u}_i \mathbf{v}_i^{\mathrm{T}} \tag{8.9}$$

where $p\ (\leq L)$ is the rank of $\mathbf{X}$.

An alternative approach estimates the covariance matrix $\mathbf{C_X}$ such that its entries $\mathbf{C_X}(i, j)$ depend only on the lag $|i - j|$, that is, as a Toeplitz matrix (Elsner and Tsonis 1996). Advantages for using each of the approaches have been discussed in literature (Vautard and Ghil 1989; Ghil et al. 2002). Without loss of generality, the SVD approach is used consistently in this chapter.

8.2.2 *Reconstruction*

Grouping of Elements

The elementary matrices in Eq. 8.9 can be split into mutually exclusive subgroups such that

$$\mathbf{G}_i = \mathbf{X}_{i_1} + \mathbf{X}_{i_2} + \cdots + \mathbf{X}_{i_q} \tag{8.10}$$

where $I_i = \{i_1, i_2, \ldots, i_q\}$ is the set of indices associated with a subgroup $\mathbf{G}_i$, $i = 1, \ldots, m$. The decomposition in Eq. 8.10 can then be expressed as

$$\mathbf{X} = \sum_{i=1}^{m} \mathbf{G}_i . \tag{8.11}$$

Diagonal Averaging

In the last step of SSA, a subgroup $\mathbf{G} \in \mathbb{R}^{K \times L}$ with $L < K$ (otherwise the matrix is first transposed, $\mathbf{G} \rightarrow \mathbf{G}^{\mathrm{T}}$) is transformed into a time series of the same length N as the original time series through diagonal averaging, that is,

$$\tilde{x}_t = \frac{1}{t} \sum_{m=1}^{t} \mathbf{G}_{m,t-m+1}, \quad 1 \leq t < L, \tag{8.12}$$

$$\tilde{x}_t = \frac{1}{L} \sum_{m=1}^{L} \mathbf{G}_{m,t-m+1}, \quad L \leq t < K + 1, \tag{8.13}$$

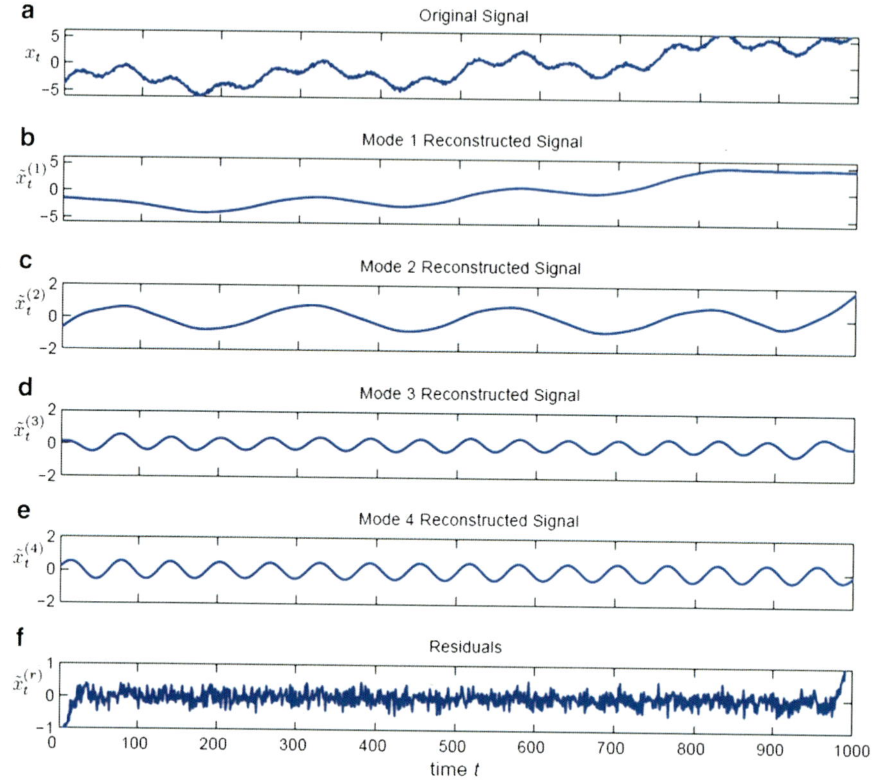

Fig. 8.2 SSA decomposition of a signal, with (**a**) the original signal, (**b**)–(**e**) different modes of the reconstructed signal and (**f**) noise

$$\tilde{x}_t = \frac{1}{N-t+1}\sum_{m=t-K+1}^{N-K+1} G_{m,t-m+1}, \quad K+1 \le t < N \tag{8.14}$$

where $\mathbf{G}_{i,j}$ is the (i,j)th entry of matrix $\mathbf{G}$. The reconstructed components $\tilde{x}_t$ recover phase information of the time series lost in the decomposition stage, and the diagonal averaging in Eq. 8.14 is an adaptive optimal filter in the least-squares sense (Vautard and Ghil 1989; Ghil et al. 2002). As no information is lost during the reconstruction, Eq. 8.14 can be written in terms of reconstructed times series $\tilde{x}_t$:

$$x_t = \sum_{k=1}^{m} \tilde{x}_t^{(k)}. \tag{8.15}$$

A typical decomposition of a signal using SSA is illustrated in Fig. 8.2. Plotted are the original signal $\left(x_t = e^{0.05t} + \sin(2t) + 2\sin(0.5t) + \varepsilon_t\right)$, reconstructed components from the first four modes as well as a residual signal. An embedding window of size $L = 100$ was used for building the trajectory matrix. In the illustration, only the residual signal can be considered as "grouped", while the first four modes are plotted "as is". It can be noted that the first mode can be associated with a trend, while modes 2–4 are oscillatory patterns, with modes 3 and 4 in quadrature.

Multichannel SSA (MSSA) generalizes the basic SSA methodology as presented above to a multivariate time series (Golyandina et al. 2001; Ghil et al. 2002). Similar to the univariate case, in MSSA two approaches for estimating the covariance matrix associated are used: (a) the Toeplitz method (Vautard and Ghil 1989; Plaut and Vautard 1994) and (b) the trajectory matrix method (Broomhead and King 1986; Allen and Smith 1996). In practice, both methods are used to verify the robustness of the MSSA results (Ghil et al. 2002).

Since SSA is principal component analysis performed on a trajectory or time-lagged matrix, all mathematical and statistical properties associated with PCA apply to SSA. It is important to note that the use of statistical concepts in the SSA framework does not require certain statistical assumptions such as stationarity or normality of residuals (Vautard and Ghil 1989; Golyandina et al. 2001). SSA belongs to a family of methods based on empirical orthogonal function (EOF) expansion whose main characteristic is the data-adaptive nature of the basis functions. The variance of the decomposition using EOF methods is defined in terms of the basis functions or mode (i.e. $\{\mathbf{u}_k\}$ in Eq. 8.9), and therefore, unlike in wavelet decomposition, the variance distribution does not imply scales or frequency content of the signal. This limitation has been cited as a critical flaw of EOF-based methods (Huang et al. 1998). However, because of the data-adaptive property, SSA and related methods have been successfully used in many applications as highlighted in Introduction. Moreover, a multiscale SSA approach has been proposed that uses a set of eigenvectors from covariance matrices defined by sliding windows of different widths across the length of a time series (Yiou et al. 2000). Multiscale SSA extends the application of SSA to nonstationary time series.

8.3 SSA-Based Statistical Process Control

As previously mentioned, SSA separates a time series into additive components that can be identified as a trend, anharmonic oscillations or noise. Because of the data-adaptive nature of the basis functions, extracted anharmonic oscillations are usually expressed in terms of a much fewer number of the basis functions than would be required when using fixed basis functions, e.g. sines and cosines in Fourier analysis. It can be expected, therefore, that separate analysis of extracted components may provide information on the underlying dynamics of the physical system that is not accessible from the analysis or use of raw measurements due to possibly confounding influences such as noise, autocorrelation or embedded features with different time–frequency localizations. This insight is key to the emergence of multiscale methods in process monitoring such as those based on wavelet decomposition (Bakshi 1998; Aradhye et al. 2003; Yoon and MacGregor 2004; Reis et al. 2008). In this case, the "scale" is related to the width of the scaling function or "mother wavelet", and wavelet coefficients at the same scale for different measurements can be considered collectively. The proposed use of SSA for process monitoring is motivated by similar reasoning but exploiting the data-adaptive properties of the obtained basis functions. The methodology, summarized in the pseudocode listing in Table 8.1 as well as schematically in Fig. 8.2, is discussed in detail below.

Table 8.1 Pseudocode – statistical process monitoring using SSA

Input: $X_t \in \mathbb{R}^{N \times d}, X'_t \in \mathbb{R}^{N' \times d}, L \in \mathbb{N}$

1. Build reference model
   - (a) SSA decomposition
     - (i) Mean centre each column of X_t and save the vector of means
     - (ii) Embed each column x_i of X_t to obtain a set of trajectory matrices $\{X_i \in \mathbb{R}^{K \times L}\}_{i=1}^{d}$, where $K = N - L + 1$.
     - (iii) Perform SVD on each trajectory matrix$X_i, i = 1, \ldots, d$
   - (b) SSA reconstruction
     - (i) For each $x_i(t)$, $i = 1, \ldots, d$, group corresponding SVD components into M groups
     - (ii) Obtain reconstructed signals $\tilde{x}_i^{(k)}$ from each set of grouped components
   - (c) Multimodal representation of original multivariate time series
     - (i) Create a new matrix $\tilde{X}^{(k)}$ by concatenating reconstructed time series $\tilde{x}_i^{(k)}$, i.e. $\tilde{X}^{(k)} = \left[\tilde{x}_1^{(k)} \tilde{x}_2^{(k)} \cdots \tilde{x}_d^{(k)}\right]$, for $k = 1, \ldots, M$.
   - (d) Determine SPC parameters (e.g. loading vectors in PCA) and control limits $\left\{\beta_{\lim}^{(k)}\right\}_{k=1}^{M}$ for the selected process monitoring method
2. Apply model to new data
   - (a) Mean centre each column of $\in \mathbb{R}^{N \times d}$ using the vector of means obtained from the training set (step 1(a)(i))
   - (b) Obtain a multimodal representation of $X_t'^{(k)} \in \mathbb{R}^{N' \times d}$ for $k = 1, \ldots, M$ through SSA decomposition and reconstruction using the same parameters as for the training data
   - (c) Compute monitoring statistics at each mode for new data, i.e. $\beta^{(k)}\left(\tilde{\mathbf{X}}'^{(k)}\right)$, for $k = 1, \ldots, M$
   - (d) At each mode k determine if new data violate detection limits of the selected monitoring method:

FOR $k = 1 \rightarrow M$ **DO**
  IF $\beta^{(k)}\left(\tilde{\mathbf{X}}'^{(k)}\right) > \beta_{\lim}^{(k)}$ **THEN**
    signal out-of-control status
  ELSE
    signal in-control status
  END IF
END FOR

8.3.1 Decomposition

Denote by X_t a multivariate time series with d variables sampled at the same N times during a period typical of normal operating conditions, and L the vector space embedding dimension. Each time series is individually expanded into an L-dimensional vector space by sliding a window of size L along each variable $x_i(t)$ to give a trajectory matrix $\widehat{\mathbf{X}}_i$ as indicated in Eq. 2.8 Selecting the optimal L to use is a design challenge, and its choice requires elaboration as it is the parameter that controls the trade-off between the amount of significant information *and* the statistical confidence in the extracted information (Vautard and Ghil 1989; Ghil et al. 2002).

Golyandina et al. (2001) have discussed in detail the interplay between choice of the window size and separability of features of interest. In general, large window sizes are preferable for a detailed decomposition of a time series, whereas small window sizes allow for as many repetitions as possible of identifiable features. Poor choice of the window length may result in mixing of interpretable components. Unfortunately, many time series are invariably disparate, and therefore, no general recommendations exist on the proper choice of the window length. In practice, L is varied within a range while noting the stable features of the decomposition (Golyandina et al. 2001).

In nonlinear time series analysis, a delay coordinate transform is used to map a univariate time series into a trajectory matrix that is assumed to represent the evolution of a nonlinear dynamic system in phase space (Broomhead and King 1986). Phase space representation is called an embedding if the mapping is *diffeomorphic*, that is, one to one and preserves differential information of the underlying attractor (Sauer et al. 1991; Kantz and Schreiber 1997). The reconstruction of the attractor in phase space requires specification of two parameters, namely, the embedding dimension L and the delay τ, both of which determine the embedding window. A number of techniques have been proposed to determine the optimal embedding parameters. Using SSA, the lag or delay is fixed at $\tau = 1$, and L is typically decided through by identifying the point of decorrelation in the autocorrelation function. Subsequently, the reconstructed attractor is validated by evaluating estimates of nonlinear invariants. It is tempting to use a similar approach in the proposed SPC method since capturing the dynamics of the time series is integral in both attractor reconstruction and time series decomposition (Golyandina et al. 2001). However, invariance of topological properties of the hypothetical dynamic system is not a desideratum in the latter.

Having obtained a set of d trajectory matrices $\left\{\widehat{\mathbf{X}}_i\right\}_{i=1}^{d}$ using a common window length L, singular value decomposition of each matrix yields the corresponding eigentriple set $\left\{\left(\mathbf{u}_i^{(k)}, \sigma_i^{(k)}, \mathbf{v}_i^{(k)}\right)\right\}_{i=1}^{L}, \quad k = 1, \ldots, d.$

8.3.2 Reconstruction

In the reconstruction phase an important design decision is determining the grouping of components. Recalling that a key objective in SSA is decomposing an observed time series as the sum of "identifiable" additive components, a successful decomposition requires that the additive components be *separable*, at least approximately (Golyandina et al. 2001). Separability can be characterized by requiring orthogonality between the rows or columns of any pair of the trajectory matrices in Eq. 8.16. Mathematically, this translates to diagonal covariance matrices between any pair of the $\mathbf{G}_i$'s defined in Eq. 8.10. An alternative necessary condition for separability can be formulated in terms of the reconstructed subseries in Eq. 8.15. More formally, denote by $X_p(t) = \big(x_p(1), \ldots, x_p(N)\big)$ and $X_q(t) = \big(x_q(1), \ldots, x_q(N)\big)$ two time series of length N. For a fixed window length L, $X_p(t)$ and $X_q(t)$ are said to be **w**-orthogonal if

$$\big(X_p(t), X_q(t)\big)_w = 0 \tag{8.16}$$

where the weighted inner product $(\cdot)_w$ is defined as

$$\big(X_p(t), X_q(t)\big)_w \overset{\text{def}}{=} \sum_{i=1}^{N} w_i x_p(i) x_q(i) \tag{8.17}$$

and the weights are defined as (see also Eqs. 8.12, 8.13, and 8.14)

$$\begin{aligned} w_i &= i, & 1 \le i \le L \\ w_i &= L, & L+1 \le i \le K \\ w_i &= N - i + 1 & K+1 \le i \le N. \end{aligned}$$

(As before, it is assumed $L < K$; otherwise, the associated trajectory matrices are transposed.) Worth to note is that separability of subseries is closely related to **w**-orthogonality, and expressing a time series as the sum of separable components is equivalent to its expansion in terms of well-defined **w**-orthogonal basis functions obtained from the observed data (Golyandina et al. 2001).

From the foregoing, two properties characterizing the quality of separability of a pair of time series have been suggested (Golyandina et al. 2001). The first quality characteristic is the maximum of the absolute value of the correlations between the rows and between the columns of a pair of trajectory matrices $\tilde{\mathbf{X}}_i$ and $\tilde{\mathbf{X}}_j$, denoted by $\rho_{\max}^{(L,K)}$. The second quality criterion is the *weighted correlation* or **w**-correlation that characterizes the deviation from **w**-orthogonality of the series $X_p(t)$ and $X_q(t)$:

$$\rho_{p,q}^{(w)} = \frac{\big(X_p(t), X_q(t)\big)}{\big\|X_p(t)\big\|_w \big\|X_p(t)\big\|_w} \tag{8.18}$$

where $\|X\| = \sqrt{(X, X)}$.

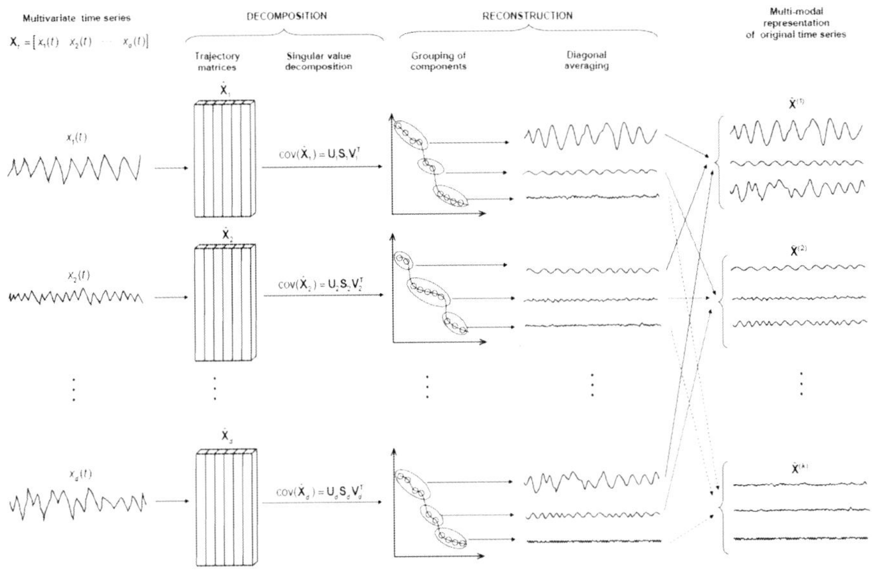

Fig. 8.3 A schematic of the SSA methodology for statistical process monitoring

Hence, a matrix of **w**-correlations of reconstructed subseries $\tilde{x}_i^{(k)}$ in Eq. 8.15 can be obtained, each subseries indexed by k in the sum corresponding to a single eigentriple. In practice, following such a prescription may prove onerous in the design of an efficient SPC procedure, particularly for the so-called phase II control (Bersimis et al. 2007). A simple approach consists of identifying the signal and noise components of the decomposition. The noise components are grouped, while the signal components can be handled collectively or individually as they generaly constitute fewer components. Inevitably, such an approach does result in redundance in certain cases as components that occur in quadrature are treated separately although they are associated with the same harmonics. As noted earlier, the spectrum of singular values only gives the proportion of variance explained by the principal directions and has no relation to the notion of scales or frequency of the signal. Therefore, the individual principal directions will be referred to as modes and the SPC method as *multimodal*. The ranking of these modes is used in reconstituting the original multidimensional time series structure at multiple views or levels. Specifically, all reconstructed components associated with the kth principal directions or group in the decomposition are collected to form a multivariate series $\tilde{X}^{(k)}, k = 1, \ldots, M$. In this sense, the method corresponds to multiscale methods based on wavelets except that the hierarchical representation is in the time domain and not wavelet domain (Fig. 8.3).

8.3.3 Statistical Process Monitoring

Once the multimodal representations are obtained after decomposing and reconstructing each variable of multivariate time series, statistical limits $\beta_{\mathrm{lim}}^{(k)}$ and other parameters for the selected monitoring method can be determined at each of the multiple levels of representation. For example, if PCA is used for monitoring, statistical limits for Hotelling's $T_{(k)}^2$- and Q-statistics can be determined for each kth approximation of the original time series. The residual limit Q_α for a significance level α is given by

$$Q_\alpha = \Theta_1\left[1 + \frac{c_\alpha\sqrt{2\Theta_2 h_0}}{\Theta_1} + \frac{\Theta_2 h_0\left(h_0 - 1\right)}{\Theta_1^!}\right]^{h_0^{-1}} \tag{8.19}$$

where

$$\Theta_i = \sum_{j=m+1}^{L}\lambda_j^i \quad i = 1,\ldots,3 \tag{8.20}$$

and

$$h_0 = 1 - \frac{2\Theta_1\Theta_3}{3\Theta_2^2}. \tag{8.21}$$

The number of principal components retained is given by m from a total of L, the λ_i's are the eigenvalues of the covariance matrix of the data, and c_α the normal deviate at the $(1-\alpha)$ percentile. Upper control limits for the T^2-statistics are given by

$$T^2 = \frac{m\left(N - 1\right)}{N - m}F_\alpha\left(m, N - m\right) \tag{8.22}$$

where $F_\alpha\left(m, N - m\right)$ is the $100\alpha\%$ critical point of the F-distribution with m and N–m degrees of freedom, with m the number of retained principal components and N the number of samples in the data (Wierda 1994; MacGregor and Kourti 1995; Wise and Gallagher 1996).

These and similar control limit estimates are based on statistical reasoning and probability theory, and conclusions derived from their use are uncertain. Therefore, the hypothesis test on which a conclusion is sought has an associated probability of type I error, or false positive, which sets the maximum acceptable probability of rejecting the null hypothesis when it is true, indicated by α in Eqs. 8.23 and 8.24. For multiple hypotheses, the number of false positives increases, and the limits need to be adjusted to maintain the probability of the type I error at α for the overall test (Kano et al. 2002). In particular, a *familywise* type I error probability is necessary when dealing with a family of tests. The following correction due to Šidàk can be used to adjust the single *testwise* type I error α_{adj} to reach a specified *familywise* type I error α for a family of n tests (Wierda 1994; Abdi 2007):

$$\alpha_{\text{adj}} = 1 - (1 - \alpha)^{1/n} \tag{8.23}$$

assuming independence of the n tests. Šidàk's correction is usually confused with Bonferroni's correction. The latter is in fact a lower bound on the Šidàk equation since it is a linear approximation from the Taylor expansion of the former and is given by

$$\alpha_{\text{adj}} \approx \frac{\alpha}{n}. \tag{8.24}$$

For non-independent tests, Šidàk's equation (Eq. 8.28) gives a lower bound on the correction.

Given new data X'_t, a time-lagged expansion is performed on each variable using a window length L. This is followed by SVD of the resulting trajectory matrices. Subsequently, a set of multilevel representation of the original data is obtained using the same parameters as for the normal data. Monitoring statistics $\beta^{(k)}$ are obtained and compared with the corresponding control limits $\beta_{\text{lim}}^{(k)}$ for each kth representation. If a sample violates a detection limit, then an out-of-control situation is signaled.

8.4 ARL Performance Analysis

In this section the performance of the proposed SSA-based approach to SPC is evaluated and compared to other SPC methods using the average run length (ARL). The run length of a process is a random variable that gives the number of samples observed before a control chart first signals an out-of-control situation. For a given process shift, a run length distribution can be defined. The expectation of the run length – *the average run length* (ARL) – measures the number of samples that, on average, are required to detect the shift. Typically, *in-control* run lengths for a process under control are large, whereas unstable processes have small ARL values that converge towards the location of shift occurrence as the magnitude of the shift increases. The inverse of the in-control run length corresponds to type I error, i.e. probability of false alarms. In most cases, theoretical evaluation of ARLs is difficult, and therefore, ARLs are determined empirically using Monte Carlo simulation. In this way, it is possible to compare the relative performance of different monitoring schemes for a fixed in-control run length by plotting ARL curves.

In the following, empirically derived ARLs for different SPC methods are compared using data generated from simulated univariate and multivariate systems with known autocorrelation structure and mean shifts, as investigated by Aradhye et al. (2003).

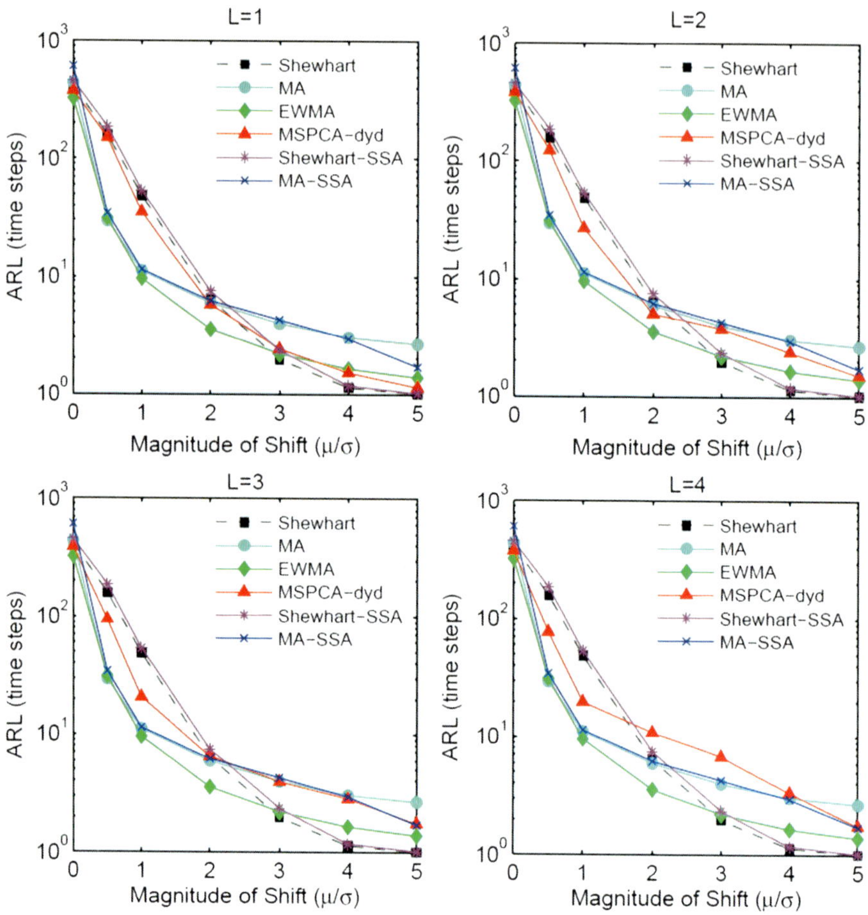

Fig. 8.4 Monte Carlo ARL curves for MSSPC-dyadic, Shewhart, MA, EWMA and SSA-based MA and Shewhart charts for a univariate IID Gaussian process. The subplots correspond to different depths of decomposition $L = \{1,2,3,4\}$ as indicated

8.4.1 Univariate SPC: Uncorrelated Gaussian Process

In the first study, 1,000 simulations of univariate data sampled from a Gaussian process with zero mean and unit variance for a specified mean shift are considered. Figure 8.4 shows the ARL curves obtained for various control charts, as indicated in the plots. The different subplots show the effect of different decomposition depths when using multiscale SPC with Haar wavelets. Time–frequency sampling over a discrete dyadic grid is used for the wavelet expansion. Similar plots are shown in Fig. 8.5 with the time–frequency sampling defined over a non-decimated or integer grid. In the case of the moving average (MA) chart, a window size of 16 is used,

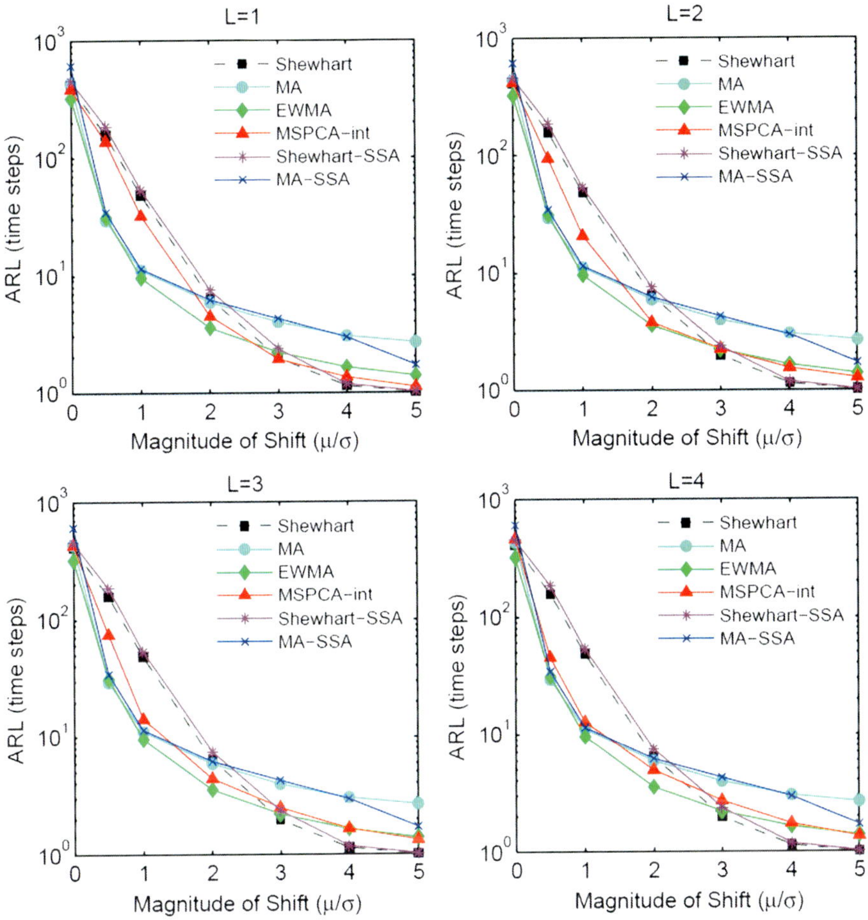

Fig. 8.5 Monte Carlo ARL curves for MSSPC-integer, Shewhart, MA, EWMA and SSA-based MA and Shewhart charts for a univariate IID Gaussian process. The subplots correspond to different depths of decomposition $L = \{1,2,3,4\}$ as indicated

while filter parameter is set at 0.2 for the EWMA control chart. SSA-based charts are based on a window length of 20 samples. For all charts, the detection limits were adjusted to achieve an in-control run length of approximately 370 samples.

The behaviour of the classical control charts conforms to well-established results in literature. More specifically, the ARL curves of Shewhart charts show the best performance in detecting large mean shifts but degrade with decreasing magnitude of shift. MA and EWMA perform best at detecting small shifts. In general, MSSPC methods show an improvement at detecting small shifts as the number of decomposition depths is increased. This has been attributed to signal-to-noise ratio enhancement arising from better separation of stochastic and deterministic effects (Aradhye et al. 2003). Also, for small shifts and high decomposition depths,

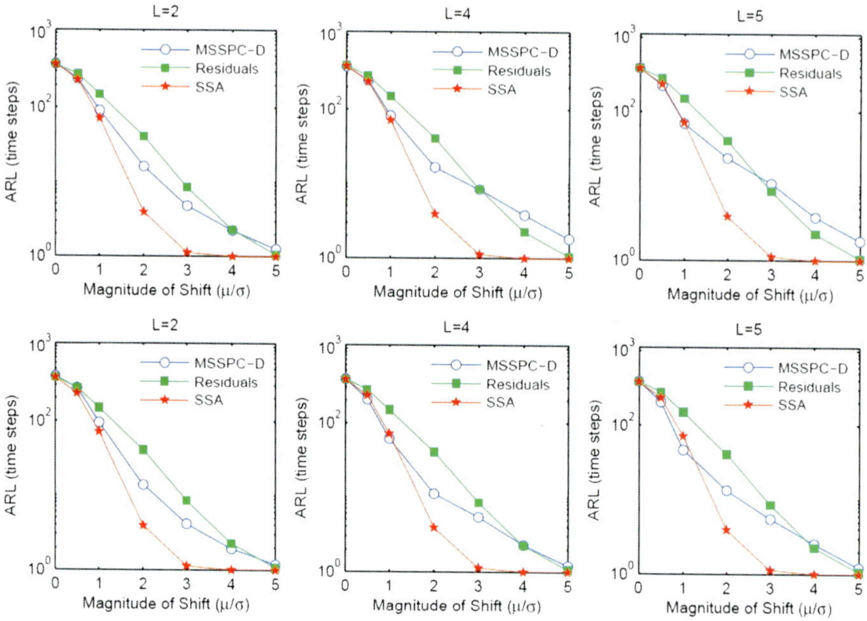

Fig. 8.6 Monte Carlo ARL curves: (*top row*) MSPCA-dyadic and (*bottom row*) MSPCA-integer for an autocorrelated process defined in Eq. 8.25 with $\beta = 0.5$ at different depths of decomposition $L = \{2,4,5\}$. Superimposed in each plot are the ARL curves for AR(1) residuals and SSA-based charts

the performance of MSSPC methods is better than Shewhart charts. On the other hand, for large shifts and high decomposition depths, MSSPC methods tend to be better than MA charts. The use of non-decimated wavelet decomposition (Fig. 8.5) enhances the performance of MSSPC, with the performance converging to that of MA or EWMA and Shewhart for small and large shifts, respectively, depending on the depth of decomposition.

The performance of SSA-based Shewhart and MA charts closely follows that of their classical counterparts. This is not unexpected since the extracted components for any mode are neither a trend nor a harmonic. In the results shown in the plots, the first two components were grouped together as signal and the rest as noise.

8.4.2 Univariate SPC: Autocorrelated Process

The ARL curves for a residuals-based control chart, multiscale SPC and SSA-based control chart for an autocorrelated process are shown in Fig. 8.6. The autocorrelated process is given by the AR(1) process

$$x(t) = \beta x(t-1) + \varepsilon(t) \tag{8.25}$$

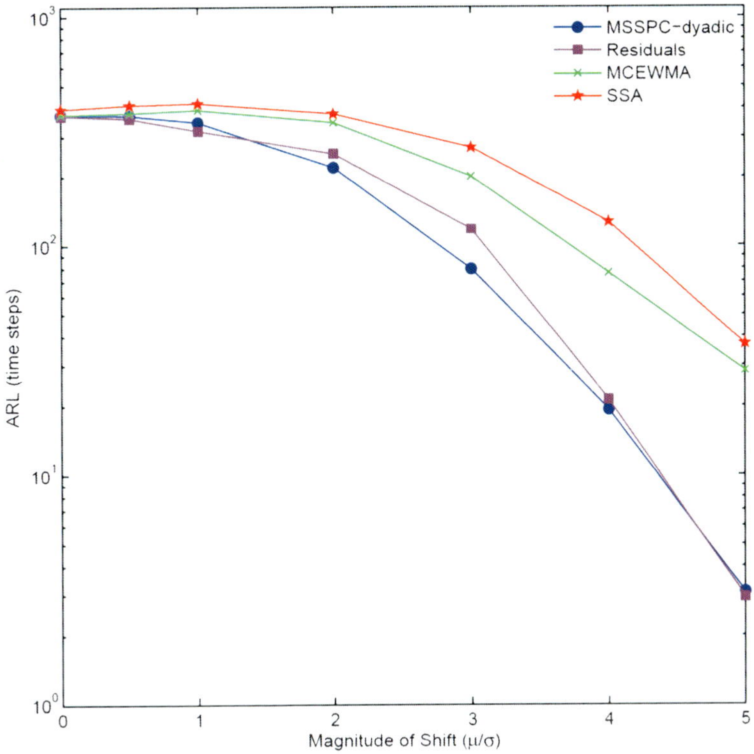

Fig. 8.7 Monte Carlo ARL curves of MSPCA-dyadic, residuals, moving centre EWMA (*MCEWMA*) and SSA control charts for a highly correlated AR(1) process, i.e. $\beta = 0.9$ in Eq. 8.25

where β is a constant coefficient and ε is Gaussian-distributed noise with zero mean and unit variance, i.e. $\mathcal{N}(a, b)$. In Fig. 8.6, $\beta = 0.5$ is used, which reflects mild autocorrelation. The plots indicate that the SSA-based Shewhart chart performs best compared to both the MSSPC and residuals SPC charts at almost all mean shifts, with the exception of small shifts at which MSSPC does at least as well as SSA for larger depths of decomposition. The time series modelling-based residuals chart had the worst performance at shifts of small to intermediate magnitudes but showed equal or better performance than MSSPC.

Increasing the degree of autocorrelation of the AR(1) process in Eq. 8.25 drastically changes the relative performance of considered SPC methods. In Fig. 8.7, with $\beta = 0.9$, the performance of all methods significantly decreases compared to Fig. 8.6, where $\beta = 0.5$ was used. Relatively, both MSSPC and residuals tend to perform better than SSA or moving centre EWMA, the latter being known to be appropriate for decorrelating integrated moving average (IMA) processes. Hence, it can be concluded that while high levels of autocorrelation have a detrimental effect on most control charts, wavelet-based methods and residuals tend to exhibit better performance than other methods. The challenge is to design an appropriate decomposition (depth and choice of basis function) in the case of wavelets and an appropriate time series model for the residuals chart.

8.4.3 Multivariate SPC: Uncorrelated Measurements

Bakshi (1998) studied a multivariate linear uncorrelated process in the context of multiscale monitoring consisting of two variables independently sampled from a Gaussian distribution with $\mathcal{N}(a,b)$, with the other two variables formed from the sum and difference of the first two, to yield a system with an intrinsic dimensionality of two:

$$\begin{aligned} x_1(t) &\sim \mathcal{N}(0,1) \\ x_2(t) &\sim \mathcal{N}(0,1) \\ x_3(t) &= x_1(t) + x_2(t) \\ x_4(t) &= x_1(t) - x_2(t) \end{aligned} \tag{8.26}$$

The observed system is affected by a random Gaussian noise of zero mean and 0.2 standard deviation that uniformly affects all measurements, i.e.

$$\mathbf{X}(t) = [x_1(t) \quad x_2(t) \quad x_3(t) \quad x_4(t)] + 0.2\varepsilon(t) \tag{8.27}$$

where $\mathcal{N}(a,b)$.

ARL curves, plotted as a function of the magnitude of shift, of SSA-based process monitoring are compared with PCA and MSPCA in Fig. 8.6. The number of modes was varied for SSA, with the last kth mode a "noise" component constituted as the sum of the remaining modes excluding the leading $(k-1)$ modes, with $k \in \{3, 5, 7, 10\}$. A window length size of 20 was used. Monitoring was performed using PCA on the reconstructed matrices corresponding to these modes, and the model subspace was defined by the first two principal directions. The same number of principal components was retained in the PCA models for the other methods. For MSPCA non-decimated 5-scale signal decomposition with Haar wavelets was used.

It can be seen that MSSPC does better in detecting small mean shifts than either PCA or SSA, but performance degrades at large shifts. PCA and SSA generally follow the same trend, a pattern also observed in the univariate uncorrelated case. However, increasing the number of modes results in deterioration in the SSA-based monitoring performance (Fig. 8.8). Note that the distinction in the relative performance of MSSPC against PCA in this case is much sharper than those reported in Aradhye et al. (2003).

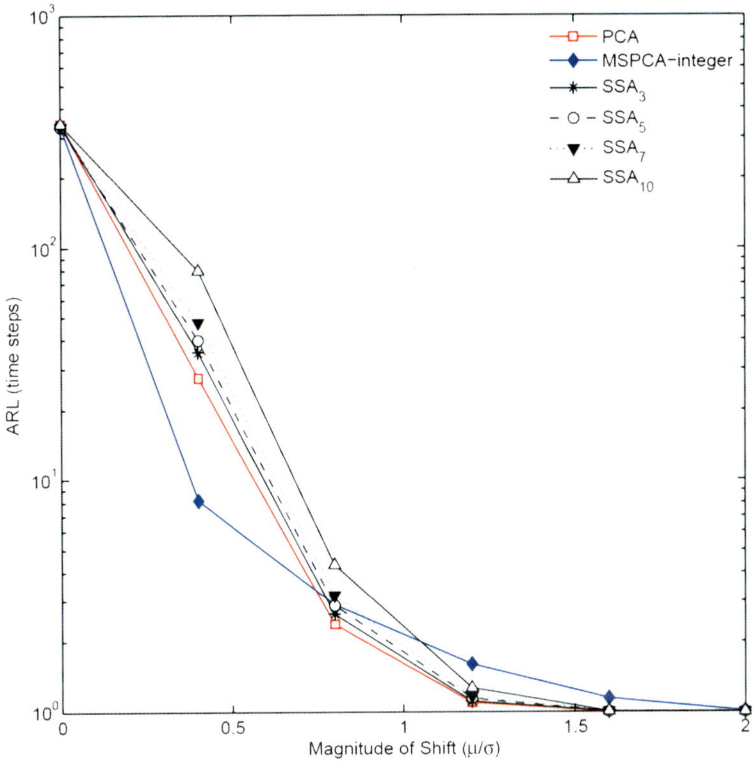

Fig. 8.8 ARL curves of PCA, MSSPC-integer and SSA statistical process monitoring for a multivariate uncorrelated process defined in Eqs. 8.26 and 8.27

8.4.4 *Multivariate SPC: Autocorrelated Measurements*

The final system considered in this section is the so-called 2×2 multivariate autocorrelated system (Ku et al. 1995; Kano et al. 2002). The system is represented by the following equation:

$$\mathbf{x}(t) = \begin{bmatrix} 0.118 & -0.191 \\ 0.847 & 0.264 \end{bmatrix} \mathbf{x}(t-1) + \begin{bmatrix} 1 & 2 \\ 3 & -4 \end{bmatrix} \mathbf{u}(t-1) \tag{8.28}$$

$$\mathbf{u}(t) = \begin{bmatrix} 0.811 & -0.226 \\ 0.477 & 0.415 \end{bmatrix} \mathbf{u}(t-1) + \begin{bmatrix} 0.193 & 0.689 \\ -0.320 & -0.749 \end{bmatrix} \mathbf{w}(t-1) \tag{8.29}$$

$$\mathbf{y}(t) = \mathbf{x}(t-1) + \mathbf{v}(t-1) \tag{8.30}$$

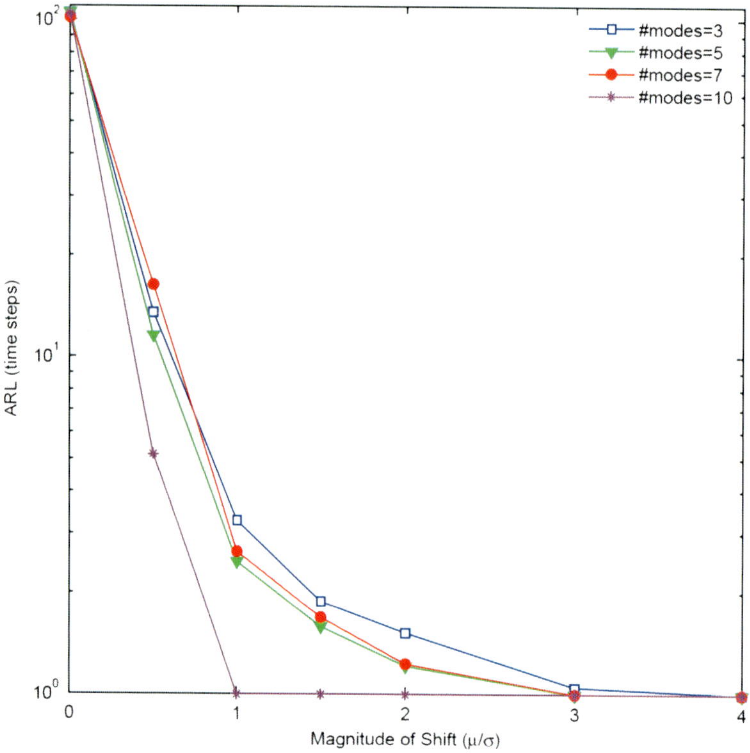

Fig. 8.9 Multivariate 2 × 2 system: variation of ARL curves as the number of modes used in the SSA model is changed

where $\mathbf{u}(t)$ is correlated input at time t, $\mathbf{y}(t)$ is a vector of measured variables and $\mathbf{v}(t)$ and $\mathbf{w}(t)$ are zero mean-centred Gaussian inputs with variances 0.01 and 0.1, respectively. The monitored variables are the input vector $\mathbf{u}(t)$ and the observation vector $\mathbf{y}(t)$. The relative performance of steady-state PCA, dynamic PCA, steady-state MSPCA, dynamic MSPCA as well as SSA is investigated. The steady-state measurement vector is $[\mathbf{u}(t)\ \mathbf{y}(t)]$, and corresponding dynamic one is $[\mathbf{u}(t)\ \mathbf{u}(t-1)\ \mathbf{y}(t)\ \mathbf{y}(t-1)]$. The mean shift disturbance is introduced in the input vector $\mathbf{u}$. The PCA and DPCA model subspaces are based on the leading two and five principal directions, respectively. Both dyadic and integer 7-level decomposition using Haar wavelets are considered. A window size of length $L = 10$ is used for SSA using the multichannel approach. Furthermore, k modes were used in the reconstruction and, as before, the last reconstructed signal being the sum of all the modes excluding the leading modes $(k-1)$ with $k \in \{3, 5, 7, 10\}$. Two principal components are selected for process monitoring at each mode.

The effect on the ARL as the number of modes k is changed is shown in Fig. 8.9, which shows a significant improvement in the ARL when $k = 10$. Lower modes do not show predictable behaviour. This may be explained by considering the

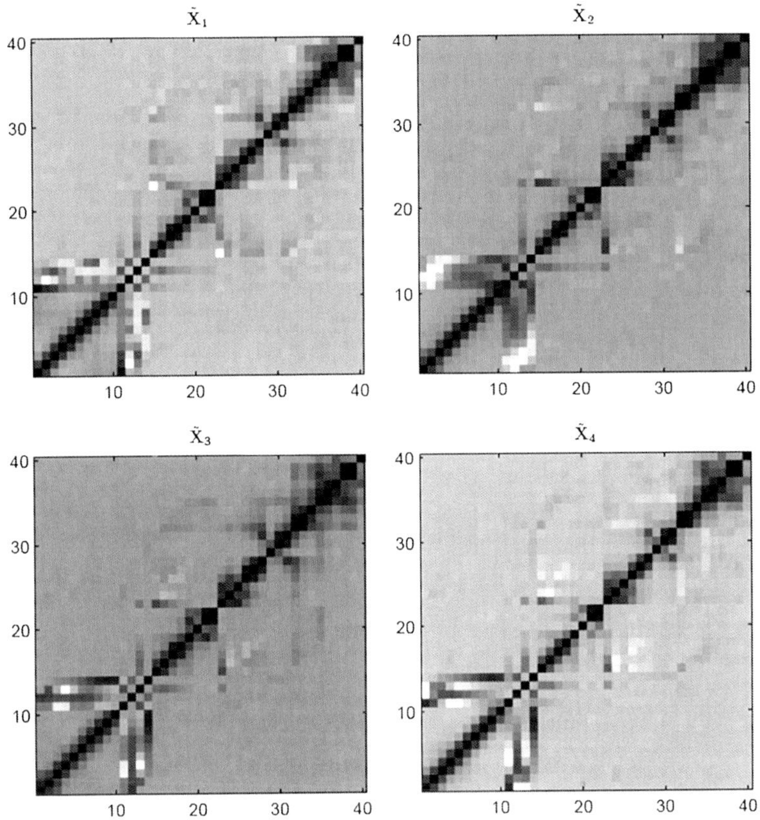

Fig. 8.10 Weighted correlation matrices for each of the variables of the 2×2 multivariate autocorrelated process defined in Eq. 8.28. The *grey* background can be attributed to the use of multichannel SSA in the decomposition, which induces coupling of the involved variables

weighted correlation matrix plots of each individual variable after reconstruction shown in Fig. 8.10. For separability, the weighted correlations between pairs of reconstructed components must be zero (white shading), non-zero for partial separability (grey shading) and unit for complete non-separability (black shading). Weighted correlation matrices of multichannel SSA have not been studied in literature, but it can be conjectured that coupling of variables precludes complete separability. Hence, the **w**-correlation matrices have a greyish background. Ignoring this background, it can be seen that reconstructed signals up to 10 show minimal scatter, whereas there is a significant scatter in the rest. Therefore, the use of the first 10 modes is likely to yield better decomposition into signal and noise components separable.

The ARL curves of the different SPC methods are shown in Fig. 8.11. MSPCA with dyadic discretization performs worst among other methods considered in detecting large shifts, but improved performance is observed at small shifts,

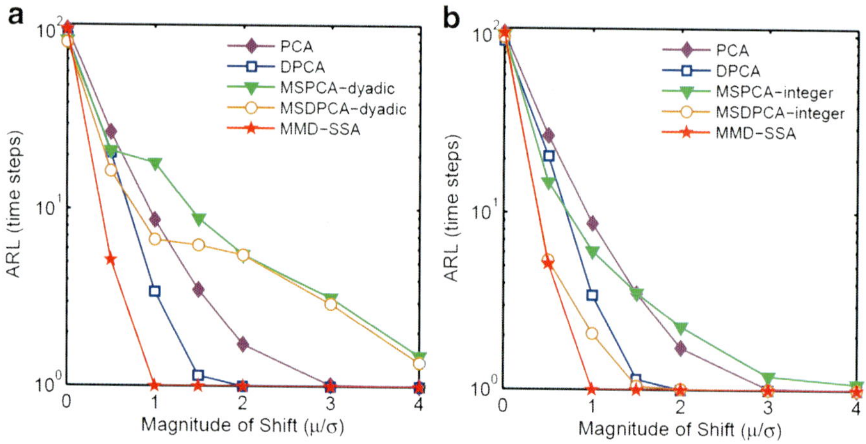

Fig. 8.11 Monte Carlo ARL curves of PCA and (**a**) MSSPC-dyadic and (**b**) MSSPC-integer for the multivariate correlated process

Fig. 8.11a. Using non-decimated wavelet decomposition improves the MSPCA performance. Dynamic PCA performs better than steady-state PCA and MSPCA. The inclusion of dynamic information in non-decimated MSSPC results in the best-performing method among PCA and MSPCA methods, Fig. 8.11b. This has been attributed to the fact that MSDPCA combines advantages of both DPCA and MSPCA, specifically autocorrelation modelling and adaptive nature of wavelets that allow timeous detection of abnormal events (Aradhye et al. 2003). SSA using 10 modes has the best performance across all shift sizes because of better separation of deterministic variation from noise effects as explained above.

8.5 Applications: Multivariate AR(1) Process

In this section SSA-based statistical process control method is applied to a multivariate AR(1) process with mean shift and parameter change. The performance of the method is compared with other existing SPC methods in terms of the reliability. Reliability gives an indication of the effectiveness of a process monitoring scheme by measuring the proportion of samples outside control limits in a window length defined from the time of occurrence of a special event. Assuming normal operating conditions, given a detection limit based on a familywise confidence limit of $100(1-\alpha)\%$, the monitored statistic is expected to be outside the control limits in only $100\alpha\%$ of samples in a suitably defined window. Abnormal conditions should reflect in a significantly higher reliability than $100\alpha\%$.

The multivariate autocorrelated process introduced earlier (Eq. 4.2) is considered first. As before, the variables used for monitoring the system are the correlated inputs $\mathbf{u}(t)$ and outputs $\mathbf{y}(t)$. Eight disturbances or fault conditions were simulated,

Table 8.2 Abnormal conditions for the multivariate autocorrelated system

Case	Type and magnitude of change
0	Normal conditions
1	Mean shift in $w_1: 0 \rightarrow 0.5$
2	Mean shift in $w_1: 0 \rightarrow 1$
3	Mean shift in $w_1: 0 \rightarrow 1.5$
4	Mean shift in $w_1: 0 \rightarrow 2$
5	Mean shift in $w_1: 0 \rightarrow 3$
6	Change in parameter mapping u_1 to x_2: $3 \rightarrow 2.5$
7	Change in parameter mapping u_1 to x_2: $3 \rightarrow 2.0$
8	Change in parameter mapping u_1 to x_2: $3 \rightarrow 1.0$

Table 8.3 Comparison of reliability measures of different SPM methods for the 2 × 2 system

		Disturbance								
Method	Index	0	1	2	3	4	5	6	7	8
PCA	T_2^2	3.0	4.8	13.8	31.4	54.7	91.6	2.7	3.0	3.3
	Q_2	3.3	3.8	5.0	7.7	13.2	37.9	3.3	3.6	4.0
DPCA	T_5^2	1.8	3.6	10.5	27.6	53.6	93.8	1.7	1.9	2.7
	Q_5	3.2	4.9	13.2	47.1	97.9	100	3.6	5.2	10.5
MSPCA-I	T_2^2	0.3	79.3	90.9	95.3	97.0	99.0	0.3	1.0	4.8
	Q_2	0.4	83.1	95.3	98.3	99.6	99.9	2.1	12.4	37.5
MSDPCA-I	T_5^2	0.4	95.1	99.0	99.9	100	99.9	3.6	22.4	53.2
	Q_5	0.5	96.5	99.8	100	100	100	4.3	29.6	64.1
SSA	T_{10}^2	0.5	19.6	98.0	100	100	100	1.1	5.9	24.2
	Q_{10}	0.0	0.0	0.7	74.5	100	100	0.0	0.1	4.7

as summarized in Table 8.2. The first five fault conditions were generated by introducing progressively larger mean shifts in w_1, similar to the ARL performance analysis. The rest of the fault conditions are induced by changing the parameter mapping u_1 to x_2, that is, the element in the second row and first row of the second coefficient matrix in Eq. 36. A total of 1,096 measurements were sampled from the system for use in constructing reference monitoring models, while 128 measurements were generated for each abnormal condition. The SPC methods applied to the data included PCA, dynamic PCA, multiscale PCA, multiscale dynamic PCA and SSA. In the reported results, non-decimated wavelet decomposition was used for the multiscale methods. Dyadic MSPCA reliability analysis results follow Kano et al. (2000, 2002) and, therefore, not shown. In all cases, the same algorithmic parameters as in the previous ARL analysis were used. The results in terms of the mean reliability from 100 simulations are shown in Table 8.3.

In general, multiscale SPC with integer discretization and SSA methods show significant improvement in reliability compared to both PCA and DPCA, particularly for mean shift changes. Moreover, the fraction of samples above detection limits in the absence of a disturbance (case 0 in Table 8.2) conforms to the significance level α for these approaches with improved mean shift detection capabilities. PCA and DPCA improve as the mean shift increases. SSA shows a T^2 reliability of 19.6 % for disturbance 1 that is much worse than for MSPCA and

MSDPCA, although all the three approaches had a comparable ARL, as indicated in Fig. 8.11. With the exception of multiscale methods, all the other methods fared poorly in detecting parametric changes, that is, disturbances 6–8.

8.6 Concluding Remarks

SSA is a method used for the analysis of time series structure, and in the context of the diagnostic framework of this book, it can be seen as a form of preprocessing of the data prior to feature extraction. The main purpose of SSA is decomposition of a time series into additive components that can be associated with a trend, oscillatory patterns that are possibly amplitude- or phase-modulated as well as an aperiodic or noise component. An important advantage of SSA compared to other methods is its adaptive nature. Specifically, the basis functions used for time series decomposition are obtained from the data themselves. This allows for a better and more compact representation of certain features in the data, such as nonlinear harmonics, that can be obtained using fixed basis functions such as sinusoids in Fourier analysis or dilated and translated mother wavelets in wavelet analysis. Process monitoring using SSA is based on obtaining a multimodal representation of a multivariate time series and subsequently applying a standard statistical process control scheme to this representation.

As demonstrated in this chapter, an SSA approach to process monitoring can compare favourably with existing methods such as PCA, dynamic PCA and multiscale PCA. The performance of these was compared using average run length, as well as reliability analysis on data generated from simulated systems. SSA could reliably detect mean shift changes in the simulated univariate systems with mild autocorrelation, while its performance degraded in the presence of excessive autocorrelation. In the case of multivariate autocorrelated systems, SSA compared favourably than MSPCA in detecting mean shifts. However, it did not perform as well in detecting parameter changes. Further investigation, which could include other variants of SSA, such as kernel SSA (Jemwa and Aldrich 2006), would be required to better establish the types of problems where SSA-based approaches to process fault diagnosis may offer advantages over other methods.

References

Abdi, H. (2007). Bonferroni and Šidàk corrections for multiple comparisons. In N. Salkind (Ed.), *Encyclopedia of measurement and statistics* (pp. 103–107). Thousand Oaks: Sage.

Allen, M., & Smith, L. (1996). Monte Carlo SSA: Detecting irregular oscillations in the presence of coloured noise. *Journal of Climate, 9*, 3373–3404.

Aradhye, H., Bakshi, B. R., Strauss, R., & Davis, J. (2003). Multiscale SPC using wavelets: Theoretical analysis and properties. *American Institution of Chemical Engineers Journal, 49*(4), 939–958.

Bakshi, B. R. (1998). Multiscale PCA with applications to multivariate statistical process monitoring. *AICHE Journal, 44*(7), 1596–1610.

Bakshi, B. R. (1999). Multiscale analysis and modeling using wavelets. *Journal of Chemometrics, 1999*, 415–434.

Bersimis, S., Psarakis, S., & Panaretos, J. (2007). Multivariate statistical process control charts: An overview. *Quality and Reliability Engineering International, 23*, 517–543.

Broomhead, D., & King, G. (1986). Extracting qualitative dynamics from experimental data. *Physica D, 20*, 217–236.

Daubechies, I. (1992). *Ten lectures on wavelets, Vol. 61 of CBMS-NSF series in Applied mathematics*. Philadelphia: SIAM.

Donoho, D., Johnstone, I., Kerkyacharian, G., & Picard, D. (1995). Wavelet shrinkage: Asymptopia? *Journal of the Royal Statistical Society, Series B, 57*, 301–369.

Elsner, J., & Tsonis, A. (1996). *Singular Spectrum Analysis – A new tool in time series analysis*. New York: Plenum Press.

Ganesan, R., Das, T., & Venkataraman, V. (2004). Wavelet-based multiscale statistical process monitoring: A literature review. *IIE Transactions, 36*, 787–806.

Ghil, M., Allen, M., Dettinger, M., Ide, K., Kondrashov, D., Mann, M., Robertson, A., Saunders, A., Tian, Y., Varadi, F., & Yiou, P. (2002). Advanced spectral methods for climatic times series. *Reviews of Geophysics, 40*(1), 3.1–3.41.

Ghil, M., Yiou, P., Hallegatte, S., Malamud, B. D., Naveau, P., Soloviev, A., Friederichs, P., Keilis-Borok, V., Kondrashov, D., Kossobokov, V., Mestre, O., Nicolis, C., Rust, H. W., Shebalin, P., Vrac, M., Witt, A., & Zaliapin, I. (2011). Extreme events: Dynamics, statistics and prediction. *Nonlinear Processes in Geophysics, 18*(3), 295–350. http://www.nonlin-processes-geophys.net/18/295/2011/

Golyandina, N., Nekrutin, V., & Zhigljavsky, A. (2001). *Analysis of time series structure: SSA and related techniques*. Boca Raton: Chapman & Hall/CRC.

Harris, T., & Ross, W. (1991). Statistical process control procedures for correlated observations. *Canadian Journal of Chemical Engineering, 69*, 48–57.

Hassani, H., & Zhigljavsky, A. (2009). Singular spectrum analysis: Methodology and application to economics data. *Journal of Systems Science and Complexity, 22*, 372–394.

Huang, N., Shen, Z., Long, S., Wu, M., Shih, H., Zheng, Q., Yen, N. C., Tung, C., & Liu, H. (1998). The empirical mode decomposition and the Hilbert spectrum for nonlinear and non-stationary time series analysis. *Proceedings of the Royal Society of London, Series A, 454*, 903–995.

Jackson, J. E. (1991). *A user's guide to principal components*. New York: Wiley.

Jemwa, G. T., & Aldrich, C. (2006). Classification of process dynamics with Monte Carlo singular spectrum analysis. *Computers and Chemical Engineering, 30*(5), 816–831.

Kano, M., Nagao, K., Hasebe, S., Hashimoto, I., Ohno, H., Strauss, R., & Bakshi, B. (2000). Comparison of statistical process monitoring methods: Application to the Eastman challenge problem. *Computers and Chemical Engineering, 24*, 175–181.

Kano, M., Nagao, K., Hasebe, S., Hashimoto, I., Ohno, H., Strauss, R., & Bakshi, B. (2002). Comparison of multivariate statistical process monitoring methods with applications to the Eastman challenge problem. *Computers and Chemical Engineering, 26*, 161–174.

Kantz, H., & Schreiber, T. (1997). *Nonlinear time series analysis*. Cambridge: Cambridge University Press.

Kresta, J., MacGregor, J. F., & Martile, T. (1991). Multivariate statistical monitoring of process operating performance. *Canadian Journal of Chemical Engineering, 69*, 35–47.

Ku, W., Storer, R. H., & Georgakis, C. (1995). Disturbance detection and isolation by dynamic principal component analysis. *Chemometrics and Intelligent Laboratory Systems, 30*, 179–196.

Lee, J. M., Yoo, C., Choi, S., Vanrolleghem, W., & Lee, I.-B. (2004). Nonlinear process monitoring using kernel principal component analysis. *Chemical Engineering Science, 59*, 223–234.

MacGregor, J. F., & Kourti, T. (1995). Statistical process control of multivariate processes. *Control Engineering Practice, 3*, 403–414.

Mallat, S. (1989). A theory for multiresolution signal decomposition: The wavelet representation. *IEEE Transactions on Pattern Analysis and Machine Intelligence, 11*(7), 674–693.

Mallat, S. (1999). *A wavelet tour of signal processing* (2nd ed.). San Diego: Academic.

Montgomery, D. C. (1996). *Introduction to statistical quality control*. New York: Wiley.

Montgomery, D. C., & Mastrangelo, C. (1991). Some statistical process control methods for autocorrelated data. *Journal of Quality Technology, 23*, 179–193.

Moskvina, V., & Zhigljavsky, A. (2003). An algorithm based on singular spectrum analysis for change-point detection. *Communications in Statistics: Simulation and Computation, 32*, 319–352.

Nomikos, P., & MacGregor, J. F. (1995a). Multivariate SPC charts for monitoring batch processes. *Technometrics, 37*(1), 41–59.

Nomikos, P., & MacGregor, J. F. (1995b). Multi-way part least squares in monitoring batch processes. *Chemometrics and Intelligent Laboratory Systems, 30*, 97–108.

Plaut, G., & Vautard, R. (1994). Spells of low-frequency oscillations and weather regimes in the Northern Hemisphere. *Journal of the Atmospheric Sciences, 51*, 210–236.

Reis, M., Saraiva, P., & Bakshi, B. R. (2008). Multiscale statistical process control using wavelet packets. *AICHE Journal, 54*(9), 2366–2378.

Runger, G. C., & Willemain, T. R. (1995). Model-based and model-free control of autocorrelated processes. *Journal of Quality Technology, 27*(4), 283–292.

Sauer, T., Yorke, J. A., & Casdagli, M. (1991). Embedology. *Journal of Statistical Physics, 65*, 579–616.

Strang, G. (2009). *Introduction to linear algebra*. Wellesley: Wellesley-Cambridge.

Tiao, G., & Box, G. (1981). Modeling multiple time series with applications. *Journal of the American Statistical Association, 76*, 802–816.

Tjostheim, D., & Paulsen, J. (1982). Empirical identification of multiple time series. *Journal of Time Series Analysis, 3*, 265–282.

Vautard, R., & Ghil, M. (1989). Singular spectrum analysis in nonlinear dynamics, with applications to paleoclimatic time series. *Physica D, 35*, 395–424.

Vautard, R., Yiou, P., & Ghil, M. (1992). Singular-spectrum analysis: A toolkit for short, noisy chaotic signals. *Physica D, 58*, 95–126.

Westerhuis, J., Kourti, T., & MacGregor, J. F. (1998). Analysis of multiblock and hierarchical PCA and PLS models. *Journal of Chemometrics, 12*, 301–321.

Wierda, S. (1994). Multivariate statistical process control – Results and directions for future research. *Statistica Neerlandica, 48*, 147–168.

Wilson, G. (1973). The estimation of parameters in multivariate time series models. *Journal of the Royal Statistical Society, Series B, 35*, 76–85.

Wise, B., & Gallagher, N. (1996). The process chemometrics approach to process monitoring and fault detection. *Journal of Process Control, 6*(6), 329–348.

Yiou, P., Sornette, D., & Ghil, M. (2000). Data-adaptive wavelets and multi-scale singular-spectrum analysis. *Physica D, 142*, 254–290.

Yoon, S., & MacGregor, J. F. (2004). Principal component analysis of multiscale data for process monitoring and fault diagnosis. *AICHE Journal, 50*(11), 2891–2903.

Nomenclature

Symbol	Description
l	Lag parameter
d	Dimensionality of data matrix, $\mathbf{X} \in \mathbb{R}^{N \times d}$
W_φ	Wavelet transform
a	Dilation parameter in wavelet transform

(continued)

(continued)

Symbol	Description
b	Translation parameter in wavelet transform
φ	Mother wavelet
L	Depth of wavelet decomposition
$\mathbf{G}_m$	mth of L detailed approximations of wavelet
$\mathbf{H}_L$	Coarse approximation of wavelet
$x_t^{(i)}$	Ith time series component resulting from decomposition of a time series with singular spectrum analysis
$\mathbf{u}_k$	kth of L left singular vector resulting from singular value decomposition of matrix
$\mathbf{v}_k$	kth of L right singular vector resulting from singular value decomposition of matrix
σ_k	kth of L singular value resulting from singular value decomposition of matrix
$\mathbf{U}$	Left singular matrix with singular vectors $\mathbf{u}_k$ $(k = 1, 2, \ldots, L)$ as columns
$\mathbf{V}$	Right singular matrix with singular vectors $\mathbf{v}_k$ $(k = 1, 2, \ldots, L)$ as columns
$\mathbf{G}_i$	Matrix consisting of the sum of the ith subgroup of decomposed trajectory matrices in singular value decomposition
$\tilde{x}_t$	Component at time t of a time series reconstructed by diagonal averaging of the elements of a matrix
ε_t	Additive noise component of time series at time t
$\tilde{\mathrm{X}}^{(k)}$	kth matrix reconstructed with singular spectrum analysis
$\tilde{x}_i^{(k)}$	ith element of kth time series reconstructed with singular spectrum analysis
$\beta_{\mathrm{lim}}^{(k)}$	Control limit of kth of M reconstructed time series
X_t'	Matrix of time series column vectors with mean-centred columns
$\widehat{\mathbf{X}}_i$	ith of d
$\tilde{\mathbf{X}}_i$	ith trajectory matrix
$\rho_{p,q}^{(w)}$	Weighted or **w**-correlation between time series p and q
$\rho_{\max}^{(L,K)}$	Maximum of the absolute value of the correlations between the rows and between the columns of a pair of trajectory matrices $\tilde{\mathbf{X}}_i$ and $\tilde{\mathbf{X}}_j$
$\mathcal{N}(a, b)$	Normal distribution with mean a and standard deviation b
$\mathbf{u}(t)$	Input vector at time t
$\mathbf{y}(t)$	Vector of measured variables at time t
$\mathbf{v}(t)$	Gaussian noise with variance 0.01
$\mathbf{w}(t)$	Gaussian noise with variance 0.1

Index

C. Aldrich and L. Auret, *Unsupervised Process Monitoring and Fault Diagnosis with Machine Learning Methods*, Advances in Computer Vision and Pattern Recognition, DOI 10.1007/978-1-4471-5185-2, © Springer-Verlag London 2013

T

U

V

W

X

Printed by Publishers' Graphics LLC
MLSI130628.15.16.17